T0234372

Lecture Notes in Physics

Founding Editors

Wolf Beiglböck
Jürgen Ehlers
Klaus Hepp
Hans-Arwed Weidenmüller

Volume 1009

Series Editors

Roberta Citro, Salerno, Italy
Peter Hänggi, Augsburg, Germany
Morten Hjorth-Jensen, Oslo, Norway
Maciej Lewenstein, Barcelona, Spain
Angel Rubio, Hamburg, Germany
Wolfgang Schleich, Ulm, Germany
Stefan Theisen, Potsdam, Germany
James D. Wells, Ann Arbor, MI, USA
Gary P. Zank, Huntsville, AL, USA

The series Lecture Notes in Physics (LNP), founded in 1969, reports new developments in physics research and teaching - quickly and informally, but with a high quality and the explicit aim to summarize and communicate current knowledge in an accessible way. Books published in this series are conceived as bridging material between advanced graduate textbooks and the forefront of research and to serve three purposes:

- to be a compact and modern up-to-date source of reference on a well-defined topic;
- to serve as an accessible introduction to the field to postgraduate students and non-specialist researchers from related areas;
- to be a source of advanced teaching material for specialized seminars, courses and schools.

Both monographs and multi-author volumes will be considered for publication. Edited volumes should however consist of a very limited number of contributions only. Proceedings will not be considered for LNP.

Volumes published in LNP are disseminated both in print and in electronic formats, the electronic archive being available at springerlink.com. The series content is indexed, abstracted and referenced by many abstracting and information services, bibliographic networks, subscription agencies, library networks, and consortia.

Proposals should be sent to a member of the Editorial Board, or directly to the responsible editor at Springer:

Dr Lisa Scalone
Springer Nature
Physics
Tiergartenstrasse 17
69121 Heidelberg, Germany
lisa.scalone@springernature.com

Arnab Rai Choudhuri

Advanced Electromagnetic Theory

 Springer

Arnab Rai Choudhuri
Department of Physics
Indian Institute of Science
Bangalore, India

ISSN 0075-8450   ISSN 1616-6361 (electronic)
Lecture Notes in Physics
ISBN 978-981-19-5943-1   ISBN 978-981-19-5944-8 (eBook)
https://doi.org/10.1007/978-981-19-5944-8

This Springer imprint is published by the registered company Springer Nature Singapore Pte Ltd.
The registered company address is: 152 Beach Road, #21-01/04 Gateway East, Singapore 189721, Singapore

To

Opu and Tipu

Preface

This book is based on a one-semester course on electromagnetic theory at advanced undergraduate or beginning graduate level which I have taught at the Indian Institute of Science several times. A prerequisite for this course is a more elementary course at the level of Purcell's *Electricity and Magnetism*, teaching students how to formulate the basic principles of electromagnetism through vectorial equations. The advanced course on which this book is based is the last course on classical electrodynamics taken by our graduate students before they embark on research in different branches of physics. The aim of this course is to expose students to all the important basic principles of the subject which professional physicists working in any area of theoretical or experimental physics are expected to know.

Teaching a course on electromagnetic theory at the level of this book is quite challenging due to the wide diversity of topics which are expected to be covered. On the one hand, there are certain practical topics which have extensive engineering applications, such as the theory of waveguides and antennae. On the other hand, one has to cover other topics which are gateways to advanced theoretical physics, such as the relation of electromagnetic theory with special relativity and the action principle formulation in a Lorentz-invariant manner. Additionally, there are topics like the calculation of electromagnetic fields from the Liénard–Wiechert potentials, which involve mathematical derivations more complicated than probably any derivations which students at this level are likely to encounter in other basic physics courses. I have tried my best to write a balanced book which should be suitable for students going into different branches of theoretical or experimental physics. I may be slightly biased by my personal research interest in plasma astrophysics, but I believe that I have kept this bias within acceptable limits by adding only a short chapter on the basics of plasma physics at the end of this book. Apart from the growing importance of plasma physics, I firmly believe that even from a conceptual point of view it is desirable that all physics students should have some idea about the basics of plasma physics—the many-body theory of classical electrodynamics.

Although most of the basic principles of electromagnetic theory covered in a book like this were established more than a century ago and have not changed since then, there have been many changes in the last few decades in the way these principles are taught at the advanced undergraduate or the beginning graduate level. When I was going to graduate school in the 1980s, it was customary for many universities around the world (especially American universities) to teach electromagnetic theory in a leisurely manner over a year (in two semesters or three quarters). Very often, there would not be a separate course on mathematical methods of physics, and students were expected to be taught at least some basic techniques of mathematical methods in the electromagnetic theory course. One textbook which held undisputed sway in that era was Jackson's *Classical Electrodynamics*. I belong to the generation of physicists who were 'raised' on Jackson's book. With tremendous advances in different areas of physics over the last few decades, it has now become imperative to give students some exposure to these modern topics at a stage sufficiently early in their career. This can only be done by cutting out some parts of the older curriculum. While any competent physicist would agree that it is not possible to become a good physicist without a good command of the older classical areas of physics, it is also generally agreed that perhaps not every topic in these older areas of physics which used to be taught to all physics students a few decades ago needs to be taught today. One can prune the older curriculum to retain only the essentials which every physics student ought to know. Many universities around the world now offer a one-semester course of electromagnetic theory at the advanced undergraduate or the beginning graduate level, rather than a year-long sequence of courses.

Many of the older textbooks of electromagnetic theory were written with the aim of teaching the subject over a year (the third edition of Jackson's book runs to about 800 pages). In contrast, my aim has been to write a lean and thin advanced electromagnetic theory textbook from a modern perspective which covers all the important topics that a professional physicist needs to know and which can be covered comfortably in a one-semester course. There being many textbooks with a more complete coverage of the subject, I have not felt the need to strive for completeness. This book essentially covers the materials which I would manage to teach in a one-semester course at the Indian Institute of Science, lecturing three hours per week. Although I am considered a reasonably fast teacher, I trust that a teacher proceeding at a moderate pace will be able to cover a large part of this book. While designing this course, I had to be ruthlessly selective in deciding which topics of electromagnetic theory a professional physicist must know from the last course on the subject. Although not everybody can be expected to be in 100% agreement with me, I do believe that any professor who is compelled to teach the essentials of advanced electromagnetic theory in one semester in a balanced course suitable for all physics students will have to end up with a course not very different from what is presented in this book. While deciding which topics to include and which topics to leave out, I had one guiding principle. I gave a higher priority to topics which introduce important new concepts rather than topics which are more like detailed working out of already established concepts. For example, although I discussed two-dimensional boundary value problems in electrostatics in considerable detail, I left out three-dimensional problems and thereby saved some precious time which could be used for teaching other topics which I considered more important.

A time-honoured practice followed in many of the standard textbooks of electro-magnetic theory had been to provide an extensive set of exercise problems. Jackson's book is famous (or notorious, depending on your point of view!) for its collection of difficult problems. In keeping with the overall spirit of the book, I have included a limited number of exercise problems specifically selected for advancing the student's understanding of the materials discussed in the text. I have avoided problems for the sole purpose of testing a student's capacity for intellectual gymnastics at its limit. There is another tricky issue connected with exercise problems: ascribing the correct credit to the persons who might have first invented some of these problems. Since this is impossible to do, I hereby declare that I collected many of the problems from different sources over the years (I cannot even recall now which problem might have been taken from where) and I make no claim of my intellectual ownership over the exercise problems.

While the thought of writing this book has been on my mind for several years, I at last got the forced opportunity of working on this book during the several depressing months when much of the world—especially my country India—was under almost complete lockdown due to the COVID-19 pandemic. Working on this book helped me to maintain my mental sanity at this difficult time. I have naturally been influenced by many authors who wrote books on this subject before me. In particular, I express my indebtedness to Feynman, Griffiths, Jackson, Landau and Lifshitz, Panofsky and Phillips. I am grateful to many students who have taken this course from me over the years and encouraged me through their questions and feedback. I have also had fruitful discussions with many departmental colleagues about the teaching of this course. I would particularly like to mention two younger colleagues who had taught this course in our department many times like me and who were very enthusiastic about my idea of writing this book, but who are no longer among us to see this book: Vasant Natarajan and V. Venkataraman. I thank Bibhuti Kumar Jha for preparing most of the figures in this book. Finally, the book would not have been possible without the continuous support of my wife Mahua.

Bangalore, India   Arnab Rai Choudhuri

Contents

Chapter 1
Introduction

1.1 Why Electromagnetic Theory Again?

We shall assume that, for most of the readers approaching this book as a textbook, this is probably the third exposure to electromagnetic theory. The first exposure is likely to have been in high school, when students are taught the basic phenomenology of electromagnetism along with the laws which govern them. This first exposure usually happens without the use of much sophisticated mathematics—often without even using calculus. The second exposure takes place at the college level, when students learn how to formulate the basic principles of electromagnetism elegantly by using vector analysis. This second exposure usually ends with a discussion of Maxwell's equations, showing that they lead to the prediction of electromagnetic waves. The practical subject of electrical circuits is usually an important part of an electromagnetism course at this level.

Students who have had these two previous exposures to electromagnetism may legitimately ask the question: what is the need of another course on the same subject? We believe that students who wish to take up physics as a career indeed should take another advanced course on electromagnetic theory. There are several reasons behind this. First of all, physicists now know that the physical universe is governed by only four fundamental interactions: strong, electromagnetic, weak, gravitational. Of these four interactions, two (strong and weak) are confined within the atomic nucleus. Electromagnetism is one of the only two other fundamental interactions which can work over long ranges. Certainly a student of physics needs a thorough understanding of such an important interaction. The earlier exposures to electromagnetism usually do not cover many important topics—the relation between special relativity and electromagnetism, and emission of electromagnetic radiation from accelerated charges. One aim of an advanced course (or textbook) is to introduce students to this core of important topics in electromagnetism which every professional physicist should know.

There is also a secondary practical aim behind an advanced course on electromagnetic theory. In such courses as classical mechanics and statistical mechanics,

© Springer Nature Singapore Pte Ltd. 2022
A. R. Choudhuri, *Advanced Electromagnetic Theory*, Lecture Notes in Physics 1009,
https://doi.org/10.1007/978-981-19-5944-8_1

students typically have to deal with a collection of particles (sometimes with rigid bodies also). Now, a state of such a system can be prescribed by a finite number of coordinates—usually the position and momentum coordinates of the particles. On the other hand, to prescribe a state of the electromagnetic field, we have to provide values of the electric field $\mathbf{E}(\mathbf{x}, t)$ and the magnetic field $\mathbf{B}(\mathbf{x}, t)$ at all points $\mathbf{x}$ of space. As indicated by the t dependence, these values can also change with time. Physicists have developed sophisticated mathematical techniques for handling such fields. One aim of an advanced electromagnetic theory course is to teach these mathematical techniques, which turn out to be extremely useful in many different areas of physics where one may have to deal with different kinds of fields. Since students can learn these mathematical techniques most easily by applying them to systems of which they know the governing principles, a course of electromagnetic theory is usually considered the most suitable place for introducing physics students to these mathematical techniques for dealing with fields.

This book thus has the following aim. We shall look at the whole of electromagnetic theory from an advanced vantage point, teaching students mathematical techniques for handling fields and also applying the basic principles to important new situations which students are not likely to have studied before.

1.2   A Possible Axiomatic Formulation

Ever since Euclid showed that the whole of plane geometry can be derived from a set of axioms, this Euclidean model of axiomatic formulation is often considered to be the ultimate goal of a mature branch of science. As it happens, there are very few branches of physics which can be profitably taught to students by following a strictly axiomatic approach. One can start even the very first course of mechanics with Newton's laws of motion. Perhaps a course of thermodynamics can also be developed around the laws of thermodynamics. However, in a subject like electromagnetic theory, the basic laws (Maxwell's equations) involve sophisticated concepts which only a student who had already learned the subject at sufficient depth can understand. One cannot teach the first course of electromagnetism to high school students by following a fully axiomatic approach. However, since readers of this book are expected to be familiar with the basic concepts of electromagnetic theory and Maxwell's equations, it is in principle possible at this advanced level to present an axiomatic development of electromagnetic theory. We are indeed going to show that the mathematical formalism of electromagnetic theory can be built up from a few fundamental principles. However, this is a subject with a rich phenomenology. While a fully axiomatic approach of the subject may have some logical and intellectual appeal, we shall not follow such an approach very strictly and shall continuously point out the connections of the subject with phenomena of the real world. We believe that this mixed approach—a judicious combination of the axiomatic and phenomenological approaches—is the best way of teaching electromagnetic theory.

Let us now enlist the basic principles of the subject. The basic idea of electromagnetic theory is that the charge density $\rho(\mathbf{x}, t)$ and the current density $\mathbf{j}(\mathbf{x}, t)$ act as sources of the electric field $\mathbf{E}(\mathbf{x}, t)$ and the magnetic field $\mathbf{B}(\mathbf{x}, t)$. The equations connecting these various quantities are the celebrated Maxwell's equations:

$$\nabla \cdot \mathbf{E} = \frac{\rho}{\epsilon_0}, \tag{1.1}$$

$$\nabla \cdot \mathbf{B} = 0, \tag{1.2}$$

$$\nabla \times \mathbf{B} = \mu_0 \mathbf{j} + \epsilon_0 \mu_0 \frac{\partial \mathbf{E}}{\partial t}, \tag{1.3}$$

$$\nabla \times \mathbf{E} = -\frac{\partial \mathbf{B}}{\partial t}, \tag{1.4}$$

where ϵ_0 and μ_0 are respectively known as the permittivity and the permeability of free space. Many of you may know that there can be two vectors $\mathbf{E}$ and $\mathbf{D}$ associated with the electric field as well as two vectors $\mathbf{B}$ and $\mathbf{H}$ associated with the magnetic field. You may also be familiar with the version of Maxwell's equations in a material medium involving all these vectors. The equations of the form (1.1)–(1.4) are often referred to as Maxwell's equations in a vacuum or in free space. So you may wonder whether these equations can be taken as the completely general basic equations of electromagnetism. A full discussion of this question will be presented later when we discuss electric and magnetic fields in the material medium (in Sects. 2.14 and 3.6). Here we give only a brief answer. Sometimes, it is useful to distinguish the charges and currents in the conductors from the charges and currents induced in the medium. Only when we make these distinctions, it becomes useful to consider two electric vectors $\mathbf{E}$ and $\mathbf{D}$ and two magnetic vectors $\mathbf{B}$ and $\mathbf{H}$. On the other hand, if the charges and currents induced in the medium are included in ρ and $\mathbf{j}$ along with the charges and currents in the conductors, then (1.1)–(1.4) are the *complete* equations valid in all macroscopic situations.

Maxwell's equations tell us how electric and magnetic fields arise from charges and currents. To complete the formulation of the basic principles, we also need to know how the electromagnetic field acts upon charges and currents. This is given by the Lorentz force equation. In this book, we shall have to deal with two different versions of the Lorentz force equation: giving the force on a moving charged particle and giving the force on a continuous distribution of charges and currents. We now write down both these versions. The first version, giving the force $\mathbf{F}$ on a discrete point charge q moving with velocity $\mathbf{v}$, is

$$\mathbf{F} = q \left(\mathbf{E} + \mathbf{v} \times \mathbf{B} \right). \tag{1.5}$$

The second version gives the force *per unit volume* $\mathbf{F}_v$ on a distribution of charges and currents:

$$\mathbf{F}_v = \rho\,\mathbf{E} + \mathbf{j} \times \mathbf{B}. \tag{1.6}$$

To show that (1.5) and (1.6) give essentially the same physics, let us consider the case of many charged particles in a region of space. If N is the number density of the charged particles moving with velocity $\mathbf{v}$, it is easy to see that

$$\rho = Nq, \quad \mathbf{j} = Nq\mathbf{v}. \tag{1.7}$$

Multiplying (1.5) by N and keeping in mind that $\mathbf{F}_v = N\mathbf{F}$, we readily arrive at (1.6).

A complete formulation of electromagnetic theory in a particular situation is obtained by combining (1.1)–(1.4) with either (1.5) or (1.6). Much of this book will be devoted to working out various consequences of this combined set of equations. However, we shall point out a curious fact to the reader. As we shall discuss in Sect. 7.7, this combination of equations *violates* one of the most sacred principles of physics: the conservation of energy. We shall see that one remarkable consequence of the basic principles of electromagnetism is that an accelerated charge will emit electromagnetic radiation. If that is the case, then we would expect that this acceler- ated charge will keep losing energy (giving rise to what is called *radiation damping*). The basic equations of electromagnetic theory which we have written down do not account for this loss of energy! We shall discuss in Sect. 7.7 the efforts of treating radiation damping in an ad hoc manner. However, a fully satisfactory formulation of the radiation damping has so far eluded physicists. This shows that sometimes we encounter unexpected surprises even when dealing with what we consider to be well- understood basic principles of physics. As it happens, radiation damping becomes really important when we deal with electromagnetic phenomena having time scales less than 10^{-23} s in the case of an electron. Keep in mind that the period of visible light is of order 10^{-15} s. This means that we can safely neglect radiation damping in most of the situations we deal with. But it is certainly a bothersome fact that our formulation of basic electromagnetism is not fully consistent at a conceptual level.

Concerning the equations of electromagnetism such as Maxwell's equations (1.1)– (1.4) and the force equation (1.5) or (1.6), one important question is the following: in which frames of reference do these equations hold? Especially, since (1.5) involves the velocity $\mathbf{v}$ of the moving charge, we need to know the frame with respect to which this velocity is specified. As we shall discuss in Chap. 5, the issue of frame of reference in electromagnetic theory is deeply connected with the concepts of special relativity. We shall show in Sects. 5.9 and 5.11 that the basic equations of electromagnetism hold in all inertial frames. Till Chap. 4, we shall proceed merely by assuming that there are some frames in which the basic equations of electromagnetism hold, and any consequences of these equations should be applicable in these frames. Especially, we shall assume that our laboratory provides a frame of reference in which the basic laws of electromagnetism hold to a very high degree of precision.

1.3 Electrostatics and Magnetostatics

It may be noted that Eqs. (1.3) and (1.4) involve both the electric field $\mathbf{E}$ and the magnetic field $\mathbf{B}$. In each of these equations, one of the fields appears in a time derivative term. In other words, when there are variations with time, the fields $\mathbf{E}$ and $\mathbf{B}$ are coupled to each other, and it is not meaningful to consider $\mathbf{E}$ alone or $\mathbf{B}$ alone. In a static situation, however, the electric field $\mathbf{E}$ and the magnetic field $\mathbf{B}$ separate out neatly. On putting the time derivative terms equal to zero, (1.1) and (1.4) give

$$\nabla . \mathbf{E} = \frac{\rho}{\epsilon_0}, \tag{1.8}$$

$$\nabla \times \mathbf{E} = 0 \tag{1.9}$$

which involve the electric field $\mathbf{E}$ alone. On the other hand, (1.2) and (1.3) give

$$\nabla . \mathbf{B} = 0, \tag{1.10}$$

$$\nabla \times \mathbf{B} = \mu_0 \mathbf{j} \tag{1.11}$$

involving $\mathbf{B}$ alone. In a static situation, we thus see that the electric field $\mathbf{E}$ and the magnetic field $\mathbf{B}$ satisfy two completely separated sets of equations which are not coupled to each other. If we have a static distribution of charges, then we can solve (1.8) and (1.9) to find the electric field $\mathbf{E}$. How to do this is the subject of electrostatics. On the other hand, if we have a static distribution of currents, then we can solve (1.10) and (1.11) to find the magnetic field $\mathbf{B}$. This is the subject of magnetostatics. Chapters 2 and 3 will respectively deal with electrostatics and magnetostatics.

We shall now present a general discussion which encompasses both electrostatics and magnetostatics as special cases. Let us consider the vector field $\mathbf{G}$ of which the divergence and the curl are given the equations

$$\nabla . \mathbf{G} = s, \tag{1.12}$$

$$\nabla \times \mathbf{G} = \mathbf{c}. \tag{1.13}$$

Suppose $s(\mathbf{x})$ and $\mathbf{c}(\mathbf{x})$ are given to us. We are assuming a static situation in which $s(\mathbf{x})$ and $\mathbf{c}(\mathbf{x})$ do not vary with time. We now want to solve (1.12) and (1.13) to find out $\mathbf{G}$ for a given distribution of $s(\mathbf{x})$ and $\mathbf{c}(\mathbf{x})$. We shall now show that it is possible to obtain a general solution of this problem. One fact is obvious. If we are able to find the general solution of (1.12)–(1.13), then the solution of the electrostatics equations (1.8)–(1.9) will follow as a special case of that, whereas the solution of the magnetostatics equations (1.10)–(1.11) will follow as a different special case. We need a mathematical tool to discuss the solution of (1.12)–(1.13). This very useful mathematical tool is presented in the next section, after which we shall discuss the solution of (1.12)–(1.13) in Sect. 1.5.

1.4   A Useful Representation of the Dirac δ-Function

We assume that the readers are familiar with the Dirac δ-function. For the sake of recapitulation, let us mention that the Dirac δ-function defined in three-dimensional space satisfies the following mathematical properties:

$$\delta(\mathbf{x} - \mathbf{x}') = 0 \ \text{ if } \ \mathbf{x} \neq \mathbf{x}', \tag{1.14}$$

$$\int f(\mathbf{x}') \, \delta(\mathbf{x} - \mathbf{x}') \, dV' = f(\mathbf{x}), \tag{1.15}$$

where $f(\mathbf{x})$ is an arbitrary function of space, and the integral is a volume integral over all space.

We now want to show that the following is a representation of the Dirac δ-function:

$$-\frac{1}{4\pi} \nabla^2 \left(\frac{1}{|\mathbf{x} - \mathbf{x}'|} \right) = \delta(\mathbf{x} - \mathbf{x}'). \tag{1.16}$$

To establish that this is true in accordance with (1.14) and (1.15), we have to demonstrate that the left-hand side (LHS) of (1.16) is zero at all points except $\mathbf{x} = \mathbf{x}'$ and that the integral of the LHS over all space gives 1.

Since we shall have to consider the gradient and the Laplacian of $1/|\mathbf{x} - \mathbf{x}'|$ throughout this book, it will be useful to work them out properly. If (x, y, z) and (x', y', z') are respectively the coordinates of $\mathbf{x}$ and $\mathbf{x}'$, then

$$|\mathbf{x} - \mathbf{x}'| = [(x - x')^2 + (y - y')^2 + (z - z')^2]^{1/2} \tag{1.17}$$

so that

$$\nabla \left(\frac{1}{|\mathbf{x} - \mathbf{x}'|} \right) = \left(\hat{\mathbf{e}}_x \frac{\partial}{\partial x} + \hat{\mathbf{e}}_y \frac{\partial}{\partial y} + \hat{\mathbf{e}}_z \frac{\partial}{\partial z} \right) \left[\frac{1}{\sqrt{(x - x')^2 + (y - y')^2 + (z - z')}} \right].$$

Noting that differentiation with respect to x, y or z operates only on the unprimed position coordinates (x, y, z) and not on the primed position coordinates (x', y', z'), we easily find that

$$\nabla \left(\frac{1}{|\mathbf{x} - \mathbf{x}'|} \right) = -\frac{\mathbf{x} - \mathbf{x}'}{|\mathbf{x} - \mathbf{x}'|^3}. \tag{1.18}$$

The Laplacian of $1/|\mathbf{x} - \mathbf{x}'|$ is given by the divergence of this, i.e.

$$\nabla^2 \left(\frac{1}{|\mathbf{x} - \mathbf{x}'|} \right) = \frac{\partial}{\partial x} \left(\frac{-(x-x')}{[(x-x')^2+(y-y')^2+(z-z')^2]^{3/2}} \right)$$

$$+ \frac{\partial}{\partial y} \left(\frac{-(y-y')}{[(x-x')^2+(y-y')^2+(z-z')^2]^{3/2}} \right)$$

$$+ \frac{\partial}{\partial z} \left(\frac{-(z-z')}{[(x-x')^2+(y-y')^2+(z-z')^2]^{3/2}} \right).$$

The first term in the right-hand side (RHS) is easily seen to be

$$\frac{\partial}{\partial x} \left(\frac{-(x-x')}{[(x-x')^2+(y-y')^2+(z-z')^2]^{3/2}} \right)$$
$$= \frac{-[(x-x')^2+(y-y')^2+(z-z')^2]+3(x-x')^2}{[(x-x')^2+(y-y')^2+(z-z')^2]^{5/2}}$$

Evaluating the other two terms similarly and adding, we get

$$\nabla^2 \left(\frac{1}{|\mathbf{x} - \mathbf{x}'|} \right) = \frac{0}{[(x-x')^2+(y-y')^2+(z-z')^2]^{5/2}}. \tag{1.19}$$

Since the denominator of (1.19) is non-zero at all points except $\mathbf{x} = \mathbf{x}'$, it easily follows that the LHS of (1.16) is equal to zero at all points except $\mathbf{x} = \mathbf{x}'$—which is one requirement (1.14) for its being identified as the δ-function. Only at $\mathbf{x} = \mathbf{x}'$ the RHS of (1.19) is of the form 0/0 and there is a possibility that it may blow up. To understand the significance of this, we consider the other requirement following from (1.15) that the volume integral of the LHS of (1.16) should give 1. Let us now check if this is the case. Applying Gauss's theorem for converting volume integrals to surface integrals, we have

$$-\frac{1}{4\pi} \int \nabla^2 \left(\frac{1}{|\mathbf{x} - \mathbf{x}'|} \right) dV = -\frac{1}{4\pi} \int \nabla \cdot \left[\nabla \left(\frac{1}{|\mathbf{x} - \mathbf{x}'|} \right) \right] dV$$
$$= -\frac{1}{4\pi} \oint \nabla \left(\frac{1}{|\mathbf{x} - \mathbf{x}'|} \right) \cdot d\mathbf{S},$$

where the surface integral is over a surface bounding the volume. Using (1.18), we get

$$-\frac{1}{4\pi} \int \nabla^2 \left(\frac{1}{|\mathbf{x} - \mathbf{x}'|} \right) dV = \frac{1}{4\pi} \oint \frac{(\mathbf{x} - \mathbf{x}') \cdot d\mathbf{S}}{|\mathbf{x} - \mathbf{x}'|^3}.$$

Now, the quantity under the surface integral is nothing but the solid angle $d\Omega$ subtended by the surface element $d\mathbf{S}$ at the point $\mathbf{x}'$. We thus have

$$-\frac{1}{4\pi} \int \nabla^2 \left(\frac{1}{|\mathbf{x} - \mathbf{x}'|} \right) dV = \frac{1}{4\pi} \oint d\Omega = 1,$$

since the total solid angle subtended at a point by a surface surrounding it is 4π. This completes our proof that the LHS of (1.16) is indeed a representation of the Dirac δ-function.

1.5   General Solution of a Vector Field with Given Divergence and Curl

We are now ready to write down the solution of (1.12)–(1.13). We shall first write down the solution and then show that it satisfies (1.12)–(1.13). A vector field $\mathbf{G}$ satisfying (1.12)–(1.13) is given by

$$\mathbf{G} = -\nabla\Phi + \nabla \times \mathbf{A}, \tag{1.20}$$

where Φ and $\mathbf{A}$ are respectively

$$\Phi(\mathbf{x}) = \frac{1}{4\pi} \int \frac{s(\mathbf{x}')}{|\mathbf{x} - \mathbf{x}'|}\, dV', \tag{1.21}$$

$$\mathbf{A}(\mathbf{x}) = \frac{1}{4\pi} \int \frac{\mathbf{c}(\mathbf{x}')}{|\mathbf{x} - \mathbf{x}'|}\, dV'. \tag{1.22}$$

The fact that (1.20) with (1.21)–(1.22) satisfies (1.12)–(1.13) is often referred to as the *Helmholtz theorem*. Now our job is to show that the divergence and the curl of the vector field $\mathbf{G}$ as given by (1.20)–(1.22) are indeed $s(\mathbf{x})$ and $\mathbf{c}(\mathbf{x})$ as we have in (1.12)–(1.13). Even if this is the case, one may raise the question if this is a unique solution of (1.12)–(1.13) or if other solutions are possible. We shall not get into a detailed analysis of this question here. Let us say only this: If $s(\mathbf{x}')$ and $\mathbf{j}(\mathbf{x}')$ have non-zero values only within a finite region of space and if we demand the boundary condition that $\Phi(\mathbf{x})$ and $\mathbf{A}(\mathbf{x})$ should both go to zero at infinity, then the solution of (1.12)–(1.13) is unique and is given by (1.20)–(1.22).

Keeping in mind that the divergence of a curl is zero, we get on taking the divergence of (1.20):

$$\nabla.\mathbf{G} = -\nabla^2\Phi = -\frac{1}{4\pi} \int \nabla^2\left(\frac{s(\mathbf{x}')}{|\mathbf{x} - \mathbf{x}'|}\right) dV'$$

on making use of (1.21). Note that the operation with ∇^2 involves differentiations with respect to only the unprimed variables x, y, or z. So $s(\mathbf{x}')$ can be treated like a constant when carrying on this differentiation. As a result, we have

$$\nabla.\mathbf{G} = -\frac{1}{4\pi} \int s(\mathbf{x}')\, \nabla^2\left(\frac{1}{|\mathbf{x} - \mathbf{x}'|}\right) dV'.$$

On making use of (1.16), we get

$$\nabla . \mathbf{G} = \int s(\mathbf{x}') \, \delta(\mathbf{x} - \mathbf{x}') \, dV' = s(\mathbf{x})$$

according to (1.15). We thus see that $\mathbf{G}$ defined through (1.20)–(1.22) satisfies (1.12).

Now our job is to take the curl of $\mathbf{G}$ and show that it satisfies (1.13). Keeping in mind that the curl of a gradient is zero, we have from (1.20)

$$\nabla \times \mathbf{G} = \nabla \times (\nabla \times \mathbf{A}) = \nabla(\nabla . \mathbf{A}) - \nabla^2 \mathbf{A} \qquad (1.23)$$

according to the standard vector identity (B.12). Making use of (1.22), the second term in (1.23) can be treated exactly as in the discussion above and can easily be shown to be $\mathbf{c}(\mathbf{x})$. In order to arrive at (1.13), we now merely have to show that the first term $\nabla(\nabla . \mathbf{A})$ in the RHS of (1.23) with $\mathbf{A}$ given by (1.22) is zero. We leave it for the readers to show this (Exercise 1.1). Readers may also look at Panofsky and Phillips [1, p. 5].

We thus complete our discussion that $\mathbf{G}$ defined by (1.20)–(1.22) satisfies the Eqs. (1.12)–(1.13). If the divergence and the curl of a vector field are given, we can write down its solution in this manner. We are now ready to write down the solution of the electrostatic equations (1.8)–(1.9) and the solution of the magnetostatic equations (1.10)–(1.11). This is done in the next two chapters.

1.6 Concluding Remarks

As we already pointed out, Maxwell's equations (1.1)–(1.4) along with the force equation (1.5) or (1.6) constitute the basics of electromagnetism. Discussions in the remainder of the book will be based on these equations. In static situations, we have seen that the equations of the electric field $\mathbf{E}$ and the magnetic field $\mathbf{B}$ neatly separate out, giving the Eqs. (1.8)–(1.9) of electrostatics and the Eqs. (1.10)–(1.11) of magnetostatics. We find that both these sets of equations are of the nature that the divergence and the curl of a vector field are given. We have discussed how to write down the solution of such a vector field of which the divergence and the curl are given. This is going to be our starting point in the next two chapters devoted to electrostatics and magnetostatics, respectively. After this analysis of the static situation in Chaps. 2 and 3, we shall start discussing time variations from Chap. 4. In order to first consider some of the simplest consequences of the time-varying situation, we shall assume $\rho = 0$ and $\mathbf{j} = 0$ in several sections of Chap. 4 (such as Sects. 4.4 and 4.6–4.8). We shall show that Maxwell's equations with $\rho = 0$ and $\mathbf{j} = 0$ lead to the famous prediction of electromagnetic waves propagating in free space with speed $c = (\epsilon_0 \mu_0)^{-1/2}$. Since one of the basic postulates of special relativity is that light (or rather an electromagnetic wave) propagates with the same speed c in all inertial frames, it is clear that there has to be some deep connection between electromagnetism and special relativity. This is discussed in Chap. 5, deriving the formulae for the transformation of electromagnetic fields from one inertial frame of reference to another. Only in Chap. 6 shall we consider in some detail the situation where no term in Maxwell's equations (1.1)–(1.4) is zero and shall show that even this completely general situation can be treated by elegant mathematical methods. One important result in this general situation with no term zero in Maxwell's equations is the emission of electromagnetic radiation, which will be the main theme of Chap. 7. The last chapter will be devoted to a very important modern application of electromagnetic theory: plasma physics—a plasma being a system consisting of negatively charged electrons and positively charged ions.

The ancient Greeks are credited with the discovery of both magnetism and electricity. They noticed that lodestones attract iron, leading them to conclude the existence of magnetism. They also noticed that when amber is rubbed against fur, they attract each other, providing the first proof of the existence of electricity. Until the beginning of the nineteenth century, magnetism and electricity were thought to be two completely disconnected phenomena. After Volta made the first primitive version of a battery around 1800 and succeeded in making electric currents flow through conducting wires, the connection between electricity and magnetism was just waiting to be discovered. Oersted found in 1820 that a current-carrying wire deflected a magnetic needle placed near it, leading to the conclusion that electric currents cause magnetic fields. This led to the first great synthesis in this field that electricity and magnetism, which were earlier regarded as distinct entities, were combined together into the science of electromagnetism. Since currents involve moving charges, Oersted's discovery basically showed that moving charges give rise to magnetic fields. Then, in 1831, Faraday discovered electromagnetic induction, which implies that moving magnets also give rise to electric fields. Finally Maxwell put the known laws of electromagnetism in the set of elegant mathematical equations known after him. His own original contribution was the addition of the displacement current term for the sake of consistency, leading to the famous result in 1865 that light is an electromagnetic wave. This is the second great synthesis in this field that combined electromagnetism and optics. Finally, various conceptual issues connected with electromagnetic theory led Einstein to formulate the special theory of relativity in 1905. In the very first paragraph of Einstein's first paper on relativity, titled 'On the electrodynamics of moving bodies', some of these conceptual issues were pointed out (Lorentz, Einstein, Minkowski and Weyl [2], p. 37). The formulation of special relativity finally settled the issue of finding the appropriate frames of reference in which the basic equations of electromagnetic theory hold—a vexing question which had bothered physicists till that time. The basic equations of electromagnetic theory were shown to hold in all inertial frames of reference.

Out of the two long-range fundamental interactions in physics—gravity and electromagnetism—physicists recognized gravity to have a universal character quite early. All material substances take part in gravitational interaction. On the other hand, only a few substances display obvious electromagnetic characteristics. Many substances around us—like wood or mud—seem to lack any electromagnetic properties. Till the end of the nineteenth century, physicists did not think of electromagnetism as something universal. Only after Rutherford established the nuclear model of the atom in 1911, and all elementary particles were subsequently discovered to have tiny magnetic moments, was it realized that electromagnetism is also universal and is the fundamental interaction responsible for the formation of atoms. Substances which do not show any electromagnetic characteristics at the macroscopic level merely have their atomic level electric charges and magnetic moments fantastically balanced inside them. It is a measure of the genius of the nineteenth-century physicists that they discovered almost all the important laws of electromagnetism before its universal character was realized. When twentieth-century physicists realized the universal nature of electromagnetism, they had all the principles and equations in their hands ready for application.

Large bodies like planets and stars are electrically neutral so that their electromagnetic effects are largely screened off (though some planets and stars have large-scale magnetic fields). Although gravity is the weakest of the fundamental interactions, it becomes the dominant interaction at the scale of stars and galaxies because it cannot be screened off. On the one hand, we have two fundamental interactions—strong and weak—dominating the physics of the atomic nucleus. On the other hand, gravity dominates the physics of very large systems. It is electromagnetism—the remaining of the only four known fundamental interactions—which is responsible for various phenomena at the intermediate scale of atoms, molecules and solid bodies. However, to study structures of atoms and molecules, we have to combine electromagnetism with quantum mechanics, which often leads to surprising results which we might not have anticipated from classical considerations. For example, in order to understand why two hydrogen atoms combine to form a hydrogen molecule, we have to consider the combined wave function of the two electrons inside the molecule. A quantum mechanical treatment of this system shows that electromagnetism can give rise to what is called the exchange interaction that provides the valency inside the hydrogen molecule (Pauling and Wilson [3], pp. 340–345). In this book, we shall not address such issues and shall restrict ourselves completely to the classical aspects of electromagnetic theory. Classical electromagnetic theory can be pushed to microscopic levels to some extent, as long as wave functions of the constituent particles do not overlap with each other. This is the case with a dilute plasma made up of negatively charged electrons and positively charged ions. Since the wave functions of electrons and ions would not overlap in a dilute plasma, we can treat them as classical particles. The last chapter of this book is devoted to the basics of plasma physics. We shall also present brief discussions of the microscopic theories of dielectric materials (Sect. 2.18), magnetic materials (Sect. 3.8) and optical dispersion (Sect. 4.9) based on classical considerations.

Exercises

1.1 Show that the term $\nabla(\nabla.\mathbf{A})$ appearing in (1.23) is zero if $\mathbf{A}$ is given by (1.22). *Hint.* You will have to make use of the fact

$$\nabla\left(\frac{1}{|\mathbf{x} - \mathbf{x}'|}\right) = -\nabla'\left(\frac{1}{|\mathbf{x} - \mathbf{x}'|}\right).$$

1.2 Show that the following is a representation of the Dirac δ-function:

$$\delta(x - x') = \frac{1}{2\pi} \int_{-\infty}^{\infty} e^{ik(x-x')} dk.$$

References

1. Panofsky, W.K.H., Phillips, M.: Classical Electricity and Magnetism, 2nd edn. Addison–Wesley (reprinted by Dover) (1962)
2. Lorentz, H.A., Einstein, A., Minkowski, H., Weyl, H.: The Principle of Relativity. Methuen and Co. (Reprinted by Dover) (1923)
3. Pauling, L., Wilson, E.B.: Introduction to Quantum Mechanics with Applications to Chemistry. McGraw Hill (reprinted by Dover) (1935)

Chapter 2
Electrostatics

2.1 Coulomb's Law

One of the central concerns of electrostatics is to find the electric field $\mathbf{E}(\mathbf{x})$ due to a static charge distribution $\rho(\mathbf{x})$. We have already pointed out that a static electric field satisfies the Eqs. (1.8) and (1.9). We also have discussed in Sect. 1.5 how we can write down the solution of the vector field $\mathbf{G}$ of which the divergence and curl are respectively s and $\mathbf{c}$, as given by (1.12)–(1.13). Substituting $\mathbf{G} = \mathbf{E}$, $s = \rho/\epsilon_0$, and $\mathbf{c} = 0$ (implying that $\mathbf{A} = 0$) in (1.20)–(1.22), we get

$$\mathbf{E} = -\nabla\Phi, \tag{2.1}$$

where Φ known as the *electrostatic potential* is given by

$$\Phi(\mathbf{x}) = \frac{1}{4\pi\epsilon_0} \int \frac{\rho(\mathbf{x}')}{|\mathbf{x} - \mathbf{x}'|} \, dV'. \tag{2.2}$$

Throughout this book, we shall use the following convention. The point in space where we consider a field is denoted by the unprimed position coordinate $\mathbf{x}$, whereas the position of the source is denoted by the primed coordinate $\mathbf{x}'$. In order to obtain the electric field, we have to take the gradient of (2.2). Keeping in mind that this operation will work only on $\mathbf{x}$ (and not on $\mathbf{x}'$), and making use of (1.18), we find from (2.1) and (2.2) that

$$\mathbf{E}(\mathbf{x}) = \frac{1}{4\pi\epsilon_0} \int \rho(\mathbf{x}') \frac{\mathbf{x} - \mathbf{x}'}{|\mathbf{x} - \mathbf{x}'|^3} \, dV'. \tag{2.3}$$

We often have to consider a point charge. The charge density corresponding to a point charge q_1 at a point $\mathbf{x}_1$ can be written as

$$\rho(\mathbf{x}') = q_1 \, \delta(\mathbf{x}' - \mathbf{x}_1). \tag{2.4}$$

© Springer Nature Singapore Pte Ltd. 2022
A. R. Choudhuri, *Advanced Electromagnetic Theory*, Lecture Notes in Physics 1009,
https://doi.org/10.1007/978-981-19-5944-8_2

On substituting this in (2.2) and (2.3), we get

$$\Phi(\mathbf{x}) = \frac{1}{4\pi\epsilon_0} \frac{q_1}{|\mathbf{x} - \mathbf{x}_1|} \tag{2.5}$$

and

$$\mathbf{E}(\mathbf{x}) = \frac{1}{4\pi\epsilon_0} q_1 \frac{\mathbf{x} - \mathbf{x}_1}{|\mathbf{x} - \mathbf{x}_1|^3} \tag{2.6}$$

giving the electric field at $\mathbf{x}$ produced by the charge q_1 at $\mathbf{x}_1$. Let us now figure out the force $\mathbf{F}_{12}$ which will be exerted by this charge q_1 at $\mathbf{x}_1$ on another charge q_2 located at $\mathbf{x}_2$. We conclude from (1.5) that in an electrostatic situation we shall have

$$\mathbf{F}_{12} = q_2 \, \mathbf{E}(\mathbf{x}_2).$$

Substituting from (2.6), we get

$$\mathbf{F}_{12} = \frac{1}{4\pi\epsilon_0} q_1 q_2 \frac{\mathbf{x}_2 - \mathbf{x}_1}{|\mathbf{x}_2 - \mathbf{x}_1|^3}. \tag{2.7}$$

This is the celebrated *Coulomb's law*—the central law in electrostatics. The constant ϵ_0 is called the *permittivity of free space*. In elementary textbooks on electromagnetic theory, the discussion on electrostatics usually begins with this equation, and then Maxwell's equations are introduced at a later stage. Here we follow the opposite approach. We see that Maxwell's equations lead to (1.8)–(1.9) in a static situation. Then we show that these equations combined with (1.5) give rise to Coulomb's law. Thus, in our approach, Coulomb's law is an outcome of the basic principles as encapsulated in Maxwell's equations along with (1.5).

It is easy to check from (2.7) that the force between two charges is repulsive if q_1 and q_2 are of the same sign, whereas the force is attractive if the two charges are of the opposite sign. Throughout this book, we shall be using the SI units in which the charge is denoted in coulomb (C), length in metre (m) and force in newton (N). We may point out that ampere (A), the unit of current, is taken as the fundamental electromagnetic unit in the SI system. How ampere is defined will be pointed out in Sect. 3.2. Once ampere is defined, we can define coulomb as the amount of electric charge passing in 1 s through a cross section of a wire carrying current of 1 A. With this definition of coulomb, the constant factor appearing in (2.7) can be experimentally found. The value of ϵ_0 can also be found from a measurement of the speed of light c, as we discuss in Sect. 4.4. In fact, using the accurate value of c is the standard convention for introducing ϵ_0 in the SI system. It is found that

$$\frac{1}{4\pi\epsilon_0} = 8.99 \times 10^9 \, \mathrm{N \, m^2 \, C^{-2}}. \tag{2.8}$$

It may be worthwhile to keep in mind that coulomb is a rather large unit compared to the typical charges we may encounter in a laboratory situation. Two charges of 1 C kept 1 m away would repel each other with an enormous force of nearly 9×10^9 N, which is about 10^5 times the force exerted on a mass of 1 kg in the earth's gravitational field. We do not come across such large electrostatic forces in the laboratory because we deal with much smaller charges. It is easy to check that a conducting sphere of radius 9 cm kept at a voltage of 100 V would have a charge of only 10^{-9} C on it. We encounter charges of the order of C only in extreme natural phenomena occurring over scales much larger than the size of a typical laboratory. A lightning strike in a thunderstorm involves a transfer of 20–30 C between a cloud and the ground.

The SI unit system was introduced with the aim of ensuring that the various basic units are of the order of the physical quantities we encounter in the laboratory. Ampere, the unit of current, is indeed defined in such a manner that it is of the order of currents we typically come across around us. If the unit of current is of the order of currents we encounter, then why does the corresponding unit of charge turn out to be so much larger than the charges we encounter around us? It may superficially seem that this question has nothing to do with special relativity. Surprisingly, we need relativistic considerations to answer this question, which will be addressed in Sect. 5.10.1.

2.2 Electrostatic Potential as Potential Energy

If the charge distribution is confined within a certain region of space, then it follows easily from (2.2) that the electrostatic potential will be zero at infinity. In such a situation, one can interpret the electrostatic potential in terms of the potential energy. If we have to bring a charge q from infinity to a point $\mathbf{x}$ inside an electrostatic field, the work by the electrostatic field in this process is clearly

$$W = q \int_\infty^{\mathbf{x}} \mathbf{E} \cdot d\mathbf{x} = -q \int_\infty^{\mathbf{x}} \nabla\Phi \cdot d\mathbf{x} = -q \left[\Phi(\mathbf{x}) - \Phi(\infty) \right].$$

If $\Phi(\infty) = 0$, then the work done is clearly $W = -q\,\Phi(\mathbf{x})$.

As a result of this work done, the electrostatic field would lose energy W. This means that the potential energy of the system after the charge has been brought to $\mathbf{x}$ is

$$U = -W = q\,\Phi(\mathbf{x}). \tag{2.9}$$

This simple argument implies that a charge kept at a point $\mathbf{x}$ has the potential energy $q\Phi(\mathbf{x})$, provided Φ is defined in such a way that it goes to zero at infinity. One can think of the electrostatic potential as the potential energy per unit charge. It should be clear from this discussion that the SI unit of electrostatic potential is J C^{-1}, which is also named volt (V). It follows from (2.1) that the unit of electric field is V m^{-1}, which is equivalent to N C^{-1}.

2.3 Poisson's and Laplace's Equations

We may point out that on substituting (2.1) in (1.8), we get

$$\nabla^2 \Phi = -\frac{\rho}{\epsilon_0}, \tag{2.10}$$

which is *Poisson's equation*. When we apply this to a region of space where the charge density ρ is zero, we get *Laplace's equation*

$$\nabla^2 \Phi = 0. \tag{2.11}$$

Much of this chapter will be devoted to discussing situations in which electric charges are confined within limited regions—usually on the surfaces of conductors. In the surrounding space, Laplace's equation (2.11) holds. Solving Laplace's equation is of central concern in electrostatics. We shall discuss how Laplace's equation (2.11) can be solved by various mathematical techniques which prove extremely useful in many other areas of physics as well.

Even when we expand ∇^2 in any appropriate coordinate system, each term in Laplace's equation (2.11) is going to be linear in Φ. Such an equation is called a *homogeneous* partial differential equation. On the other hand, Poisson's equation (2.10) has the term $-\rho/\epsilon_0$, which does not involve Φ. In fact, we can think of ρ as the source which gives rise to Φ. Poisson's equation is an example of an *inhomogeneous* partial differential equation. A standard method of solving inhomogeneous differential equations is through Green's function (see, for example, Arfken, Weber and Harris [1], Chap. 10). We shall discuss this method in Sect. 6.2. From our discussion so far, we see that (2.2) gives the solution of (2.10). Although we have not introduced the term Green's function, the Green's function method would indeed give (2.2) as the solution of (2.10), which will become clear from the discussion of Sect. 6.2.

2.4 Electric Field Due to a Dipole and a Surface Dipole Layer

We sometimes have to find the electric field due to a static charge distribution. We can do this by using (2.3), which requires first calculating electric fields due to different parts of the charge distribution and then summing them up. This requires a vector summation (or integration), since the electric field produced by any part of the charge distribution is a vector. Sometimes, it is easier to handle such problems by making use of (2.2), which requires that we first calculate electrostatic potentials due to different parts of the charge distribution. Since the electrostatic potential is a scalar, we have to do a scalar summation (or integration) to find out the electrostatic potential due to

Fig. 2.1 A sketch indicating
an electric dipole with
charges q and $-q$ separated
by distance **s**. We wish to find
the electric field at a point at
distance **r** from the charge q

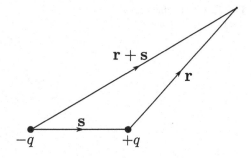

the whole charge distribution. This is usually easier to do than a vector summation
(or integration). Once we have found the resultant electrostatic potential, it is easy
to find the electric field by using (2.1). We shall now illustrate this with some simple
examples.

Let us consider an electric dipole consisting of a positive charge q and a negative
charge $-q$ separated by **s**. We use the convention that the vector **s** points from the
negative charge towards the positive charge. We now want to find the electric field at
a point at distance **r** away from the positive charge q, as shown in Fig. 2.1. This point
is clearly at a distance $\mathbf{r} + \mathbf{s}$ away from the negative charge $-q$. As pointed out above,
the first step in finding the electric field due to this dipole will be the determination
of the electrostatic potential due to the dipole. The electrostatic potential at the field
point due to the positive charge q clearly is

$$\Phi_q(\mathbf{r}) = \frac{1}{4\pi\epsilon_0}\frac{q}{|\mathbf{r}|} = \frac{1}{4\pi\epsilon_0}\frac{q}{r} \tag{2.12}$$

according to (2.5). The electrostatic potential due to the negative charge $-q$ will
obviously be $-\Phi_q(\mathbf{r} + \mathbf{s})$ so that the total electrostatic potential due to the dipole is

$$\Phi_{\text{dipole}} = \Phi_q(\mathbf{r}) - \Phi_q(\mathbf{r} + \mathbf{s}) = -\mathbf{s}.\nabla\Phi_q.$$

On substituting for Φ_q from (2.12) and making use of the expression of gradient in
spherical coordinates, which will be discussed in Sect. 2.7 and is also listed as (C.5),

$$\nabla\Phi = \frac{\partial\Phi}{\partial r}\hat{\mathbf{e}}_r + \frac{1}{r}\frac{\partial\Phi}{\partial\theta}\hat{\mathbf{e}}_\theta + \frac{1}{r\sin\theta}\frac{\partial\Phi}{\partial\phi}\hat{\mathbf{e}}_\phi, \tag{2.13}$$

we get

$$\Phi_{\text{dipole}} = \frac{1}{4\pi\epsilon_0}\mathbf{s}.\frac{q\mathbf{r}}{r^3} = \frac{1}{4\pi\epsilon_0}\frac{\mathbf{p}.\mathbf{r}}{r^3}, \tag{2.14}$$

where $\mathbf{p} = q\mathbf{s}$ is the electric dipole moment.

When $|\mathbf{r}| \gg |\mathbf{s}|$ (which is usually the case we consider), we can think of the
whole dipole of strength **p** located at the origin of the spherical coordinates, and the

Fig. 2.2 Electric field lines
due to an electric dipole

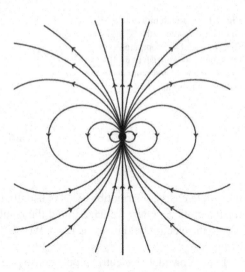

electrostatic potential due to this dipole at some point in space is given by (2.14).
Taking the direction of **p** to be the polar axis, we have $\mathbf{p}.\mathbf{r} = pr \cos\theta$ so that (2.14)
becomes

$$\Phi_{\text{dipole}} = \frac{1}{4\pi\epsilon_0} \frac{p \cos\theta}{r^2}. \tag{2.15}$$

To get the electric field due to an electric dipole **p** located at the origin, we have to
substitute (2.15) into (2.1). On making use of (2.13), we have

$$\mathbf{E}_{\text{dipole}} = \frac{1}{4\pi\epsilon_0} \left[\frac{2p \cos\theta}{r^3} \hat{\mathbf{e}}_r + \frac{p \sin\theta}{r^3} \hat{\mathbf{e}}_\theta \right]. \tag{2.16}$$

The field lines of such an electric field are sketched in Fig. 2.2.

Next, we consider the problem of finding out the electrostatic potential and the
electric field due to a surface layer with uniform electric dipole density τ per unit
area. At first sight, this may appear a little bit of a contrived and artificial problem.
However, when we discuss magnetostatics, we shall point out in Sect. 3.1 that this
problem of electric dipole layer has interesting connections with some key issues in
magnetostatics. Let us consider an area element ds of this surface, which will have
dipole moment $\tau \, ds$ and will produce the electrostatic potential at a point at distance
r from it:

$$d\Phi = \frac{1}{4\pi\epsilon_0} \tau \frac{\mathbf{r}.d\mathbf{s}}{r^3}$$

according to (2.14). Keeping in mind that τ is constant over the surface, the total
electrostatic potential due to the whole surface must be given by

Fig. 2.3 Calculating electric
field due to a uniform
electric dipole surface

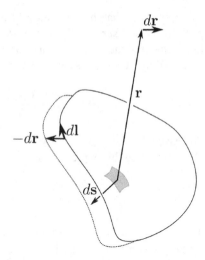

$$\Phi = \frac{1}{4\pi\epsilon_0} \, \tau \int \frac{\mathbf{r}.d\mathbf{s}}{r^3}.$$

When we use the convention that $d\mathbf{s}$ is taken in the outward direction with respect
to the field point (which implies that the dipole moment is also in this direction for
positive τ) and $\mathbf{r}$ is taken from the area element to the field point, it is easy to see that
$\mathbf{r}.d\mathbf{s}/r^3$ should be the negative of the solid angle $d\Omega$ submitted by the area element
to the field point (see Fig. 2.3). Therefore,

$$\Phi = -\frac{\tau}{4\pi\epsilon_0} \int d\Omega = -\frac{\tau}{4\pi\epsilon_0} \Omega, \qquad (2.17)$$

which is a very elegant result, showing that the electrostatic potential Φ produced
by an electric dipole surface at a field point is proportional to the solid angle Ω
subtended by the surface at that point. If τ in (2.17) is positive, then the negatively
charged side of the dipole surface is towards the field point, making the electrostatic
potential there negative.

After obtaining this elegant expression of the electrostatic potential Φ, now our
job is to the find out the electric field. For this purpose, we have to consider the
difference in electrostatic potentials between the points $\mathbf{r} + d\mathbf{r}$ and $\mathbf{r}$, which must be
$d\Phi = -\mathbf{E}.d\mathbf{r}$. Now, $d\Phi$ should also be equal to the difference between electrostatic
potentials due to the given dipole surface and a dipole surface produced by displacing
the given dipole surface by an amount $-d\mathbf{r}$, both at the field point $\mathbf{r}$, as sketched
in Fig. 2.3. If $d\Omega$ is the difference between the solid angles subtended by these two
surfaces at the field point $\mathbf{r}$, then it follows from (2.17) that

$$d\Phi = -\frac{\tau}{4\pi\epsilon_0} d\Omega = -\mathbf{E}.d\mathbf{r}. \qquad (2.18)$$

It is easy to see that the $-d\Omega$ must be equal to the solid angle subtended at the field point by the side surface of which the two dipole surfaces are the two edges. If $d\mathbf{l}$ is a line element of the bounding curve of the dipole surface, then the area of an element produced by $d\mathbf{r}$ and $d\mathbf{l}$ is $d\mathbf{r} \times d\mathbf{l}$ so that

$$d\Omega = \oint \frac{(d\mathbf{r} \times d\mathbf{l}) \cdot \mathbf{r}}{r^3} = \oint \frac{(d\mathbf{l} \times \mathbf{r}) \cdot d\mathbf{r}}{r^3}.$$

We leave it for the reader to check that all the standard sign conventions are followed. On substituting this expression for $d\Omega$ in (2.18), we conclude

$$\mathbf{E} = \frac{\tau}{4\pi\epsilon_0} \oint \frac{d\mathbf{l} \times \mathbf{r}}{r^3}. \tag{2.19}$$

This means that the electric field due to an electric dipole surface can be obtained by carrying on a line integral along the boundary curve of the dipole surface. As we already mentioned, we shall see in Sect. 3.1 that this result has a striking similarity to an important result in magnetostatics giving the magnetic field due to a current-carrying circuit.

2.5 Dipoles in Electromagnetic Theory

In various discussions in this book, we shall encounter dipoles (both electric and magnetic) repeatedly. One well-known implication of (1.10) is that isolated magnetic monopoles cannot exist. As a consequence of this, a magnetic dipole is the simplest kind of magnetic entity that can occur in nature. Even though electric monopoles (i.e. electric charges) can exist by themselves, the concept of an electric dipole is still very important in many practical situations. When an electrically neutral atom or molecule is kept in a strong electric field, it can become distorted and acquire an electric dipole moment.

Let us consider the case of an electric dipole moment $\mathbf{p}$ kept in an electric field $\mathbf{E}$. Based on the fact that a dipole is made up of a positive charge and a negative charge separated by a small distance and that the potential energy of a charge in an electric field is given by (2.9), we leave it for the reader to justify that the potential energy of an electric dipole in an electric field is given by

$$U = -\mathbf{p} \cdot \mathbf{E}. \tag{2.20}$$

Then the force on the dipole would be

$$\mathbf{F} = -\nabla(-\mathbf{p} \cdot \mathbf{E}) = (\mathbf{p} \cdot \nabla)\mathbf{E} + \mathbf{p} \times (\nabla \times \mathbf{E})$$

on making use of the vector identity (B.6) and keeping in mind that $\mathbf{p}$ is not a field variable so that its spatial derivatives would vanish. In view of (1.9) for an electrostatic field, the force on an electric dipole in an electric field is given by

$$\mathbf{F} = (\mathbf{p}.\nabla)\mathbf{E}. \tag{2.21}$$

An electric dipole kept in an electric field experiences a torque also. Again, considering the forces acting on the positive charge and the negative charge making up the dipole, the reader is asked to justify that the torque on an electric dipole in an electric field is given by

$$\mathbf{L} = \mathbf{p} \times \mathbf{E}. \tag{2.22}$$

On the ground of similarity, we expect that a magnetic dipole $\mathbf{p}_m$ in a magnetic field $\mathbf{B}$ will have the potential energy, the force and the torque given by expressions similar to (2.20)–(2.22)—with $\mathbf{p}_m$ replacing $\mathbf{p}$ and $\mathbf{B}$ replacing $\mathbf{E}$.

2.6 Gauss's Law in Electrostatics and Applications

Let us now integrate both sides of (1.8) over a certain volume:

$$\int (\nabla.\mathbf{E})\,dV = \frac{1}{\epsilon_0} \int \rho\,dV.$$

We can use Gauss's theorem in vector analysis (B.13) to convert the LHS into a surface integral over the surface bounding this volume and also note that $\int \rho\,dV$ appearing on the RHS is the total charge Q enclosed within this bounding surface. Hence,

$$\oint \mathbf{E}.d\mathbf{S} = \frac{1}{\epsilon_0}\,Q. \tag{2.23}$$

The LHS of this equation is the total outward electric flux across the closed surface bounding the volume. According to (2.23), this outward electric flux is essentially equal to the total charge enclosed within this surface (divided by ϵ_0). This result (2.23) is often referred to as *Gauss's law in electrostatics*.

This law often provides a quick method for calculating the electric field in highly symmetric situations. To illustrate how it is used, we consider an infinite plane surface having electric charge σ per unit area with free space on both sides, as shown in Fig. 2.4a. From symmetry, we expect the electric field lines to go out on both sides perpendicular to the plane surface. Let us consider a cylindrical box of which the two end surfaces are parallel to the plane surface with the charge. If A is the cross-sectional area of this cylinder, it is obvious that the total electric flux $\oint \mathbf{E}.d\mathbf{S}$ out of the bounding surfaces of the cylindrical box is $2EA$. The total charge included within this volume is $Q = \sigma A$. According to (2.23), we have

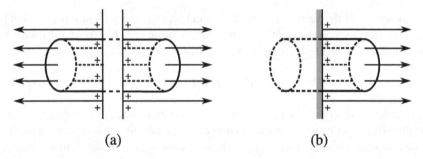

Fig. 2.4 An infinite plane surface having surface charge density σ, shown with a hypothetical box-like volume used for calculating the electric field due to the charged surface. **a** There is free space on both sides of the charged surface. **b** The charged surface is the surface of a conductor with free space on one side

$$2EA = \frac{1}{\epsilon_0}\sigma A,$$

from which

$$E = \frac{\sigma}{2\epsilon_0}. \tag{2.24}$$

Apart from a point charge and an electric dipole, one other kind of system that we encounter repeatedly in our study of electrostatics is an electrical conductor. If there is an unbalanced electric field inside a conductor, that would lead to a current involving the movement of electric charges and would cause a non-static situation. So, while considering electrostatics, we require $\mathbf{E} = 0$ inside a conductor. It then follows from (1.8) that the electric charge density ρ has to be zero inside a conductor in a static situation. However, there can be a surface charge density at the surface of a conductor. Since it follows from (2.1) that the electrostatic potential Φ inside a conductor must be constant in order to make the electric field zero, the surface of a conductor should be a surface of constant electrostatic potential Φ. One implication of (2.1) is that the electric field just outside the surface of the conductor should be normal to the surface. We now want to relate the electric field just outside a conducting surface with the charge density σ on the surface. If we enlarge the region near the surface of a conductor, then the surface would look like a plane, and the electric field just outside it would be what we expect outside a plane conductor with surface charge density σ. As shown in Fig. 2.4b, we can again consider a cylindrical box and apply Gauss's law in electrostatics. Now we have the electric field only on one side of the surface, and the total electrical flux out of the bounding surfaces of this cylindrical box would be EA rather than $2EA$ as in the case discussed earlier. Substituting this in Gauss's law (2.23), we get

$$E = \frac{\sigma}{\epsilon_0}. \tag{2.25}$$

We conclude that the electric field just outside the surface of a conductor is normal to the surface having the value given by (2.25).

2.7 Cylindrical and Spherical Coordinates

In many problems of electromagnetism, it is often convenient to use cylindrical or spherical coordinates depending on the symmetry of the problem. We now present a discussion of these coordinate systems—in particular we discuss how to write down the Laplacian ∇^2 in these systems, which will be very useful in treating many topics. The cylindrical and spherical coordinates are examples of what are called orthogonal coordinate systems. If ds is a small line element, its square in such a coordinate system is given by

$$ds^2 = h_1^2 \, dq_1^2 + h_2^2 \, dq_2^2 + h_3^2 \, dq_3^2, \tag{2.26}$$

where (q_1, q_2, q_3) are the three coordinates in three-dimensional space. If the coordinate system is orthogonal, then there are no cross-terms like $dq_1 dq_2$.

Readers would know that ds^2 in cylindrical coordinates (r, θ, z) is given by

$$ds^2 = dr^2 + r^2 d\theta^2 + dz^2, \tag{2.27}$$

which implies that

$$h_r = 1, \ h_\theta = r, \ h_z = 1 \tag{2.28}$$

for cylindrical coordinates. On the other hand, the line element in spherical coordinates (r, θ, ϕ) is

$$ds^2 = dr^2 + r^2 d\theta^2 + r^2 \sin^2\theta \, d\phi^2, \tag{2.29}$$

from which it follows that

$$h_r = 1, \ h_\theta = r, \ h_\phi = r \sin\theta \tag{2.30}$$

for spherical coordinates.

Since $\nabla^2 \Phi = \nabla(\nabla \Phi)$, we first need to figure out how to express the gradient in curvilinear coordinates and then to figure out how to express the divergence in curvilinear coordinates. That will enable us to write down the Laplacian in curvilinear coordinates. We know that the component of $\nabla \Phi$ in a particular direction is merely the spatial derivative of Φ in that direction. Let us consider the direction in which only one coordinate q_i is varying. According to (2.26), the spatial separation between q_i and $q_i + dq_i$ in that direction is $h_i \, dq_i$. Hence, the component of $\nabla \Phi$ in that direction is

$$(\nabla \Phi)_i = \lim_{dq_i \to 0} \frac{\Phi(q_i + dq_i) - \Phi(q_i)}{h_i dq_i} = \frac{1}{h_i} \frac{\partial \Phi}{\partial q_i}. \tag{2.31}$$

Fig. 2.5 A small volume element in an orthogonal curvilinear coordinate system chosen such that one coordinate is constant over any surface

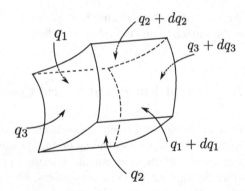

It now clear from (2.30) that (2.13) gives the expression of the gradient in spherical coordinates. The expression of the gradient in cylindrical coordinates follows from (2.28) and is

$$\nabla \Phi = \frac{\partial \Phi}{\partial r} \hat{\mathbf{e}}_r + \frac{1}{r} \frac{\partial \Phi}{\partial \theta} \hat{\mathbf{e}}_\theta + \frac{\partial \Phi}{\partial z} \hat{\mathbf{e}}_z. \tag{2.32}$$

Our next job is to find the divergence of an arbitrary vector field $\mathbf{A}$ in curvilinear coordinates. Let us consider a small volume element, as shown in Fig. 2.5, such that one coordinate is constant over any side surface. We focus our attention on the side surfaces over which the first coordinate has values q_1 and $q_1 + dq_1$, whereas the edges have the lengths $h_2 \, dq_2$ and $h_3 \, dq_3$ so that the area of the surface is $h_2 h_3 \, dq_2 \, dq_3$. An application of Gauss's theorem in vector analysis over this volume element gives

$$\nabla . \mathbf{A} \, dV = \mathbf{A} . \, d\mathbf{S}, \tag{2.33}$$

where $dV = h_1 h_2 h_3 \, dq_1 \, dq_2 \, dq_3$ is the volume element, and $\mathbf{A} . \, d\mathbf{S}$ has to be evaluated over the six side surfaces. The contribution to $\mathbf{A} . \, d\mathbf{S}$ from the two side surfaces we are focusing on is

$$(A_1 h_2 h_3 \, dq_2 \, dq_3)_{q_1 + dq_1} - (A_1 h_2 h_3 \, dq_2 \, dq_3)_{q_1} = \frac{\partial}{\partial q_1} (h_2 h_3 A_1) dq_1 \, dq_2 \, dq_3$$

$$= \frac{1}{h_1 h_2 h_3} \frac{\partial}{\partial q_1} (h_2 h_3 A_1) \, dV.$$

Adding the similar contributions to $\mathbf{A} . \, d\mathbf{S}$ from the other two pairs of side surfaces, we have from (2.33)

$$\nabla . \mathbf{A} = \frac{1}{h_1 \, h_2 \, h_3} \left[\frac{\partial}{\partial q_1} (h_2 h_3 A_1) + \frac{\partial}{\partial q_2} (h_3 h_1 A_2) + \frac{\partial}{\partial q_3} (h_1 h_2 A_3) \right]. \tag{2.34}$$

Finally, let us now take $\mathbf{A}$ to be $\nabla\Phi$ of which the ith component is given by (2.31). We then get

$$\nabla.(\nabla\Phi) = \nabla^2\Phi = \frac{1}{h_1h_2h_3}\left[\frac{\partial}{\partial q_1}\left(\frac{h_2h_3}{h_1}\frac{\partial\Phi}{\partial q_1}\right) + \frac{\partial}{\partial q_2}\left(\frac{h_3h_1}{h_2}\frac{\partial\Phi}{\partial q_2}\right) + \frac{\partial}{\partial q_3}\left(\frac{h_1h_2}{h_3}\frac{\partial\Phi}{\partial q_3}\right)\right].$$
(2.35)

Making use of (2.28), we conclude that the Laplacian in cylindrical coordinates is given by

$$\nabla^2\Phi = \frac{1}{r}\frac{\partial}{\partial r}\left(r\frac{\partial\Phi}{\partial r}\right) + \frac{1}{r^2}\frac{\partial^2\Phi}{\partial\theta^2} + \frac{\partial^2\Phi}{\partial z^2}.$$
(2.36)

On the other hand, (2.30) implies that the Laplacian in spherical coordinates is given by

$$\nabla^2\Phi = \frac{1}{r^2}\frac{\partial}{\partial r}\left(r^2\frac{\partial\Phi}{\partial r}\right) + \frac{1}{r^2\sin\theta}\frac{\partial}{\partial\theta}\left(\sin\theta\frac{\partial\Phi}{\partial\theta}\right) + \frac{1}{r^2\sin^2\theta}\frac{\partial^2\Phi}{\partial\phi^2}.$$
(2.37)

In some of the later sections of this chapter, we shall make use of (2.36) and (2.37).

2.8   Boundary Value Problems and Uniqueness Theorem

We have already discussed a few cases of finding the electrostatic field due to a given distribution of charges. We have considered the following methods of solving this problem.

1. One can consider the electrostatic fields produced by different parts of the charge distribution and then sum them up, in accordance with (2.3). This method involves a vector summation or integration.
2. If one wants to avoid the vector summation, then one can use (2.2) to first calculate the electrostatic potential and then use (2.1) to find the electrostatic field from it. Illustrations are given in Sect. 2.4.
3. If the charge distribution is highly symmetric, then one can use Gauss's law in electrostatics to quickly find the electrostatic field. Illustrations are given in Sect. 2.6.

Now we want to discuss a class of problems known as boundary value problems. Suppose we have a few conductors lying in free space. We have pointed out in Sect. 2.6 that the electric field has to be zero inside the conductors and that electric charges can reside only on their surfaces. If there are electric charges on the surfaces, then electric fields would be produced in the regions of free space where Laplace's equation (2.11) must be satisfied. We can try to find these electric fields by solving Laplace's equation. This equation has to be solved only in the regions of free space, since we know that the electric field is zero inside conductors. The surfaces of the conductors, therefore, provide the boundaries of the region in which we want to

solve Laplace's equation. We naturally need some appropriate boundary conditions
to solve this problem. Sometimes the electrostatic potential on a boundary surface
may be given (if the boundary surface is a continuous conductor, then the potential
over it has to be constant). This type of boundary condition is called the *Dirichlet
condition*. One other possible boundary condition is that surface charge densities on
the surfaces may have been given. We know that the electric field has to be normal
to a conducting surface and that this normal electric field

$$E_n = -\frac{\partial \Phi}{\partial n}. \tag{2.38}$$

is related to the surface charge density by (2.25). In (2.38), n is the distance from
the conducting surface measured in the direction normal to the surface. Specifying
the surface charge density on a conducting surface essentially implies specifying the
normal derivative of the electrostatic potential over the surface. This type of boundary
condition is called the *Neumann condition*.

We can now mathematically state what a boundary value problem is. Suppose
we have a region of free space within which the electrostatic potential Φ satisfies
Laplace's equation, and the region is bounded by surfaces over which either the
Dirichlet type boundary condition (Φ given) or the Neumann type boundary condition
(the normal derivative $\partial \Phi / \partial n$ given) is specified. We want to find the electrostatic
potential Φ within the region of free space by solving Laplace's equation subject to
the given boundary condition. The next few sections of this chapter will be devoted
to discussing techniques for solving boundary value problems. As we shall see, this
is a mathematically elegant topic. However, it may seem at first sight that this is
a somewhat artificial topic. One of the reasons why the boundary value problem
is given such prominence in an advanced course of electrodynamics is that, while
learning to solve this problem, students learn some immensely useful mathematical
techniques which are used in many areas of mathematical physics. The first thing
we shall do is to prove the theorem that the solution of a boundary value problem is
unique. In other words, it is not possible to have two distinctly different solutions of
a boundary value problem.

To prove the uniqueness theorem, let us begin by assuming the opposite—that two
different solutions are possible of a boundary value problem—and see where that
leads us to. Let Φ_1 and Φ_2 be the two solutions of the electrostatic potential which
satisfy Laplace's equation in a region of space bounded by surfaces over which either
Φ (Dirichlet boundary condition) or $\partial \Phi / \partial n$ (Neumann boundary condition) is given.
In a portion of the surface over which the Dirichlet boundary condition is given, the
two solutions must have the same value so that

$$D: \quad \Phi_1 - \Phi_2 = 0 \tag{2.39}$$

in that portion of the surface. Similarly, in a portion of the surface over which the
Neumann boundary condition is satisfied, we must have

$$N : \frac{\partial}{\partial n}(\Phi_1 - \Phi_2) = 0 \qquad (2.40)$$

In the region of space where both the solutions satisfy Laplace's equation, we have

$$\nabla^2 (\Phi_1 - \Phi_2) = 0 \qquad (2.41)$$

We now consider an arbitrary scalar field Φ in the region of space where we are trying to solve the boundary value problem and focus our attention on the surface integral of $\Phi \nabla\Phi$ on the bounding surfaces. By Gauss's theorem (B.13),

$$\oint \Phi \nabla\Phi . \, d\mathbf{S} = \int \nabla.(\Phi \nabla\Phi) \, dV. \qquad (2.42)$$

Due to the vector identity (B.4), we have

$$\nabla.(\Phi \nabla\Phi) = (\nabla\Phi)^2 + \Phi \nabla^2\Phi$$

so that (2.42) becomes

$$\oint \Phi \nabla\Phi . \, d\mathbf{S} = \int [(\nabla\Phi)^2 + \Phi \nabla^2\Phi] \, dV. \qquad (2.43)$$

Since this is an identity which should be satisfied by any scalar field Φ, we can now substitute $\Phi = \Phi_1 - \Phi_2$ so that (2.43) becomes

$$\oint (\Phi_1 - \Phi_2) \nabla(\Phi_1 - \Phi_2). \, d\mathbf{S} = \int [\{\nabla(\Phi_1 - \Phi_2)\}^2 + (\Phi_1 - \Phi_2) \nabla^2(\Phi_1 - \Phi_2)] \, dV. \qquad (2.44)$$

From (2.41), the second term on the RHS is zero. It is easy to see that the surface integral on the LHS also must be zero because either (2.39) or (2.40) is satisfied in any portion of the surface. So we are left with

$$\int [\nabla(\Phi_1 - \Phi_2)]^2 \, dV = 0.$$

Since the integrand is a square quantity, it cannot be negative anywhere so that the integral can be zero only if the integrand is identically zero everywhere. This implies

$$\nabla(\Phi_1 - \Phi_2) = 0$$

so that

$$\mathbf{E}_1 - \mathbf{E}_2 = 0,$$

where $\mathbf{E}_1$ and $\mathbf{E}_2$ are the electric fields obtained from Φ_1 and Φ_2. This completes our proof that we cannot have two different electric fields as the solution of a boundary value problem.

There is an important corollary of this uniqueness theorem. We emphasize that we have to specify **either** Φ (Dirichlet boundary condition) **or** $\partial\Phi/\partial n$ (Neumann boundary condition) in any portion of the bounding surface. If we specify both, the problem becomes over-determined and, in general, may not permit a solution. Suppose both types of boundary conditions are given in a boundary value problem. We first find a solution by using the boundary condition that Φ is given on the bounding surfaces. According to the uniqueness theorem, this is the only possible solution. We can calculate the $\partial\Phi/\partial n$ at the boundaries from this solution. If this turns out to be different from the $\partial\Phi/\partial n$ given as boundary condition, then there will also be no way of satisfying that boundary condition.

We close our discussion of the uniqueness theorem by pointing out a variation of this theorem. In realistic boundary value problems, we often have a few isolated conductors for each of which the total amount of charge is given, with the understanding that the charges will distribute themselves over the conductors in such a manner that the surface of any conductor is at a constant electrostatic potential. Is the solution of this boundary value problem unique? It can be shown that the solution of this problem is also unique. We shall not discuss the proof of this version of uniqueness theorem here. The interested reader may try to prove this theorem or look up its proof given by Griffiths [2, Sect. 3.1.6].

2.9 Method of Images

One important consequence of the uniqueness theorem of boundary value problems is this: Suppose by some trick we are able to write down a solution of a boundary value problem which satisfies Laplace's equation in the appropriate region of space and which also satisfies the boundary conditions on the surrounding boundaries. Since the uniqueness theorem tells us that there can be only one solution, we can safely conclude that this solution obtained through the trick is the actual—and, in fact, the only possible—solution. We do not have to bother that the trick might have misled us by giving a spurious solution and the real solution could be different!

We now consider one trick known as the *method of images*. There are only a few problems which can be solved with this trick. But it is great when it works. Let us explain this method considering the following problem. Suppose there is a charge q in front of a semi-infinite conductor kept at zero potential, i.e. $\Phi = 0$ on this conductor. We also can take $\Phi = 0$ at infinity. Figure 2.6 shows the conductor on the left side of the surface $x = 0$, along with the charge q at $(a, 0)$. We are interested in finding a solution of Laplace's equation only in the free space on the right side of the conductor, taking the existence of the charge q into consideration, with the boundary condition that Φ is zero both on the conductor and at infinity. We now apply the trick of considering a hypothetical image charge $-q$ at $(-a, 0)$ inside the

Fig. 2.6 A charge q in front
of a plane conductor along
with an image charge $-q$
inside the conductor

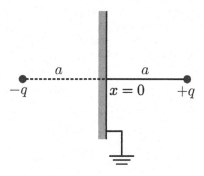

semi-infinite conductor which is outside the region where we are trying to solve our
problem. The electrostatic potential at a point (x, y) of free space due to the actual
charge q at $(a, 0)$ and the image charge $-q$ at $(-a, 0)$ is

$$\Phi(x, y) = \frac{q}{4\pi\epsilon_0} \left[\frac{1}{\sqrt{(x-a)^2 + y^2}} - \frac{1}{\sqrt{(x+a)^2 + y^2}} \right]. \tag{2.45}$$

It is easy to check that this expression satisfies all the requirements of the boundary
value problem. It is clear from the discussion of Sect. 1.4—especially from (1.19)—
that (2.45) satisfies Laplace's equation in the region of free space to the right of the
conductor, except at the point $(a, 0)$ where the charge is located. It is trivial to see
that Φ as given by (2.45) is zero at the surface of the conductor $x = 0$ and at infinity.
We thus conclude that (2.45) is the solution of our boundary value problem. We
could readily write down this solution by the trick of considering the electrostatic
potential produced by a hypothetical image charge kept somewhere outside the region
of interest where we want to solve the boundary value problem. Once we have found
the potential, we can calculate the normal electric field at the conducting surface
by using (2.38), and then (2.25) would give us the charge density induced on the
conducting surface. This is given as Exercise 2.8.

One other problem that can be handled with the method of images is finding the
electric field due to a charge q placed at a distance d from the centre of an earthed
conducting sphere having radius a ($d > a$). This is shown in Fig. 2.7. The boundary
value problem has to be solved in the free space outside the conductor with the
boundary conditions that Φ should be zero on the surface of the sphere ($r = a$) and
at infinity. We now want to argue that this problem can be solved by placing an image
charge $-qa/d$ inside the conductor at a distance a^2/d from the centre on the line
connecting the centre with the outside charge q. It is trivial to see that the combined
potential Φ due to the actual charge and the image charge will satisfy Laplace's
equation in the free space outside the sphere (except at the location of the charge q)
and will be zero at infinity. The only remaining thing is to show that Φ should also
be zero on the surface of the sphere ($r = a$). Let $\hat{\mathbf{n}}$ be the unit vector from the centre
of the sphere towards a point on the surface where we now evaluate the potential.

Fig. 2.7 A charge q in front of a spherical conductor along with an image charge $-qa/d$ inside this sphere

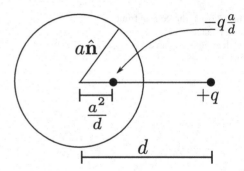

This point is at a location $a\,\hat{\mathbf{n}}$ if we take the centre of the sphere as our origin. If $\hat{\mathbf{n}}'$ is the unit vector towards the charge q, then the location of this charge is $d\,\hat{\mathbf{n}}'$, and the location of the image charge is $(a^2/d)\hat{\mathbf{n}}'$. The electrostatic potential at the point $a\hat{\mathbf{n}}$ on the spherical surface of the conductor due to the actual charge q at $d\,\hat{\mathbf{n}}'$ and the image charge $-qa/d$ at $(a^2/d)\hat{\mathbf{n}}'$ is

$$\Phi(a\,\hat{\mathbf{n}}) = \frac{q}{4\pi\epsilon_0}\left[\frac{1}{|a\,\hat{\mathbf{n}} - d\,\hat{\mathbf{n}}'|} - \frac{a/d}{|a\,\hat{\mathbf{n}} - (a^2/d)\,\hat{\mathbf{n}}'|}\right]. \tag{2.46}$$

Dividing both the numerator and the denominator of the second term by a/d, we get

$$\Phi(a\,\hat{\mathbf{n}}) = \frac{q}{4\pi\epsilon_0}\left[\frac{1}{|a\,\hat{\mathbf{n}} - d\,\hat{\mathbf{n}}'|} - \frac{1}{|d\,\hat{\mathbf{n}} - a\,\hat{\mathbf{n}}'|}\right],$$

which is easily seen to be zero. This completes our proof that the combined electrostatic potential due to the actual charge and the image charge we have considered gives the solution of our boundary value problem.

2.10 Boundary Value Problems in Two-Dimensional Cartesian Coordinates

In this section and the next two sections, we shall discuss how to solve boundary value problems in different coordinate systems. If the boundary condition happens to be given on a rectangular, cylindrical or spherical boundary, then it becomes particularly convenient to use the Cartesian, cylindrical or spherical coordinate system. First, we have to solve Laplace's equation in this coordinate system. The solution will be valid in the region of free space and will, in general, have several coefficients. These coefficients are then evaluated by applying the boundary conditions. The method for solving Laplace's equation which we shall use repeatedly is the method of *separation of variables*, as we shall explain below.

In this book, we shall restrict our discussion to problems in which one coordinate direction in three-dimensional space will be assumed to be a direction of symmetry. In other words, we shall assume that no quantity varies with one coordinate variable. This will make the problem two-dimensional. We can illustrate the basic principles and methodology by considering two-dimensional problems. Three-dimensional problems are solved with the same methodology, but the mathematical calculations are much more complicated and involved. Readers desirous of learning about three-dimensional boundary value problems should consult other more complete books on electromagnetic theory (Panofsky and Phillips [3], Chap. 5; Jackson [4], Chap. 3). Finally, we shall mention that the techniques of solving boundary value problems which we are going to discuss now turn out to be useful for solving different kinds of problems in different areas of physics.

Let us first consider Cartesian coordinates, by assuming that nothing varies in the z direction. Then Laplace's equation (2.11) becomes

$$\frac{\partial^2 \Phi}{\partial x^2} + \frac{\partial^2 \Phi}{\partial y^2} = 0. \tag{2.47}$$

In the method of separation of variables, we assume that the solution can be written as a product of functions—each of which is a function of one coordinate variable alone. In this particular situation, we assume the solution to be of form

$$\Phi(x, y) = X(x)Y(y). \tag{2.48}$$

Substituting this in (2.47) and dividing by $X(x)\, Y(y)$, we get

$$\frac{1}{X}\frac{d^2 X}{dx^2} = -\frac{1}{Y}\frac{d^2 Y}{dy^2}. \tag{2.49}$$

The LHS of this equation can be a function of x only, whereas the RHS can be a function of y only. The demand of consistency requires that both sides must be equal to a quantity independent of x or y. Denoting this quantity by $-k^2$, we get the following two equations for X and Y from (2.49):

$$\frac{d^2 X}{dx^2} + k^2 X = 0, \tag{2.50}$$

$$\frac{d^2 Y}{dy^2} - k^2 Y = 0. \tag{2.51}$$

The solutions of these two equations are

$$X(x) = A_k \cos kx + B_k \sin x,$$

$$Y(y) = C_k e^{-ky} + D_k e^{ky},$$

Fig. 2.8 A region of space
bounded by three plane
conductors on three sides.
We have to find the
electrostatic potential Φ in
this region of space with Φ
specified to be zero on the
side conductors and V on the
bottom conductor

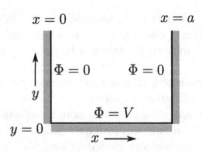

where A_k, B_k, C_k and D_k are constant coefficients. According to (2.48), $\Phi(x, y)$ should be given by a product of these two functions. Also, k can have different values. Therefore, the general solution is given by

$$\Phi(x, y) = \int (A_k \cos kx + B_k \sin kx)(C_k e^{-ky} + D_k e^{ky}) \, dk. \tag{2.52}$$

The significance of this solution is this: if the potential varies periodically in one direction, then it has to vary exponentially in the perpendicular direction. Here we have chosen the x direction as the direction in which the potential varies periodically. Had we assumed that both the sides of (2.49) are equal to a positive constant k^2, we would have a solution varying periodically in the y direction and exponentially in the x direction. Which is the more appropriate solution to consider depends on the boundary conditions. Now, we shall consider two examples to illustrate how the constant coefficients appearing in the general solution (2.52) can be evaluated.

Example 1. Let us consider a region of space bounded by three conductors as shown in Fig. 2.8. The vertical conductors at $x = 0$ and $x = a$ are kept at potential $\Phi = 0$, whereas the horizontal conductor at $y = 0$ has the potential $\Phi = V$. Obviously, one has to ensure that the three conductors do not touch each other and can be at different potentials. The potential at the upper boundary $y \to \infty$ also is assumed to go to zero. The electrostatic potential would satisfy Laplace's equation in the rectangular region $0 < x < a, 0 < y < \infty$.

If we want to make $\Phi \to 0$ at the upper boundary, then the e^{ky} term should not be there, and we conclude that $D_k = 0$ in (2.52). Additionally, in order to make $\Phi = 0$ at $x = 0$, we need $A_k = 0$. Then, from (2.52), we have

$$\Phi(x, y) = \int F_k \sin kx \, e^{-ky} \, dk,$$

where we have written $F_k = B_k C_k$. Now, in order to satisfy the boundary condition $\Phi = 0$ at $x = a$, we must have $\sin ka = 0$, which implies that k can have only discrete values given by $k = n\pi/a$, where n is an integer. Then we can write

$$\Phi(x, y) = \sum_n F_n \sin \frac{n\pi x}{a} e^{-\frac{n\pi y}{a}}. \tag{2.53}$$

It may be noted that we have already made use of all the other boundary conditions except that $\Phi = V$ at $y = 0$. We can now use this remaining boundary condition to evaluate the coefficients F_k. We have

$$\Phi(x, y = 0) = \sum_n F_n \sin \frac{n\pi x}{a} = V.$$

We assume that the readers know how to evaluate coefficients of such Fourier series. We write

$$F_n = \frac{2}{a} \int_0^a V \sin \frac{n\pi x}{a} dx.$$

This is a straightforward integration which gives

$$F_n = \frac{4V}{n\pi}$$

for odd n and zero for even n. Substituting this in (2.53), we get the final solution of the boundary value problem:

$$\Phi(x, y) = \frac{4V}{\pi} \sum_{\text{odd } n} \frac{1}{n} \sin \frac{n\pi x}{a} e^{-\frac{n\pi y}{a}}, \tag{2.54}$$

which obeys Laplace's equation in the rectangular region of interest and satisfies all the boundary conditions.

Example 2. To illustrate how charges are handled in boundary value problems, we now consider a charge q at the position $(s, 0)$ between two plane conductors at $x = 0$ and $x = a$ at zero potential, as shown in Fig. 2.9 ($s < a$). Note that a point charge in a two-dimensional problem implies a line charge in three dimensions, the line being perpendicular to the two-dimensional plane in which we are considering the problem. As in the previous example, we should have all A_k-s zero and $k = n\pi/a$ in (2.52) in order to make $\Phi = 0$ at $x = 0$ and $x = a$. Further, in order to allow for a possible discontinuity at the $y = 0$ plane due to the presence of the charge q, we write separate solutions for Φ above and below this plane. Requiring that $\Phi \to 0$ both as $y \to \infty$ and as $y \to -\infty$, the appropriate solutions for Φ above and below the $y = 0$ plane are

$$\Phi_1(x, y > 0) = \sum_n C_n \sin \frac{n\pi x}{a} e^{-\frac{n\pi y}{a}}, \tag{2.55}$$

$$\Phi_2(x, y < 0) = \sum_n D_n \sin \frac{n\pi x}{a} e^{\frac{n\pi y}{a}}. \tag{2.56}$$

Fig. 2.9 An electric charge in the region of free space between two plane conductors kept at zero potential

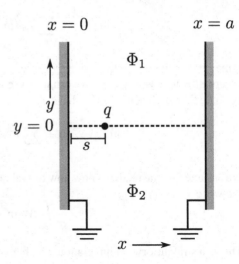

Since these two solutions must be continuous across the $y = 0$ plane at all points except the point $x = s$ where the charge is located, we must have

$$C_n = D_n. \tag{2.57}$$

To handle the expected discontinuity at $x = s$ in this plane, we have to take account of the charge q which corresponds to a charge density $q\delta(x - s)\delta(y)$ so that Poisson's equation around the charge is

$$\frac{\partial^2 \Phi}{\partial x^2} + \frac{\partial^2 \Phi}{\partial y^2} = -\frac{q\,\delta(x - s)\,\delta(y)}{\epsilon_0}$$

according to (2.10). We now integrate this equation from a point $(s, -\epsilon)$ slightly below the charge to the point (s, ϵ) slightly above the charge. This gives

$$\int_{-\epsilon}^{\epsilon} \frac{\partial^2 \Phi}{\partial y^2} dy = -\frac{q\,\delta(x - s)}{\epsilon_0}, \tag{2.58}$$

since the term $\partial^2 \Phi / \partial y^2$ is expected to make a significant contribution to the integration over this infinitesimal range due to the discontinuity of Φ along y, but that is not the case for the other term $\partial^2 \Phi / \partial x^2$. From (2.58), we get

$$\frac{\partial \Phi}{\partial y}\bigg|_{y=\epsilon} - \frac{\partial \Phi}{\partial y}\bigg|_{y=-\epsilon} = \frac{\partial \Phi_1}{\partial y}\bigg|_{y=0} - \frac{\partial \Phi_2}{\partial y}\bigg|_{y=0} = -\frac{q\,\delta(x - s)}{\epsilon_0}.$$

On substituting from (2.55)–(2.57) for Φ_1 and Φ_2, we get

$$-\sum_{n} C_n \frac{2n\pi}{a} \sin \frac{n\pi x}{a} = -\frac{q\,\delta(x-s)}{\epsilon_0}.$$

The coefficients C_n can now be found by multiplying both sides of this equation by $\sin(m\pi x/a)$ and integrating from $x = 0$ to $x = a$. The readers can check that this procedure gives

$$C_n = \frac{q}{\epsilon_0 n\pi} \sin \frac{n\pi s}{a}.$$

Keeping (2.57) in mind, we can now write down (2.55)–(2.56) in the final forms

$$\Phi_1(x, y > 0) = \frac{q}{\epsilon_0 \pi} \sum_{n} \frac{1}{n} \sin \frac{n\pi s}{a} \sin \frac{n\pi x}{a} e^{-\frac{n\pi y}{a}}, \tag{2.59}$$

$$\Phi_2(x, y < 0) = \frac{q}{\epsilon_0 \pi} \sum_{n} \frac{1}{n} \sin \frac{n\pi s}{a} \sin \frac{n\pi x}{a} e^{\frac{n\pi y}{a}}. \tag{2.60}$$

This is the solution of our boundary value problem, in which we have illustrated how we can handle a charge (which is a line charge in three dimensions) in a two-dimensional boundary value problem.

2.11 Boundary Value Problems in Polar Coordinates

After this discussion of boundary value problems in two-dimensional rectangular geometry, let us now turn our attention to cylindrical geometry. If we assume that there are no variations in the z direction, then the problem becomes two-dimensional and the cylindrical coordinates become planar polar coordinates. From (2.36), we can write down Laplace's equation in polar coordinates:

$$r\frac{\partial}{\partial r}\left(r\frac{\partial \Phi}{\partial r}\right) + \frac{\partial^2 \Phi}{\partial \theta^2} = 0. \tag{2.61}$$

We again solve this equation by separation of variables, by trying out a solution of the form

$$\Phi(r, \theta) = R(r)\,\Theta(\theta). \tag{2.62}$$

On substituting this in (2.61) and dividing by $R(r)\Theta(\theta)$, we get

$$\frac{1}{R} r \frac{d}{dr}\left(r\frac{dR}{dr}\right) = -\frac{1}{\Theta}\frac{d^2\Theta}{d\theta^2}.$$

Since the LHS can be a function of r only and RHS a function of θ only, both the sides must be equal to a quantity independent of r and θ. Writing this quantity as k_n^2, we get the following equations for $\Theta(\theta)$ and $R(r)$:

$$\frac{d^2\Theta}{d\theta^2} + k_n^2\Theta = 0, \tag{2.63}$$

$$r\frac{d}{dr}\left(r\frac{dR}{dr}\right) - k_n^2 R = 0. \tag{2.64}$$

To solve these equations, we have to handle the cases $k_n \neq 0$ and $k_n = 0$ separately. First, for the case $k_n \neq 0$, the solutions of (2.63) and (2.64) are given by

$$\Theta_n = A_n \cos k_n\theta + B_n \sin k_n\theta, \tag{2.65}$$

$$R_n = C_n r^{k_n} + D_n r^{-k_n}. \tag{2.66}$$

For the case $k_n = 0$, the solutions are simpler

$$\Theta_0 = E + F\theta, \quad R_0 = G + H \ln r. \tag{2.67}$$

In a boundary value problem which encompasses all values of θ from 0 to 2π, the requirement that Θ_n given by (2.65) should not be multiple valued dictates that k_n can only be equal to an integer n, and we shall also have $F = 0$ in (2.67). The solution of Laplace's equation for a particular $k_n = n$ will be given by a product of (2.65) and (2.66), which then will have to be summed up over all possible values on n. In other words, the general solution of Laplace's equation in polar coordinates is

$$\Phi(r, \theta) = \sum_{n=1}^{\infty}(A_n \cos n\theta + B_n \sin n\theta)(C_n r^n + D_n r^{-n}) + E(G + H \ln r). \tag{2.68}$$

Readers should be able to convince themselves easily that it is not necessary to include terms for negative n in this series.

2.11.1 Conducting Cylinder in a Uniform Electric Field

Let us now work out the example of a conducting cylinder kept in a uniform electric field with its axis perpendicular to the electric field. The boundary value problem will clearly be a two-dimensional problem in the plane perpendicular to the axis of the cylinder. In the absence of the cylinder, we would have a uniform electric field. It is obvious that the presence of the cylinder will distort the electric field, since the surface of the cylinder has to be an equipotential surface on which the field

Fig. 2.10 A conducting
cylinder in a region of space
where there would have been
a uniform electric field in the
absence of this conductor.
The figure shows a view
taken along the axis of the
cylinder so that the surface
of the cylinder appears as a
circle

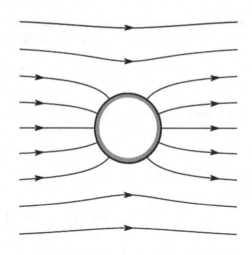

lines will have to be normal. This is shown in Fig. 2.10. Taking the origin of the
polar coordinates at the centre of the cylinder, we consider the case in which the
electrostatic potential on the surface of the cylinder at $r = a$ is zero, that is,

$$\Phi(r=a,\ \theta) = 0, \tag{2.69}$$

which is one of our boundary conditions. Since we have an electric field extending
to infinity, it is not possible to make the $\Phi \to 0$ at infinity. Rather, the influence of
the cylinder will increasingly become insignificant as we go to larger r, and we shall
have an electrostatic potential at infinity corresponding to a uniform electric field

$$\mathbf{E}_\infty = E_0\,\hat{\mathbf{e}}_x,$$

where we have chosen the x axis in the direction of the electric field. The corresponding electrostatic potential is

$$\Phi_\infty = -E_0\,x = -E_0\,r\cos\theta. \tag{2.70}$$

Now, our boundary value problem is to solve Laplace's equation in the region $a <
r < \infty$ subject to the boundary conditions (2.69) and (2.70).

We expect Φ in our region of interest to be given by $-E_0 r\cos\theta$ (which is its
value at infinity) plus some other terms from (2.68) arising out of the presence of
the cylinder. We can keep only such terms among these other terms which go to zero
at infinity. It is not difficult to check that the most general solution satisfying the
boundary condition (2.70) is

$$\Phi(r,\theta) = \sum_{n=1}^{\infty}(A_n\cos n\theta + B_n\sin n\theta)D_n\,r^{-n} - E_0\,r\cos\theta.$$

Now, when we want to apply the boundary condition (2.69), it is easy to see that we can keep only the term going as $\cos\theta$ in the above series, i.e. we should have

$$\Phi(r, \theta) = A_1 D_1 \frac{\cos\theta}{r} - E_0 r \cos\theta. \qquad (2.71)$$

Applying the boundary condition (2.69), we find that

$$A_1 D_1 = E_0 a^2$$

so that (2.71) becomes

$$\Phi(r, \theta) = E_0 \cos\theta \left(\frac{a^2}{r} - r\right). \qquad (2.72)$$

This is the solution of the boundary value problem in our region of interest $a < r < \infty$. The field lines shown in Fig. 2.10 result from the electrostatic potential given by (2.72).

2.11.2 Wedge-Shaped Region Between Conductors

Let us consider a wedge-shaped region of angle α between two plane conductors connected to each other, as shown in Fig. 2.11a. If some charge is put on the conductors, it will get distributed on the surfaces of the conductors in such a way that two conductors should have the same potential, which we can take to be zero. We should be able to find out the distribution of charge density on the conductors by solving the boundary value problem. The electrostatic potential within the wedge-shaped region can be written from (2.62), (2.65) and (2.66). Now we have to select possible values of k_n in such a way that Φ is zero at both $\theta = 0$ and $\theta = \alpha$. It is easy to see that the solution should be of the form

$$\Phi = \sum_m A_m \sin\frac{m\pi\theta}{\alpha} r^{\frac{m\pi}{\alpha}}. \qquad (2.73)$$

Fig. 2.11 a A wedge-shaped region between two plane conductors making an α between them. **b** The special case of $\alpha = 3\pi/2$

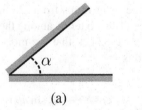

(a)

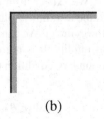

(b)

The surface charge density σ is given by (2.25), which takes the following form for the conductor at $\theta = 0$:

$$\sigma = -\frac{\epsilon_0}{r}\frac{\partial \Phi}{\partial \theta}.$$

It is clear that the most significant contribution near the origin will come from the term with the lowest power of r. Taking

$$\Phi \propto \sin \frac{\pi \theta}{\alpha} \, r^{\frac{\pi}{\alpha}}$$

near the origin, we get

$$\sigma \propto \cos \frac{\pi \theta}{\alpha} \, r^{\frac{\pi}{\alpha} - 1}. \tag{2.74}$$

One interesting thing to note is that σ becomes independent of r when $\alpha = \pi$, which certainly should be the case because $\alpha = \pi$ implies a plane conductor. Let us take note of the situation when α is larger than π. In such a situation, (2.74) shows that σ goes as a negative power of r, meaning that there would be a concentration of charge at the apex point of the wedge. Figure 2.10b shows the configuration with $\alpha = 3\pi/2$ for which σ would go as $r^{-1/3}$. The significance is obvious. If a conductor with a surface charge distribution has a sharp corner, charges tend to get concentrated there.

2.12 Boundary Value Problems in Spherical Coordinates

At last, we come to a discussion of spherical coordinates. In this book, we shall restrict our discussion only to systems which are symmetric around a certain axis. If that axis is taken as the polar axis of the coordinate system, then certainly all quantities are independent of ϕ. We need to find a solution $\Phi(r, \theta)$ of Laplace's equation. From (2.37), we write down the equation we need to solve:

$$\frac{\partial}{\partial r}\left(r^2 \frac{\partial \Phi}{\partial r}\right) + \frac{1}{\sin \theta}\frac{\partial}{\partial \theta}\left(\sin \theta \frac{\partial \Phi}{\partial \theta}\right) = 0. \tag{2.75}$$

Again, we try the method of separation of variables by writing

$$\Phi(r, \theta) = R(r)\,\Theta(\theta). \tag{2.76}$$

Substituting in (2.75) and dividing by $R(r)\Theta(\theta)$, we get

$$\frac{1}{R}\frac{d}{dr}\left(r^2 \frac{dR}{dr}\right) = -\frac{1}{\Theta}\frac{1}{\sin \theta}\frac{d}{d\theta}\left(\sin \theta \frac{d\Theta}{d\theta}\right).$$

Both sides should be independent of r or θ, and we take them to be equal to $n(n + 1)$. This leads to the following equations for $R(r)$ and $\Theta(\theta)$:

$$\frac{d}{dr}\left(r^2\frac{dR}{dr}\right) - n(n + 1)R = 0, \tag{2.77}$$

$$\frac{1}{\sin\theta}\frac{d}{d\theta}\left(\sin\theta\frac{d\Theta}{d\theta}\right) + n(n + 1)\Theta = 0. \tag{2.78}$$

Let us substitute $\mu = \cos\theta$. Noting that

$$\frac{d}{d\theta} = \frac{d\mu}{d\theta}\frac{d}{d\mu} = -\sin\theta\frac{d}{d\mu},$$

we easily find that (2.78) becomes

$$\frac{d}{d\mu}\left[(1 - \mu^2)\frac{d\Theta}{d\mu}\right] + n(n + 1)\,\Theta = 0. \tag{2.79}$$

This is the celebrated Legendre equation which is discussed in all standard textbooks of mathematical physics (see, for example, Arfken, Weber and Harris [1], Sect. 15.1). If we require that the solution should be regular for all possible values of θ from 0 to $\pi/2$, then n can only have integral values 0, 1, 2, 3, ..., and the solutions of (2.78) are the Legendre polynomials of different order n, that is,

$$\Theta(\theta) = P_n(\cos\theta). \tag{2.80}$$

Assuming $R(r)$ to go as r^l and substituting in (2.77), we get

$$l(l + 1) - n(n + 1) = 0.$$

This equation is satisfied by two possible values of l: either $l = n$ or $l = -(n + 1)$. Clearly, $R(r)$ can be a superposition of two terms following from these two values of l. Then it follows from (2.76) and (2.80) that

$$\Phi(r, \theta) = \sum_{n=0}^{\infty}\left(A_n r^n + \frac{B_n}{r^{n+1}}\right) P_n(\cos\theta). \tag{2.81}$$

This is the most general axisymmetric (i.e. independent of the ϕ) solution of Laplace's equation in spherical coordinates. If axisymmetry is not assumed, then we shall have spherical harmonics in the place of Legendre polynomials. Several advanced textbooks on electromagnetic theory discuss fully three-dimensional solutions of Laplace's equation without assuming axisymmetry (see, for example, Jackson [4], Sects. 3.5–6). We shall refrain from discussing this mathematically involved subject.

The basic principles used in solving such three-dimensional boundary value problems are very much like extensions of the principles used in two-dimensional problems.

2.12.1 Some Properties of Legendre Polynomials

In order to solve boundary value problems in spherical geometry by making use of the solution (2.81) of Laplace's equation, we need to know some properties of Legendre polynomials. We here list those properties without derivations. Readers will have to consult textbooks of mathematical physics for their derivations (see, for example, Arfken, Weber and Harris [1], Sects. 15.1–2)). The important properties we may need to know for solving problems in spherical geometry are the following:

1. Legendre polynomials can be obtained from the following generating function by the mathematical relation

$$\frac{1}{\sqrt{1 - 2\mu t + t^2}} = \sum_{n=0}^{\infty} P_n(\mu) \, t^n. \quad (|t| \le 1) \tag{2.82}$$

2. The Legendre polynomial $P_n(\mu)$ satisfies the differential equation

$$\frac{d}{d\mu} \left[(1 - \mu^2) \frac{dP_n}{d\mu} \right] + n(n + 1)P_n = 0. \tag{2.83}$$

3. Legendre polynomials $P_n(\mu)$ and $P_l(\mu)$ obey the orthogonality relation

$$\int_{-1}^{1} P_n(\mu) \, P_l(\mu) \, d\mu = \frac{2}{2n + 1} \, \delta_{nl}. \tag{2.84}$$

One can easily work out the first few Legendre polynomials from the generating function (2.82). For future reference, we write down the first few here:

$$P_0(\mu) = 1, \quad P_1(\mu) = \mu, \quad P_2(\mu) = \frac{1}{2}(3\mu^2 - 1), \ldots. \tag{2.85}$$

Note that we always have $P_n(1) = 1$ when $\mu = 1$, i.e. when $\theta = 0$.

2.12.2 Boundary Value Problem Around a Sphere

To illustrate how boundary value problems in spherical geometry are handled, let us consider a sphere of radius $r = a$ over which the electrostatic potential has the axisymmetric distribution $\Phi(a, \theta)$. Obviously, the spherical surface could not be a conductor. Otherwise, the electrostatic potential could not vary with θ. We now want to find out the electrostatic potential in the free space around the sphere. If there is no source at infinity, then the electrostatic potential has to be zero at infinity, which means that we must have $A_n = 0$ in the solution (2.81) of Laplace's equation. Hence, the electrostatic potential outside the spherical surface must be given by

$$\Phi(r, \theta) = \sum_{n=0}^{\infty} \frac{B_n}{r^{n+1}} P_n(\cos \theta). \tag{2.86}$$

We have to satisfy the following boundary condition on the spherical surface $r = a$:

$$\Phi(a, \theta) = \sum_{n=0}^{\infty} \frac{B_n}{a^{n+1}} P_n(\cos \theta).$$

To determine the coefficients B_n from the given electrostatic potential $\Phi(a, \theta)$ over the spherical surface, we need to multiply both sides of this equation by $P_l(\cos \theta) d(\cos \theta)$ and integrate from $\theta = 0$ to $\theta = \pi/2$ (which is equivalent to integrating $\cos \theta$ from -1 to $+1$). On making use of the orthogonality relation (2.84), we easily find

$$B_n = \frac{2n + 1}{2} a^{n+1} \int_{-1}^{1} \Phi(a, \theta) P_n(\cos \theta) \, d(\cos \theta). \tag{2.87}$$

It is clear that (2.86) with the coefficients B_n given by (2.87) would be the solution of our boundary value problem.

2.13 Multipole Expansion

We shall now discuss a kind of series expansion which is useful in many areas of physics when we deal with charge distributions. Keeping up with the spirit of restricting our discussions to axisymmetric situations, let us consider an axisymmetric charge distribution $\rho(r', \theta')$ in a finite region around the origin. We assume that outside this finite region, we have free space where Laplace's equation should hold, and the electrostatic potential should be given by (2.86). Keeping in mind that $P_n(\mu) = 1$ at a point $(r, 0)$ on the symmetry axis, we must have

$$\Phi(r, 0) = \sum_{n=0}^{\infty} \frac{B_n}{r^{n+1}} = \frac{B_0}{r} + \frac{B_1}{r^2} + \frac{B_2}{r^3} + \cdots. \tag{2.88}$$

Let us now find out the electrostatic potential at this point $(r, 0)$ due to the charge distribution $\rho(r', \theta')$. As we shall see, this will enable us to express $B_1, B_2, B_3, \ldots$ in terms of this charge distribution $\rho(r', \theta')$.

The distance between the source point (r', θ') and the field point $(r, 0)$ is given by

$$|\mathbf{x} - \mathbf{x}'| = \sqrt{r^2 + r'^2 - 2rr'\cos\theta'} = r\sqrt{1 + \left(\frac{r'}{r}\right)^2 - 2\frac{r'}{r}\cos\theta'}.$$

Considering field points sufficiently far away such that $r \gg r'$ and by making use of (2.82), we are led from this to

$$\frac{1}{|\mathbf{x} - \mathbf{x}'|} = \frac{1}{r}\sum_{n=0}^{\infty} P_n(\cos\theta')\left(\frac{r'}{r}\right)^n. \tag{2.89}$$

Although we have derived (2.89) by considering the field point on the symmetry axis, a little reflection should convince you that it is a more general expression valid for any two arbitrary position vectors $\mathbf{x}$ and $\mathbf{x}'$ (Exercise 2.15). Substituting (2.89) in (2.2), we get

$$\Phi(r, 0) = \frac{1}{4\pi\epsilon_0}\frac{1}{r}\sum_{n=0}^{\infty}\int \rho(r', \theta')\, P_n(\cos\theta')\left(\frac{r'}{r}\right)^n dV'. \tag{2.90}$$

Comparing this with (2.88), we conclude

$$B_n = \frac{1}{4\pi\epsilon_0}\int \rho(r', \theta')\, P_n(\cos\theta')\, r'^n\, dV'. \tag{2.91}$$

We write the first three coefficients B_n in the following way:

$$B_0 = \frac{q}{4\pi\epsilon_0}, \quad B_1 = \frac{p}{4\pi\epsilon_0}, \quad B_2 = \frac{1}{4\pi\epsilon_0}\frac{Q}{2}. \tag{2.92}$$

Making use of (2.85) and writing $z' = r'\cos\theta'$, we easily find from (2.91) that

$$q = \int \rho(r', \theta')\, dV', \tag{2.93}$$

$$p = \int \rho(r', \theta')\, z'\, dV', \tag{2.94}$$

$$Q = \int \rho(r', \theta')(3z'^2 - r'^2)\, dV'. \tag{2.95}$$

We readily recognize that q given by (2.93) is the total charge and p given by (2.94) is the electric dipole moment of the charge distribution. The quantity Q given by (2.95) is called the *electric quadrupole moment*.

By considering the electrostatic potential at a point on the symmetry axis, we have evaluated B_n-s, of which the first three are given by (2.92)–(2.95). Once we know B_n-s, we can write down the electrostatic potential at any point in space outside the charge distribution satisfying $r \gg r'$ (i.e. not only on the symmetry axis) by using (2.86). On using the notation (2.92), the fist few terms in the expression of the electrostatic potential at points away from the charge distribution are

$$\Phi(r, \theta) = \frac{1}{4\pi\epsilon_0} \left[\frac{q}{r} + \frac{p}{r^2} P_1(\cos\theta) + \frac{1}{2}\frac{Q}{r^3} P_2(\cos\theta) + \cdots \right] \tag{2.96}$$

with q, p and Q given by (2.93)–(2.95). This series, of which we have written down the first three terms explicitly, is called the *multipole expansion* of the electrostatic potential.

It may be noted that successive terms in the series (2.96) fall off at increasingly higher power of r and will consequently decrease more and more rapidly at large r. If the total charge q is non-zero, then the term involving q is the dominant term at large r and (2.96) tends to the electrostatic potential due to a charge q. If $q = 0$, then the next term will be the dominant term at large r. It is the electrostatic potential due to an electric dipole p and is the same as what we have worked out in (2.15). If both $q = 0$ and $p = 0$, then it is the next quadrupole term which will be dominant at large r, and so on. We point out one other interesting fact. Any point on the symmetry axis within the region of charge distribution can be taken as the origin of our coordinate system. Obviously, the total charge q is independent of the choice of the origin, but higher moments p, Q, ..., in general, depend on where we take the origin. However, readers should be able to easily prove that if q is zero, then p should be independent of the choice of origin. Again, if q and p are both zero, then it can be easily shown the Q is independent of the origin (Exercise 2.14). In other words, the lowest non-zero multipole moment is independent of the choice of the origin, and the term in the multipole expansion involving this moment is the dominant term in the expansion at large r.

Note that we have worked out the multipole expansion only for the axisymmetric case. Even if the charge distribution is not axisymmetric, it is still possible to do a more complicated multipole expansion. In that case, spherical harmonics appear in the place of Legendre polynomials (see, for example, Jackson [4], Sect. 4.1). The monopole term turns out to be exactly the same for the non-axisymmetric charge distribution as for the axisymmetric distribution. For the non-axisymmetric charge distribution, even the dipole term comes out quite simple (Exercise 2.15). However, from the quadrupole term onwards, the mathematical expressions are much more complicated in the non-axisymmetric case.

2.14 Polarization in Dielectric Medium

So far we have considered charges and conductors distributed in free space. Sometimes we have space filled with a dielectric medium. We shall consider this situation now.

In the presence of an electric field, a dielectric medium gets polarized, which means that a small volume element dV' of this medium located at the position $\mathbf{x}'$ will have an electric dipole moment $\mathbf{P}\,dV'$, where $\mathbf{P}$ is the *polarization density*. We can apply (2.14) to find the electrostatic potential due to this volume element dV' at the field point $\mathbf{x}$. Integrating over the dielectric medium, the total electrostatic potential at the field point due to the polarization in the medium is

$$\Phi(\mathbf{x}) = \frac{1}{4\pi\epsilon_0} \int \frac{\mathbf{P}\cdot(\mathbf{x} - \mathbf{x}')}{|\mathbf{x} - \mathbf{x}'|^3}\, dV'.$$

Note that we are again following the convention of using primes to indicate spatial variables associated with the source region. By making use of (1.18) and taking into consideration that an operation with ∇' will give rise to a minus sign with respect to what we get in an operation with ∇, we write

$$\Phi(\mathbf{x}) = \frac{1}{4\pi\epsilon_0} \int \mathbf{P}\cdot\nabla'\left(\frac{1}{|\mathbf{x} - \mathbf{x}'|}\right) dV'.$$

By making use of the vector identity (B.4), we get

$$\Phi(\mathbf{x}) = \frac{1}{4\pi\epsilon_0} \int \left[\nabla'\cdot\left(\frac{\mathbf{P}}{|\mathbf{x} - \mathbf{x}'|}\right) - \frac{\nabla'\cdot\mathbf{P}}{|\mathbf{x} - \mathbf{x}'|}\right] dV'.$$

We can use Gauss's theorem in vector analysis (B.13) to convert the first term above into a surface integral, which should go to zero if the surface is taken at infinity. We can then finally write

$$\Phi(\mathbf{x}) = \frac{1}{4\pi\epsilon_0} \int \frac{-\nabla'\cdot\mathbf{P}}{|\mathbf{x} - \mathbf{x}'|}\, dV'. \tag{2.97}$$

Comparing with (2.2), we at once conclude that any region in the interior of a dielectric medium has an effective charge density $-\nabla'\cdot\mathbf{P}$. This implies that a region of uniform polarization density $\mathbf{P}$ does not have an effective charge. We leave it for the readers to argue why only a spatially varying $\mathbf{P}$ would give rise to an effective electric charge.

In the discussions in the remaining part of this section, we shall not distinguish source points from field points, and a spatial derivative will simply mean a derivative in our region of interest. So we no longer use primes, and the polarization charge inside the dielectric medium is taken to be $-\nabla\cdot\mathbf{P}$. Now, we often use the convention that the charge density ρ refers to only the charges on the conductors (or point charges). Taking account of the polarization charge $-\nabla\cdot\mathbf{P}$, the total charge density should be $\rho - \nabla.\mathbf{P}$, and ρ in (1.8) should be replaced by this when dealing with a dielectric medium:

$$\nabla.\mathbf{E} = \frac{\rho - \nabla.\mathbf{P}}{\epsilon_0}.$$

Now, in equations of mathematical physics, we often like to keep sources on one side and the effects produced by the sources on the other side. In this particular situation, it is often useful to view ρ as the only source, which produces both the electric field $\mathbf{E}$ and the polarization density $\mathbf{P}$ in the dielectric medium. So, if we keep only ρ on one side, the above equation can be written as

$$\nabla.(\epsilon_0\mathbf{E} + \mathbf{P}) = \rho. \tag{2.98}$$

We now introduce the *electric displacement* vector $\mathbf{D}$ defined in the following way:

$$\mathbf{D} = \epsilon_0\mathbf{E} + \mathbf{P}. \tag{2.99}$$

Substituting this in (2.98), we have

$$\nabla.\mathbf{D} = \rho. \tag{2.100}$$

When we consider electrostatics with a dielectric medium, we normally replace (1.8) by (2.100), whereas (1.9) remains unchanged.

Since the polarization density $\mathbf{P}$ is caused by the electric field $\mathbf{E}$, we expect $\mathbf{P}$ to be proportional to $\mathbf{E}$ in a simple isotropic dielectric medium, and we can write

$$\mathbf{P} = \epsilon_0\chi\,\mathbf{E}, \tag{2.101}$$

where χ is called the *susceptibility* of the medium. In an anisotropic medium (such as a crystalline solid), the polarization density $\mathbf{P}$ and the electric field $\mathbf{E}$ are, in general, related by a more complicated susceptibility tensor (see, for example, Panofsky and Phillips [3], p. 30, pp. 99–100; Born and Wolf [5], Chap. XIV). We shall not discuss such media in this book. We also write

$$\mathbf{D} = \epsilon\mathbf{E} = \epsilon_0\kappa\,\mathbf{E}, \tag{2.102}$$

where κ is the *dielectric constant* given by

$$\kappa = \frac{\epsilon}{\epsilon_0} = 1 + \chi \tag{2.103}$$

according to (2.99), (2.101) and (2.102).

To summarize the discussions in this short but very important section, there are two possible ways of handling electrostatic problems involving dielectric media. If we include polarization charges induced in the medium in ρ along with charges on conductors, then we can use the basic equation (1.8), which remains unchanged. On the other hand, if we want to include only the charges on the conductors in ρ and treat the induced polarization charges along with the electric field as effects, then we have to use (2.100). This is often the preferred and more logical approach. The other basic equation of electrostatics (1.9) remains unchanged in the presence of a dielectric medium. So, when solving electrostatic problems with a dielectric medium, we have to make use of (2.100) and (1.9), connecting **D** to **E** through the relation (2.102).

2.15 Boundary Conditions Between Dielectric Media

As an illustration of how we deal with electrostatic problems with dielectric media, let us work out the boundary conditions across the interface between two media with dielectric constants κ_1 and κ_2. As we pointed out at the end of the previous section, we have to make use of (2.100) and (1.9), along with (2.102). If there is no surface charge at the interface between the two media, then (2.100) becomes

$$\nabla . \mathbf{D} = 0. \tag{2.104}$$

As shown in Fig. 2.12a, we consider a pillbox-like volume chosen such that the interface between the two media lies at the middle of the box. Integrating $\nabla.\mathbf{D}$ over this volume, we have

$$\int (\nabla . \mathbf{D}) \, dV = \oint \mathbf{D} . d\mathbf{S} = 0$$

according to (2.104). The surface integral is clearly over the bounding surfaces of this pillbox-like volume. If we make the thickness of this pillbox very small, then this integral will be only over the surfaces parallel to the interface of which we can take the area to be S. Then $\oint \mathbf{D}.d\mathbf{S} = 0$ implies

$$D_{2n} S - D_{1n} S = 0,$$

where D_{1n} and D_{2n} are the normal components of **D** on the two sides of the interface. It is clear that

$$D_{1n} = D_{2n}, \tag{2.105}$$

Fig. 2.12 a A pillbox and **b** a rectangular path at the interface between two dielectric media with dielectric constants κ_1 and κ_2

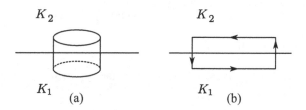

which means that the normal component of the displacement vector $\mathbf{D}$ has to be continuous across an interface between two dielectric media. Making use of (2.102), we can write (2.105) as

$$\kappa_1 E_{1n} = \kappa_2 E_{2n}. \tag{2.106}$$

Next, we consider a rectangular path with the two long sides of the path parallel to the interface on its two sides, as shown in Fig. 2.12b. We now consider the line integral of $\mathbf{E}$ over this rectangular path:

$$\oint \mathbf{E} \cdot d\mathbf{l} = \int (\nabla \times \mathbf{E}) \cdot d\mathbf{S} = 0$$

on making use of Stokes's theorem and then the electrostatic Eq. (1.9). If the short sides of the rectangular path are made infinitesimally small, then the only contributions to the line integral come from the two long sides of length L, and we have

$$E_{1t}L - E_{2t}L = 0,$$

where E_{1t} and E_{2t} are the tangential components of the electric field $\mathbf{E}$ on two sides of the interface. We thus end up with the boundary condition

$$E_{1t} = E_{2t}. \tag{2.107}$$

Readers should able to argue that the same boundary condition will follow from (1.4) even when we allow time variations.

An electric field line will certainly change its direction as it crosses an interface between two dielectric media. This effect, often called the *refraction of electric field lines*, can easily be worked out from (2.106) and (2.107). Suppose θ_1 and θ_2 are the two angles which the electric field lines make with the vertical on the two sides of the interface. Certainly (2.106) implies

$$\kappa_1 E_1 \cos \theta_1 = \kappa_2 E_2 \cos \theta_2,$$

whereas (2.107) gives

$$E_1 \sin \theta_1 = E_2 \sin \theta_2.$$

Dividing the first relation by the second, we get

$$\kappa_1 \cot \theta_1 = \kappa_2 \cot \theta_2,$$

from which

$$\frac{\tan \theta_1}{\tan \theta_2} = \frac{\kappa_1}{\kappa_2}. \tag{2.108}$$

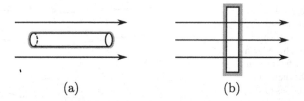

Fig. 2.13 Cavities inside a medium of dielectric constant κ with uniform electric field. **a** A needle-shaped cavity parallel to the electric field. **b** A disk-shaped cavity perpendicular to the electric field

As a straightforward application of these boundary conditions, let us consider an electrostatic field **E** inside a dielectric medium having a cavity. Figure 2.13a shows a needle-shaped cavity having its axis parallel to **E**. From the condition (2.107) that the tangential component of the electric field has to be continuous across the side surfaces of this needle-shaped cavity, we conclude that the electric field $\mathbf{E}_c$ in the cavity should be

$$\mathbf{E}_c = \mathbf{E}. \tag{2.109}$$

Next, we consider a disk-shaped cavity with the plane of the disk being perpendicular to **E**, as shown in Fig. 2.13b. From the continuity of the normal component of **D** across the surface, as given by (2.105), we easily conclude that the displacement vector $\mathbf{D}_c$ inside the disk-shaped cavity should be

$$\mathbf{D}_c = \mathbf{D}.$$

Using (2.102), we can write $\mathbf{D} = \epsilon_0 \kappa \mathbf{E}$ and $\mathbf{D}_c = \epsilon_0 \mathbf{E}_c$ so that

$$\mathbf{E}_c = \kappa\, \mathbf{E} \tag{2.110}$$

in the case of a disk-shaped cavity.

2.16 Dielectric Sphere Inside a Uniform Electric Field

Sometimes we have to solve boundary value problems involving dielectric media. To give an idea how such problems are handled, we consider the case of a dielectric sphere kept in an electric field which would have been uniform with value $\mathbf{E}_0$ in the absence of the sphere. This dielectric sphere of radius $r = a$ (the origin being taken at the centre of the sphere) is shown in Fig. 2.14.

From (2.100) and (2.102), we conclude that, inside a uniform dielectric medium, we shall have

$$\epsilon_0 \kappa\, \nabla . \mathbf{E} = \rho.$$

Fig. 2.14 A sphere of
dielectric constant κ and
radius a in a region of space
where there would have been
a uniform electric field $\mathbf{E}_0$ in
the absence of the dielectric
sphere. The field lines of the
electric displacement vector
$\mathbf{D}$ in the presence of the
dielectric sphere are shown
(assuming $\kappa = 1.5$)

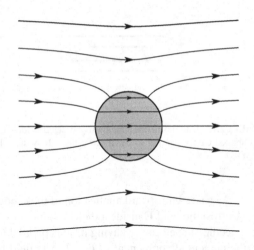

If $\rho = 0$ inside this medium, then we clearly have $\nabla.\mathbf{E} = 0$. On combining with
(2.1), we see that Laplace's equation (2.11) is valid inside the uniform dielectric
medium. If an electrostatics problem involves several regions inside each of which
we have a uniform dielectric medium, then we have Laplace's equation valid inside
each of the regions. However, we have to be careful about boundaries between these
regions of uniform dielectric media because Laplace's equation would not be valid
across the boundaries. We have to apply boundary conditions (2.106) and (2.107) to
connect solutions of Laplace's equation on two sides of a boundary.

Now we come to the problem of the dielectric sphere inside the uniform electric
field. We must have Laplace's equation valid both inside the sphere ($r < a$) and in the
free space around it ($r > a$) with the uniform electric field at infinity. But we have to
be careful about the boundary at $r = a$. Let us first consider the region $r > a$ outside
the sphere. As in the case of the conducting cylinder in a uniform electric field, as
discussed in Sect. 2.11.1, the electrostatic potential at infinity should be given by
(2.70). It may be noted that we were using cylindrical coordinates in Sect. 2.11.1 and
are using spherical coordinates now, although the electrostatic potential corresponding to a uniform electric field happens to be the same in both these systems. Making
use of the general solution of Laplace's equation given by (2.81), we can write the
electrostatic potential outside the dielectric sphere:

$$\Phi_{r>a}(r, \theta) = -E_0 r \cos\theta + \sum_{n=0}^{\infty} \frac{B_n}{r^{n+1}} P_n(\cos\theta). \qquad (2.111)$$

It is easy to see that the electrostatic potential inside the sphere should be given by

$$\Phi_{r<a}(r, \theta) = \sum_{n=0}^{\infty} A_n r^n P_n(\cos\theta) \qquad (2.112)$$

if we demand that the electrostatic potential should not blow up at the origin.

Now we come to the boundary conditions. The condition (2.107) implies that

$$\left.\frac{\partial \Phi_{r>a}}{\partial \theta}\right|_{r=a} = \left.\frac{\partial \Phi_{r<a}}{\partial \theta}\right|_{r=a}, \tag{2.113}$$

whereas the other boundary condition (2.106) requires

$$\left.\frac{\partial \Phi_{r>a}}{\partial r}\right|_{r=a} = \kappa \left.\frac{\partial \Phi_{r<a}}{\partial r}\right|_{r=a}. \tag{2.114}$$

It is not difficult to check that these boundary conditions can be satisfied if we keep only the terms involving $P_1(\cos\theta) = \cos\theta$ in (2.111) and (2.112). Retaining only these terms, (2.111) and (2.112) become

$$\Phi_{r>a}(r, \theta) = -E_0 r \cos\theta + \frac{B_1}{r^2} \cos\theta, \tag{2.115}$$

$$\Phi_{r<a}(r, \theta) = A_1 r \cos\theta. \tag{2.116}$$

Substituting these solutions (2.115) and (2.116) in the boundary conditions (2.113) and (2.114), we get

$$-E_0 a + \frac{B_1}{a^2} = A_1 a,$$

$$-E_0 - \frac{2B_1}{a^3} = \kappa A_1.$$

It easily follows from these equations that

$$A_1 = -\frac{3E_0}{\kappa + 2}, \tag{2.117}$$

$$B_1 = \frac{\kappa - 1}{\kappa + 2} a^3 E_0. \tag{2.118}$$

Once the values of the coefficient A_1 and B_1 in the solution (2.115) and (2.116) are given by (2.117) and (2.118), we have the required solution of the boundary value problem in terms of the given parameters, such as κ, a and E_0. The field lines shown in Fig. 2.14 are as given by this solution. We should point out that the field lines shown in Fig. 2.14 are actually field lines of the electric displacement **D**, in conformity with the fact that the field lines of **D** should be continuous according to (2.100) when $\rho = 0$, that is, when there is no surface charge on the surface of the dielectric sphere. It is easy to conclude from (2.116) and (2.117) that the electric field inside the dielectric sphere is given by

$$E_{r<a} = \frac{3}{\kappa + 2} E_0. \tag{2.119}$$

Inside the dielectric sphere, we shall have uniform polarization. From (2.101), (2.103) and (2.119), this uniform polarization is

$$P_{r<a} = \epsilon_0(\kappa - 1) E_{r<a} = \frac{3(\kappa - 1)}{(\kappa + 2)} \epsilon_0 E_0. \tag{2.120}$$

We leave it for the reader to check that, while $E_{r<a}$ inside the dielectric sphere is less than E_0, the electric displacement D inside the sphere is higher than that in the surrounding regions, which is apparent in Fig. 2.14.

It is interesting to consider the case $\kappa = 1$, which will make the dielectric properties inside and outside the sphere indistinguishable. In such a situation, (2.120) implies that there will be no polarization inside the sphere, and (2.119) implies that the electric field inside the sphere will be E_0. Since it follows from (2.118) that B_1 will be zero, we conclude from (2.115) that in this situation we would have the undisturbed electric field E_0 everywhere.

2.17 Energy Density of an Electrostatic Field

An electric field has an energy associated with it. We can estimate the energy in the following way. Suppose we consider a charge distribution $\rho(x)$ which gives rise to the electrostatic potential $\Phi(x)$. Since Poisson's equation (2.10) is a linear equation, we expect the electrostatic potential to be $f\Phi(x)$ when the charge distribution is $f\rho(x)$ (here, f is a fraction lying between 0 and 1). Suppose we bring some more charge from infinity and distribute it in such a way that charge distribution now becomes $(f + df)\rho(x)$. According to (2.9), the work done in bringing this extra charge $df\rho(x)$ from infinity is

$$dU = \int_V df \, \rho(x) \, f \, \Phi(x) \, dV. \tag{2.121}$$

Note that in Sect. 2.2 we had denoted the work done by the electric field as W, which is negative of the work U done by 'us' in building up the electric field configuration. To get the total work done in building up the field, we have to integrate (2.121) from $f = 0$ to $f = 1$. This gives

$$U = \int_{f=0}^{f=1} dU = \int_V \rho(x) \, \Phi(x) \, dV \int_{f=0}^{f=1} f \, df = \frac{1}{2} \int_V \rho(x) \, \Phi(x) \, dV.$$

On making use of (2.100), this becomes

$$U = \frac{1}{2} \int (\nabla \cdot \mathbf{D}) \, \Phi \, dV.$$

The use of the vector identity (B.4) now gives us

$$U = \frac{1}{2} \int \nabla \cdot (\Phi \mathbf{D}) \, dV - \frac{1}{2} \int \mathbf{D} \cdot \nabla \Phi \, dV. \tag{2.122}$$

The first term can be converted into a surface integral by Gauss's theorem (B.13). If this surface is taken at a large distance from the region where the charges are distributed, then this surface integral will clearly be zero. Making use of (2.1), we now have

$$U = \frac{1}{2} \int \mathbf{E} \cdot \mathbf{D} \, dV. \tag{2.123}$$

Since this is the total work done in building the electric field by bringing the charges from infinity little by little, it is natural to interpret this as the energy residing with the electric field. The expression (2.123) readily implies that we can interpret $\frac{1}{2}\mathbf{E}.\mathbf{D}$ as the energy density associated with the electric field.

Note that we have derived this result by taking the possible presence of dielectric media into account. We shall see in Sect. 4.2 that, even in situations involving evolution with time, the energy associated with the electric field is given by (2.123).

We point out one important consequence of this expression of energy for dielectric materials. Suppose we have charges put on a few conductors kept in empty space. We now consider the hypothetical situation that empty space everywhere is replaced by a uniform dielectric medium, while the charges on the conductors remain unchanged. It is easy to argue that $\mathbf{D}$ outside the conductors will remain unchanged on replacing the empty space by the dielectric medium. On the other hand, $\mathbf{E}$ will decrease by the factor κ (we are assuming $\kappa > 1$). As a result, the total energy given by (2.123) will decrease. Consider now a piece of dielectric material in an electric field. If the piece moves into a region where the electric field is stronger (i.e. where E^2 has a higher value), then it is easy to check that the overall energy of the system decreases. On this ground, we expect that there will be a force acting on a piece of dielectric material dragging it to regions where E^2 is stronger. A more detailed analysis indeed shows this to be the case. We shall not get into this detailed analysis here and would refer the interested reader to the discussion of this topic given by Panofsky and Phillips [3, Sects. 6-6 and 6-7].

2.18 Microscopic Theory of Dielectric Materials

One important topic is to understand what happens at the molecular level inside a dielectric material to give rise to its macroscopic properties. We give a very brief introduction to this subject.

In some materials, the molecules themselves intrinsically have electric dipole moments associated with them. An electric field tries to align these molecular dipoles. A molecular dipole $\mathbf{p}_0$ making an angle θ with respect to an electric field $\mathbf{E}$ would have potential energy $-p_0 E \cos \theta$ according to (2.20). We expect from elementary thermal physics that the probability of such a dipole making an angle θ is proportional to $\exp(p_0 E \cos \theta / k_B T)$, where k_B is the Boltzmann constant. Since such a molecule would contribute an electric dipole $p_0 \cos \theta$ in the direction of the electric field, the average electric dipole contributed by all the molecules can be obtained by averaging over all of them. This is

$$\overline{p} = \frac{\int d\Omega \; p_0 \cos \theta \; \exp \left(\frac{p_0 E \cos \theta}{k_B T} \right)}{\int d\Omega \; \exp \left(\frac{p_0 E \cos \theta}{k_B T} \right)}. \tag{2.124}$$

We can take

$$\exp \left(\frac{p_0 E \cos \theta}{k_B T} \right) \approx \left(1 + \frac{p_0 E \cos \theta}{k_B T} \right),$$

since $p_0 E$ is usually much smaller than $k_B T$ for temperatures of interest to us. On substituting this in (2.124), we get

$$\overline{p} \approx \frac{p_0^2}{3 k_B T} E. \tag{2.125}$$

Even molecules which are not polar may develop electric dipole moments when kept in an electric field. Let us assume that the electrons are bound to molecules in such a way there is a restoring force of $-m_e \omega_0^2 \mathbf{x}$ when an electron is displaced from its mean position by $\mathbf{x}$. From the force balance condition that the sum of this force and the electric force $-e\mathbf{E}$ should be zero, the resulting electric dipole moment is

$$\mathbf{p} = -e\,\mathbf{x} = \frac{e^2}{m_e \omega_0^2} \mathbf{E}. \tag{2.126}$$

From (2.125) and (2.126), it is clear that a molecule would make a contribution to the electric dipole moment proportional to the electric field $\mathbf{E}$ for both the cases we have considered. We can write

$$\overline{\mathbf{p}} = \alpha \epsilon_0 \mathbf{E}. \tag{2.127}$$

If there are N molecules per unit volume, then the electric polarization per unit volume is

$$\mathbf{P} = N\overline{\mathbf{p}} = N\alpha \epsilon_0 \mathbf{E}. \tag{2.128}$$

This is of the form (2.101). However, there is still a small twist in the complete theory of electrical susceptibility. The electric field experienced by a molecule need not necessarily be the macroscopic electric field $\mathbf{E}$. One can give some theoretical arguments that the effective electric field experienced by the molecules is

$$\mathbf{E}_{\text{eff}} = \mathbf{E} + \frac{1}{3\epsilon_0}\mathbf{P}, \qquad (2.129)$$

a result which appears to agree with experimental data. We refer the readers to Feynman et al. [6, Sect. 11-4] for the theoretical arguments, with the comment that one has to consider a small spherical cavity inside the dielectric material. However, (2.129) is different from what you would get by solving the boundary value problem with a spherical cavity (Exercise 2.16). To derive (2.129), one has to assume that the electric field is not disturbed by the presence of the cavity. If we replace $\mathbf{E}$ in (2.128) by $\mathbf{E}_{\text{eff}}$ as given by (2.129), we have the expression

$$\mathbf{P} = N\alpha\epsilon_0 \left(\mathbf{E} + \frac{1}{3\epsilon_0}\mathbf{P}\right). \qquad (2.130)$$

Substituting for $\mathbf{P}$ from (2.101) and using (2.103), we get

$$\epsilon_0(\kappa - 1) = N\alpha\epsilon_0 \left[1 + \frac{1}{3}(\kappa - 1)\right],$$

from which

$$\frac{N\alpha}{3} = \frac{\kappa - 1}{\kappa + 2}. \qquad (2.131)$$

This is known as the *Clausius–Mossotti relation*, which relates the microscopic molecular polarizability coefficient α to the macroscopic dielectric constant κ.

Exercises

2.1 Consider an infinite line charge with charge density Q per unit length. Find out the electric field at a radial distance r (i) by Gauss's law and (ii) by Coulomb's law. Are the results the same? What is the electrostatic potential? Is it possible to make this potential zero at infinity? If not, then how do you interpret this result?

2.2 Consider a circular disk of radius R having a uniform charge density σ per unit area.

(i) By direct application of Coulomb's law, show that the electric field at a point on the axis at a distance z from the charged surface is along the axis and is given by

$$E = \frac{\sigma}{2\epsilon_0}\left[1 - \frac{z}{\sqrt{z^2 + R^2}}\right].$$

Check the results you get in the limits $z \to 0$ and $z \to \infty$, and interpret them.

(ii) Use (2.2) to show that the electrostatic potential at the point on the axis is given by

$$\Phi = \frac{\sigma}{2\epsilon_0}[\sqrt{z^2 + R^2} - z].$$

Obtain the electric field from this. Is this the same as what you get by direct application of Coulomb's law?

2.3 Consider a force field going as

$$\mathbf{F} = \frac{k}{r^n}\hat{\mathbf{r}},$$

where n is a real number and k is a constant. Calculate $\nabla.\mathbf{F}$ and, for points other than the origin, show that this is zero only when $n = 2$. When is the divergence positive and when is it negative? Can you give a physical interpretation?

2.4 Suppose there is a charge $-2q$ at the origin and that there are two charges $+q$ each on the polar axis at distances $+l$ and $-l$ from the origin. Determine the electrostatic potential at a *distant* point. Calculate the electric field and sketch the field lines.

2.5 Two electric dipoles $\mathbf{p}_1$ and $\mathbf{p}_2$ are separated by a distance $\mathbf{r}$. Show that the electrostatic potential energy between these two dipoles is

$$U = \frac{1}{4\pi\epsilon_0}\left[\frac{\mathbf{p}_1.\mathbf{p}_2}{r^3} - \frac{3}{r^5}(\mathbf{p}_1.\mathbf{r})(\mathbf{p}_2.\mathbf{r})\right].$$

2.6 Prove that electrostatic forces alone cannot hold a charged particle in *stable* equilibrium at a point in space.

2.7 We can represent the time-averaged potential of a neutral hydrogen atom reasonably well by the expression

$$\Phi = \frac{1}{4\pi\epsilon_0}\frac{qe^{-\alpha r}}{r}\left(1 + \frac{\alpha r}{2}\right),$$

where q is the electronic charge and $\alpha^{-1} = a_0/2$, a_0 being the Bohr radius. Find the charge distribution which will give rise to this potential and interpret your result. What is the total charge?

2.8 Suppose a charge q is at a distance a from an infinite plane conductor kept at zero potential. Find out the surface charge density σ induced on the surface of the conductor and then calculate the total charge induced by integrating σ over the surface. Assuming the force per unit area on the conductor to be σE_n, find out the total force of attraction experienced by the conductor by integrating over the whole surface of the conductor and check whether you get the same result as the force on the image charge exerted by the actual charge. If there is a discrepancy, can you think of the reason behind it?

2.9 Consider a line charge placed parallelly between two grounded parallel conducting plates. This problem has been solved as Example 2 in Sect. 2.10. Write a computer programme to calculate the potential at several points and then draw equipotential surfaces.

2.10 Suppose that $E_z(x, y)$ on the plane $z = 0$ is given by

$$E_z = E_0 \sin(lx) \sin(my).$$

Find out the potential at all points above the $z = 0$ plane, assuming that the potential satisfies Laplace's equation everywhere above the plane.

Argue that a metallic mesh with an electric field on one side of it is almost as effective as a metal sheet in screening the electric field from the other side.

2.11 Consider a ring carrying a charge q placed symmetrically with respect to the polar axis and located at the point (r_0, θ_0) in spherical polar coordinates. Find out the electrostatic potential at all points in space (i.e. for both $r < r_0$ and $r > r_0$).

Hint. You first have to write down the expression of the electrostatic potential at a point on the axis and expand it in powers of r by using (2.82). Then you can use (2.81) to solve this problem in the form of a series.

2.12 Consider a thin spherical shell of electrically conducting material with radius a and suppose that a very small electric dipole **p** is kept at the centre of this spherical shell. What will be the electrostatic potentials inside and outside the spherical shell? Calculate the surface charge density induced in the inner surface of the shell and make a sketch of the electric field lines.

2.13 Suppose we want to solve Laplace's equation in a cylindrical region. Assuming the solution to be of the form
$$R(r)e^{\pm kz \pm im\phi},$$

find out the equation to be satisfied by $R(r)$. Assuming that there is a source of electric field inside a conducting cylinder of which the surface $r = a$ is kept at zero potential, write down the general solution inside this cylinder. (You will need a basic knowledge about Bessel functions to answer this question.)

2.14 If the total charge and the electric dipole moment of an axisymmetric charge distribution is zero, show that the value of the electric quadrupole moment does not depend on where you take the origin on the symmetry axis.

2.15 (a) We had arrived at (2.89) by assuming that the field point **x** is on the $\theta = 0$ axis. For any two arbitrary position vectors **x** and **x**′, argue that

$$\frac{1}{|\mathbf{x} - \mathbf{x}'|} = \frac{1}{r} \sum_{n=0}^{\infty} P_n(\cos\alpha) \left(\frac{r'}{r}\right)^n,$$

where $r = |\mathbf{x}| > r' = |\mathbf{x}'|$ and α is the angle between these position vectors such that $\cos\alpha = \mathbf{x}.\mathbf{x}'/rr'$.

(b) We have discussed the electrostatic multipole expansion in Sect. 2.13 for an axisymmetric charge distribution. Even for a *non-axisymmetric* charge distribution in a localized region, show that the electrostatic potential at a point far away from the charge distribution would have a dipole term given by

$$\frac{1}{4\pi\epsilon_0} \frac{\mathbf{p}.\mathbf{x}}{r^2},$$

where $\mathbf{p}$ is the dipole moment of the non-axisymmetric charge distribution given by

$$\mathbf{p} = \int \rho(\mathbf{x}') \, \mathbf{x}' \, dV'.$$

Note that we are using here our standard notation in which $\mathbf{x}'$ refers to source points and $\mathbf{x}$ to field points.

2.16 Consider a spherical cavity inside a material of dielectric constant κ within which there is an electric field having the constant value $\mathbf{E}$ away from the cavity. Show that the electric field inside the cavity is given by

$$\mathbf{E}_c = \mathbf{E} + \frac{1}{(1 + 2\kappa)} \frac{\mathbf{P}}{\epsilon_0},$$

where $\mathbf{P}$ is the polarization induced in the dielectric medium due the electric field $\mathbf{E}$.

2.17 A conductor is kept in a region where the surrounding medium has dielectric constant κ. If the surface charge in a place at the surface of the conductor is σ, find out the normal component of the electric displacement $\mathbf{D}$ and the normal component of the electric field $\mathbf{E}$ just outside the surface of the conductor. What is the surface charge density induced on the surface of the dielectric adjacent to the conducting surface? Verify that (2.25) is satisfied if σ appearing in (2.25) is replaced by the sum of the surface charge density on the conductor and the induced surface charge on the adjacent surface of the dielectric.

2.18 Two coaxial conducting cylinders of inner and outer radii a and b carry charges q and $-q$ per unit length. A part of the space between the cylinders with angular extent θ_0 is filled with a material of dielectric constant κ, as shown in Fig. 2.15.

(a) Find the electric field everywhere between the cylinders. (b) Calculate the surface charge distribution on the inner cylinder.

Fig. 2.15 Refer to
Exercise 2.18

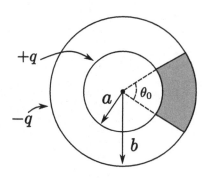

References

1. Arfken, G.B., Weber, H.J., Harris, E.E.: Mathematical Methods for Physicists, 7th edn. Elsevier, Amsterdam (2013)
2. Griffiths, D.J.: Introduction to Electrodynamics, 4th edn. Cambridge University Press, Cambridge (2017)
3. Panofsky, W.K.H., Phillips, M.: Classical Electricity and Magnetism, 2nd edn. Addison-Wesley (reprinted by Dover), Boston (1962)
4. Jackson, J.D.: Classical Electrodynamics, 3rd edn. Wiley, New York (1999)
5. Born, M., Wolf, E.: Principles of Optics, 6th edn. Pergamon Press, Oxford (1980)
6. Feynman, R.P., Leighton, R.B., Sands, M.: The Feynman Lectures on Physics, vol. II: Mainly Electromagnetism and Matter. Addison-Wesley, Boston (1964)

Chapter 3
Magnetostatics

3.1 Basic Principles

We pointed out in Sect. 1.3 that, under static circumstances, Maxwell's equations neatly split into separate equations for the electric field and for the magnetic field. After discussing electrostatics in Chap. 2, we now turn to magnetostatics. The main aim of magnetostatics is to study how static magnetic fields are produced by static electric currents. The basic equations of magnetostatics are (1.10) and (1.11). We have discussed in Sect. 1.5 how to solve a vector field $\mathbf{G}$ of which the divergence s and the curl $\mathbf{c}$ are specified. We now write down the solution by substituting $\mathbf{G} = \mathbf{B}$, $s = 0$ and $\mathbf{c} = \mu_0 \mathbf{j}$ into (1.20)–(1.22). The magnetic field $\mathbf{B}(\mathbf{x})$ produced by the static current $\mathbf{j}(\mathbf{x})$ can be specified in the following way:

$$\mathbf{B} = \nabla \times \mathbf{A}, \tag{3.1}$$

where $\mathbf{A}$ known as the vector potential of the magnetic field is given by

$$\mathbf{A}(\mathbf{x}) = \frac{\mu_0}{4\pi} \int \frac{\mathbf{j}(\mathbf{x}')}{|\mathbf{x} - \mathbf{x}'|} \, dV'. \tag{3.2}$$

It is clear that we have to take the curl of $\mathbf{A}$, as given by (3.2), in order to obtain the magnetic field $\mathbf{B}$. First note that

$$\nabla \times \left(\frac{\mathbf{j}(\mathbf{x}')}{|\mathbf{x} - \mathbf{x}'|} \right) = \nabla \left(\frac{1}{|\mathbf{x} - \mathbf{x}'|} \right) \times \mathbf{j}(\mathbf{x}') + \frac{1}{|\mathbf{x} - \mathbf{x}'|} \nabla \times \mathbf{j}(\mathbf{x}')$$

according to the vector identity (B.5). Since the operator ∇ operates only on the field coordinates $\mathbf{x}$ and not on the source coordinates $\mathbf{x}'$, it is obvious that the second term will be zero. Making use of (1.18) in the first term, we get

© Springer Nature Singapore Pte Ltd. 2022
A. R. Choudhuri, *Advanced Electromagnetic Theory*, Lecture Notes in Physics 1009,
https://doi.org/10.1007/978-981-19-5944-8_3

$$\nabla \times \left(\frac{\mathbf{j}(\mathbf{x'})}{|\mathbf{x} - \mathbf{x'}|} \right) = \frac{\mathbf{j}(\mathbf{x'}) \times (\mathbf{x} - \mathbf{x'})}{|\mathbf{x} - \mathbf{x'}|^3} \tag{3.3}$$

It follows from (3.1) to (3.3) that

$$\mathbf{B}(\mathbf{x}) = \frac{\mu_0}{4\pi} \int \frac{\mathbf{j}(\mathbf{x'}) \times (\mathbf{x} - \mathbf{x'})}{|\mathbf{x} - \mathbf{x'}|^3} \, dV'. \tag{3.4}$$

This is the expression of the magnetic field $\mathbf{B}$ due to the current distribution $\mathbf{j}$. Sometimes we have to consider currents flowing through conductors. If we consider a small length element $d\mathbf{l'}$ of a conductor through which a current I is flowing, it is obvious that we have to write $I \, d\mathbf{l'}$ in the place of $\mathbf{j} \, dV'$. In this case, (3.4) would clearly become

$$\mathbf{B}(\mathbf{x}) = \frac{\mu_0}{4\pi} I \int \frac{d\mathbf{l'} \times (\mathbf{x} - \mathbf{x'})}{|\mathbf{x} - \mathbf{x'}|^3}. \tag{3.5}$$

If we want to determine the magnetic field due to a closed circuit, we have to carry on the line integral over the circuit. This expression (3.5) is sometimes referred to as the *Biot–Savart law*.

While deriving an expression of the electric field due to a surface electric dipole layer in Sect. 2.4, we pointed out that it has a connection with magnetostatics. Now we can see that connection. In (2.19), $\mathbf{r}$ is the vectorial distance of the field point from the source point, which we are denoting here by $\mathbf{x} - \mathbf{x'}$. Keeping in mind that $\mu_0/4\pi$ appears in magnetostatics in a manner very similar to the manner in which $1/4\pi\epsilon_0$ appears in electrostatics, the magnetic field due to a surface magnetic dipole layer with uniform magnetic dipole density τ_m per unit area should be given by the following expression similar to (2.19):

$$\mathbf{B}(\mathbf{x}) = \frac{\mu_0}{4\pi} \tau_m \oint \frac{d\mathbf{l'} \times (\mathbf{x} - \mathbf{x'})}{|\mathbf{x} - \mathbf{x'}|^3}, \tag{3.6}$$

where the integration has to be carried out over the boundary curve of the magnetic dipole surface, similar to what is done in (2.19). A comparison between (3.5) and (3.6) readily leads to the following conclusion: as far as the production of magnetic field is concerned, a current carrying circuit is equivalent to a surface of uniform magnetic dipole of which the circuit is the boundary and of which the strength (i.e. the magnetic dipole density) is

$$\tau_m = I. \tag{3.7}$$

3.2 Biot–Savart Law

Suppose we have two current carrying circuits kept at some distance. We want to find the force $d\mathbf{F}_{12}$ exerted by a small element of length dl_1 of the first circuit carrying current I_1 on a small element of length dl_2 of the second circuit carrying current I_2. If $\mathbf{x}_1$ and $\mathbf{x}_2$ are the vectorial positions of the current-carrying elements dl_1 and dl_2, then we conclude from (3.5) that the magnetic field $d\mathbf{B}$ produced by dl_1 at the location of dl_2 is

$$dB = \frac{\mu_0}{4\pi} I_1 \frac{dl_1 \times (\mathbf{x}_2 - \mathbf{x}_1)}{|\mathbf{x}_2 - \mathbf{x}_1|^3}. \tag{3.8}$$

The next step is to figure out the force experienced by dl_2 due to this magnetic field $d\mathbf{B}$. For this, we need to use (1.6), according to which the force on a volume element dV carrying current density $\mathbf{j}$, where the magnetic field is $\mathbf{B}$, is given by $\mathbf{j} \times \mathbf{B} \, dV$. Writing $I \, dl$ in the place of $\mathbf{j} \, dV$ as we did in going from (3.4) to (3.5), the force on an element dl of a circuit carrying current I placed in a magnetic field $\mathbf{B}$ is given by

$$d\mathbf{F} = I \, (dl \times \mathbf{B}). \tag{3.9}$$

This is another version of the Lorentz force equation. In order to obtain the force $d\mathbf{F}_{12}$ exerted by dl_1 on dl_2, we simply have to put $I = I_2$ and $dl = dl_2$ in (3.9), while substituting for $\mathbf{B}$ the magnetic field $d\mathbf{B}$ produced by dl_1 as given by (3.8). This gives

$$d\mathbf{F}_{12} = \frac{\mu_0}{4\pi} I_1 I_2 \frac{dl_2 \times [dl_1 \times (\mathbf{x}_2 - \mathbf{x}_1)]}{|\mathbf{x}_2 - \mathbf{x}_1|^3}. \tag{3.10}$$

We shall generally refer to this as the *Biot–Savart law* in magnetostatics, although (3.5) also often goes by that name. The constant μ_0 is called the *permeability of free space*.

The Biot–Savart law, as given by (3.10), has a role in magnetostatics somewhat similar to the role of Coulomb's law (2.7) in electrostatics. Just as point charges can be regarded as basic entities in electrostatics, small elements of current-carrying conductors can be regarded as basic entities in magnetostatics. In college-level textbooks on electromagnetism, the discussion of magnetostatics often begins with the Biot–Savart law. Since the aim of this book is to show how the whole of electromagnetism can be developed from Maxwell's equations, we have obtained the Biot–Savart law from Maxwell's equations—by considering the Eqs. (1.10)–(1.11) dealing with magnetic fields in a static situation and then by combining the magnetic field obtained from them with the Lorentz force equation (3.9).

So far, we have not said anything about the unit of current or the value of μ_0. Now we are in a position to introduce the SI unit of current, which is taken as the basic electrical unit in the SI system from which the units of other electrical quantities can be obtained. We can introduce the SI unit of current in such a way that μ_0 in (3.10) has a certain specified value when the length is expressed in metre (m) and the force in newton (N). The SI unit ampere (A) of electrical current is introduced by demanding that we have

$$\frac{\mu_0}{4\pi} = 10^{-7}\,\mathrm{N\,A^{-2}}. \tag{3.11}$$

The choice of such a value for μ_0 ensures that the SI unit of current turns out to be very convenient for everyday use. For example, the typical current flowing through a wire in a modern household would be of the order of a few amperes. We have pointed out in Sect. 2.1 that the unit of charge follows from the unit of current, and we no longer have the freedom to choose the permittivity of free space ϵ_0 after we have chosen the value of permeability of free space μ_0 in accordance with (3.11).

From Coulomb's law (2.7) giving the force $\mathbf{F}_{12}$ exerted by charge q_1 on charge q_2, we can at once write down the force $\mathbf{F}_{21}$ exerted by charge q_2 on charge q_1, which is

$$\mathbf{F}_{21} = \frac{1}{4\pi\epsilon_0} q_1 q_2 \frac{\mathbf{x}_1 - \mathbf{x}_2}{|\mathbf{x}_1 - \mathbf{x}_2|^3}.$$

We see at once that

$$\mathbf{F}_{21} = -\mathbf{F}_{12}.$$

In other words, Newton's third law of motion holds for Coulomb's law. An interesting question is whether Newton's third law holds for the Biot–Savart law as well. Since the expression of $d\mathbf{F}_{12}$ given by (3.10) is somewhat complicated, the answer to this question is not at once obvious and requires some analysis. The term in the numerator of (3.10) can be written as

$$d\mathbf{l}_2 \times [d\mathbf{l}_1 \times (\mathbf{x}_2 - \mathbf{x}_1)] = [d\mathbf{l}_2.(\mathbf{x}_2 - \mathbf{x}_1)]d\mathbf{l}_1 - (d\mathbf{l}_1.d\mathbf{l}_2)(\mathbf{x}_2 - \mathbf{x}_1).$$

On making use of this, (3.10) can be written as

$$d\mathbf{F}_{12} = \frac{\mu_0}{4\pi} I_1 I_2 \left[-\left\{ d\mathbf{l}_2. \nabla_2 \left(\frac{1}{|\mathbf{x}_2 - \mathbf{x}_1|} \right) \right\} d\mathbf{l}_1 - \frac{d\mathbf{l}_1.d\mathbf{l}_2}{|\mathbf{x}_2 - \mathbf{x}_1|^3} (\mathbf{x}_2 - \mathbf{x}_1) \right], \tag{3.12}$$

where ∇_2 is the gradient with respect to the position coordinate $\mathbf{x}_2$, and we have used the fact

$$\nabla_2 \left(\frac{1}{|\mathbf{x}_2 - \mathbf{x}_1|} \right) = -\frac{(\mathbf{x}_2 - \mathbf{x}_1)}{|\mathbf{x}_1 - \mathbf{x}_2|^3},$$

which should be evident from (1.18). The second term within the square bracket in (3.12) obviously would pick up a negative sign on interchanging the subscripts 1 and 2. This means that this term obeys Newton's third law. But that is not the case for the first term within the square bracket in (3.12). Hence, the expression (3.12) for $d\mathbf{F}_{12}$ as a whole would not be equal to $-d\mathbf{F}_{21}$ and would not satisfy Newton's third law. Let us now consider the force exerted by the whole first circuit on the whole second circuit. To get this force, we have to integrate (3.12) both over $d\mathbf{l}_1$ and $d\mathbf{l}_2$. It is clear that the first term in (3.12) would give rise to the line integral over $d\mathbf{l}_2$ of a gradient, which gives zero. In other words, when we carry on the line integral, the term not consistent with Newton's third law drops outs, and only the term consistent with Newton's third law remains. The final expression of the force is

$$\mathbf{F}_{12} = -\frac{\mu_0}{4\pi} I_1 I_2 \oint_1 \oint_2 \frac{(\mathbf{x}_2 - \mathbf{x}_1)}{|\mathbf{x}_2 - \mathbf{x}_1|^3} d\mathbf{l}_1.d\mathbf{l}_2. \tag{3.13}$$

We draw the very important and enigmatic conclusion: while Newton's third law of motion does not hold if we consider the force between small elements in two circuits, this law holds when we consider the force between the two whole circuits.

3.3 Ampere's Law in Magnetostatics

In Sect. 2.6 we integrated the basic electrostatic Eq. (1.8) over a volume to obtain Gauss's law in electrostatics. In a similar spirit, we now integrate the basic magnetostatic equation (1.11) over a surface. This gives

$$\int (\nabla \times \mathbf{B}).\,d\mathbf{S} = \mu_0 \int \mathbf{j}.\,d\mathbf{S}. \tag{3.14}$$

Now, $\int \mathbf{j}.\,d\mathbf{S}$ appearing on the RHS of (3.14) is nothing but the total current flowing through the surface, which we can denote by I. We can use Stokes's theorem (B.15) to convert the LHS of (3.14) into a line integral over the boundary of the surface. We get

$$\oint \mathbf{B}.\,d\mathbf{l} = \mu_0 I. \tag{3.15}$$

This is the well-known *Ampere's law in magnetostatics*.

Just as Gauss's law in electrostatics can be used to calculate the electric field effortlessly in highly symmetric situations, Ampere's law (3.15) can similarly be used for a quick calculation of the magnetic field in some highly symmetric situations. Let us use this law to calculate the magnetic field at a distance r from a straight wire of infinite length through which a current I is flowing. We consider a circular surface of radius r perpendicular to the wire which is at the centre of this circular surface. The circumference of the circular surface is the path along which the line integral in the LHS of (3.15) has to be evaluated, as indicated in Fig. 3.1. From symmetry, we expect the magnetic field $\mathbf{B}$ to be along this path so that we should have $\oint \mathbf{B}.\,d\mathbf{l} = 2\pi r B$. The current I flowing through the wire is clearly the full current across the circular surface. Thus, (3.15) becomes

$$2\pi r B = \mu_o I,$$

Fig. 3.1 An infinite straight current-carrying wire with a circular path around it along which we have to carry out the integration for finding the magnetic field

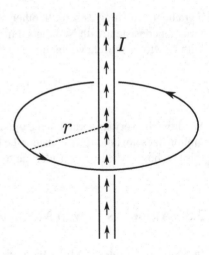

from which the magnetic field at a distance r from a straight wire carrying current I is given by

$$B = \frac{\mu_0}{2\pi} \frac{I}{r} \qquad (3.16)$$

and is along the circular path encircling the wire.

3.4 Techniques for Solving Magnetostatic Problems

It should be clear by now that the basic equations of magnetostatics are more complicated than the basic equations of electrostatics. As a result, calculating the magnetic field due to a current distribution is usually a more difficult problem than calculating the electric field due to a charge distribution. There are very few current distributions for which the magnetic field can be obtained analytically and written down in a closed form. Now we list various possible techniques for finding the magnetic field due to a current distribution. In a particular situation, one has to figure out the particular technique which will be the most suitable for handling the situation.

3.4.1 Using the Biot–Savart Law

In any arbitrary situation involving a current distribution, one can always apply (3.5), sometimes referred to as the Biot–Savart law, to calculate the magnetic field. In a few cases, the integration in (3.5) is not too difficult to perform, and we can derive an expression of the magnetic field. For example, the magnetic field due to a straight

Fig. 3.2 A sketch indicating how the magnetic field due to a circular loop on its symmetry axis can be evaluated. The magnetic field $d\mathbf{B}$ produced by an element $d\mathbf{l}'$ of the loop is indicated. One has to carry on an integration over the entire loop

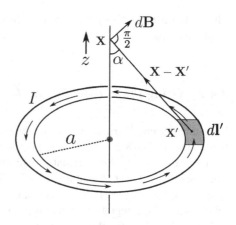

infinite current-carrying conductor can be calculated by using the Biot–Savart law. We obtained this magnetic field in Sect. 3.3 by using Ampere's law. It is given as Exercise 3.1 to prove this from the Biot–Savart law.

As an example of the application of the Biot–Savart law, we consider a circular current-carrying loop of radius a and calculate the magnetic field at a point $\mathbf{x}$ on the symmetry axis (see Fig. 3.2). The calculation of the magnetic field at off-axis points involves elliptic integrals and will not be discussed in this book. Let α be the angle subtended by the radius of the current-carrying loop at the point where we want to find the magnetic field. A small element $d\mathbf{l}'$ of the loop is obviously perpendicular to the distance $\mathbf{x} - \mathbf{x}'$ of the field point from this element. The magnetic field $d\mathbf{B}$ due to this element will be in the direction of the vector $d\mathbf{l}' \times (\mathbf{x} - \mathbf{x}')$ making an angle $\alpha + \pi/2$ with the symmetry axis. When we integrate over the whole loop, the component of the magnetic field perpendicular to the symmetry axis will certainly be zero due to symmetry. The vertical component according to (3.5) will be

$$B_z = \frac{\mu_0}{4\pi} I \int \frac{dl'}{(a^2 + z^2)} \sin\alpha, \qquad (3.17)$$

where we have taken the z axis in the upward direction, with the centre of the loop as the origin. Since the quantities inside the integrand remain constant when we do the integration over the loop, the integration of dl' simply gives $2\pi a$. On noting further that

$$\sin\alpha = \frac{a}{\sqrt{a^2 + z^2}},$$

we finally get from (3.17):

$$B_z = \frac{\mu_0}{2} \frac{I a^2}{(a^2 + z^2)^{3/2}}. \qquad (3.18)$$

This is the expression of the magnetic field due to a current-carrying loop on its symmetry axis.

3.4.2   Using the Vector Potential

For some current distributions, it may be easy to first calculate the vector potential by making use of (3.2). Then we can get the magnetic field by taking its curl.

In our discussion of electrostatics, we saw that very often we calculate the electrostatic potential first by making use of (2.2), and then we can get the electric field from (2.1). In the case of electrostatics, the calculation of the electrostatic potential involves a scalar integration, which is often much easier to do than a vector integration like (2.3). It is for this reason that the electrostatic potential is found to be so useful in many situations and is widely used. In the case of magnetostatics, however, the calculation of the vector potential (3.2) involves a vector integration exactly as in the case of the direct application of the Biot–Savart law (3.5). It is true that we can write down the three spatial components of (3.2) in the form of three scalar integrations. However, it happens only in an occasional problem that going through the intermediate step of first calculating the vector potential makes things simpler. Hence, the usefulness of the vector potential for solving practical problems of magnetostatics is somewhat limited compared to the usefulness of the scalar potential for solving practical problems of electrostatics. One particular problem in which the vector potential is found useful is the problem of finding the magnetic field at *any* point in space due to a circular current-carrying loop. We have calculated the magnetic field due to such a loop at a point on the axis in Sect. 3.4.1 by applying the Biot–Savart law. Integration of the Biot–Savart law for finding the magnetic field at points away from the axis presents formidable mathematical difficulties. The best way to proceed is to first find the vector potential due to the loop at off-axis points. Even the calculation of the vector potential involves elliptic integrals and will not be discussed here. The interested reader may look up Jackson [1, Sect. 5.5] for a discussion.

It follows from (3.1) that we would get the same magnetic field **B** if **A** were replaced by $\mathbf{A} + \nabla \psi$. In other words, there is some indefiniteness in defining **A**. We shall discuss this issue in Sect. 4.10 when we consider the vector potential in the time-varying situation. In a static situation, one thing is obvious: If we calculate **A** for a current distribution **j** by using (3.2), then we would get the correct magnetic field **B**. We do not have to bother about the indefiniteness of **A** as long as we follow this forward approach. However, when we follow the reverse approach of finding out **A** for a given field **B**, this indefiniteness can become bothersome—even for the simplest situation of a uniform magnetic field (Exercise 3.3). One important question is whether this indefiniteness of **A** has any experimentally verifiable consequences.

This brings us to the issue of whether the vector potential is "real" in the sense of giving rise to some physical consequences in addition to the consequences due to the magnetic field. A magnetic field gives rise to forces on magnets or currents. We usually measure a magnetic field by measuring such forces. Is it possible to do any experimental measurements which will directly give us information about **A** rather than **B**? Nobody has figured out any way of doing this within the framework of classical electromagnetism. As long as we limit ourselves to the classical regime, the vector potential **A** merely appears to be a mathematical trick which sometimes helps us in calculating the magnetic field—the measurable entity. However, things change when we carry on quantum mechanical calculations for a particle in a magnetic field—calculations which involve solving Schrödinger's equation with the vector potential in it (see, for example, Landau and Lifshitz [2], Chap. XV; Merzbacher [3], pp. 71–74). As you are asked to determine in Exercise 3.6, the vector potential outside an infinite solenoid is non-zero, even though there is no magnetic field there. We can make two beams of electrons pass through the two sides of a solenoid and then undergo quantum interference. Since these beams never pass through regions where the magnetic field is non-zero, you may expect that the quantum interference pattern should be independent of whether there is a current through the solenoid creating a magnetic field inside the solenoid or not. Very surprisingly, that is not the case! The quantum interference pattern changes on changing the current, showing that the existence of the non-zero vector potential in the regions through which the electron beams passed has a measurable consequence. This follows from quantum mechanical considerations and can also be measured experimentally. Further discussion of this effect—often called the *Aharanov–Bohm effect*—is beyond the scope of this book. Interested readers may look up Feynman [4, Sect. 15–5] for an elementary discussion. We shall present a very brief discussion of how charged particles in electromagnetic fields are treated in quantum mechanics in Sect. 5.13 (with Exercise 5.11 dealing with the Aharanov–Bohm effect).

3.4.3 Using Ampere's Law

We have discussed in Sect. 3.3 how the application of Ampere's law (3.15) may enable one to calculate the magnetic field in a highly symmetric situation almost effortlessly. As in the case of Gauss's law in electrostatics, while these are universally valid laws in all situations, they cut down the calculations only in a limited number of cases involving high symmetry. However, when Ampere's law can be easily applied, it works almost like magic. Apart from the problem of finding the magnetic field due an infinite straight current-carrying wire, the other well-known problem in which Amprere's law can by used for a quick result is the problem of finding the magnetic field inside an infinite current-carrying solenoid (see Exercise 3.2).

3.4.4 Using Scalar Potential

The magnetic scalar potential Φ_m introduced through the relation

$$\mathbf{B} = -\nabla \Phi_m \tag{3.19}$$

analogous to the similar electrostatic relation (2.1) implies that the current density will be zero according to (1.11). Hence, this magnetic scalar potential can be introduced only in regions where there are no currents. We shall discuss later in this chapter how the magnetic scalar potential can be introduced in problems involving magnetic media without currents, and we shall see that some techniques developed in the context of electrostatics can be readily applied.

The magnetic scalar potential is also often useful when we have a current distribution in a finite region of space and we are interested in finding the magnetic field outside this region where the magnetic scalar potential can be introduced. To find the magnetic scalar potential at a field point due to a current-carrying circuit, we can use the equivalence relation (3.7) that the circuit can be replaced by a surface of uniform magnetic dipole density. Now, we have already derived the elegant result (2.17) for the electrostatic potential due to a surface of uniform electric dipole density. By analogy with (2.17), we can write down the magnetic scalar potential due to a circuit on making use of (3.7). If Ω is the solid angle subtended at a field point by a circuit through which a current I is flowing, then the magnetic scalar potential at the field point due to this circuit is given by

$$\Phi_m = \frac{\mu_0}{4\pi} I \, \Omega, \tag{3.20}$$

if we use the convention that the area element (of which the circuit is the boundary) is in the direction of the field point—this is the opposite of the convention used in Sect. 2.4 leading to a negative sign in (2.17).

We can apply this result to obtain the magnetic field at a point on the symmetry axis of a current-carrying loop. It is easy to check that the solid angle subtended by the current-carrying loop at the field point shown in Fig. 3.2 is $2\pi(1 - \cos\alpha)$. Substituting this in (3.20), we get

$$\Phi_m = \frac{\mu_0 I}{2}(1 - \cos\alpha) = \frac{\mu_0 I}{2}\left[1 - \frac{z}{\sqrt{a^2 + z^2}}\right]. \tag{3.21}$$

It is straightforward to show that (3.19) with Φ_m given by (3.21) substituted in it gives the same magnetic field as (3.18).

It is also possible to write down the expression of the magnetic field directly by making use of the equivalence relation (3.7). We already pointed out in Sect. 3.4.1 that it is difficult to obtain the exact expression of the magnetic field at an off-axis point due to a circular current-carrying coil. However, if we consider a field point far away (i.e. its distance $r \gg a$, the radius of the coil), then it is possible to write down an approximate expression of the magnetic field at an off-axis point. By the equivalence relation (3.7), at a distant point the circuit would appear as a point magnetic dipole, with the dipole moment given by $\pi a^2 I$, since its area is πa^2, and the magnetic moment per unit area can be taken as I. Then, on replacing $1/4\pi\epsilon_0$ in (2.16) by $\mu_0/4\pi$ and substituting $\pi a^2 I$ for p, we get

$$\mathbf{B} = \frac{\mu_0 a^2 I}{4} \left[\frac{2\cos\theta}{r^3} \hat{\mathbf{e}}_r + \frac{\sin\theta}{r^3} \hat{\mathbf{e}}_\theta \right], \tag{3.22}$$

where r, θ are the polar coordinates with the current-carrying loop at the origin. The magnetic field on the symmetry axis can easily be found from (3.22) by putting $\theta = 0$ and is

$$\mathbf{B} = \frac{\mu_0 a^2 I}{2} \frac{1}{z^3} \hat{\mathbf{e}}_r,$$

which is the same as what we get from (3.18) in the limit $|z| \gg a$.

3.5   The Magnetic Dipole Moment of a Localized Current System

We have discussed multipole expansion in electrostatics in Sect. 2.13 (assuming the charge distribution to be axially symmetric and confined within a region). Due to the more complicated mathematical expressions in magnetostatics, doing a multipole expansion in magnetostatics is a more challenging task. Here we only show what the first term in such a multipole expansion would be like. In the case of electrostatics, for a symmetric but arbitrary charge distribution for which the total charge is non-zero, the lowest-order term is the monopole term which goes as $1/r$. However, in magnetostatics, there are no magnetic monopoles, as encapsulated in (1.10), and we expect the monopole term to be zero in magnetostatics. We show now that the lowest-order term in the multipole expansion in magnetostatics is the dipole term and obtain its expression for a general current distribution. In fact, one might anticipate the result that the lowest order term would be a dipole term from (3.7), according to which a closed circuit is equal to a magnetic dipole surface.

As in the case of multipole expansion in electrostatics discussed in Sect. 2.13, we consider a current distribution (instead of a charge distribution) in a finite region around the origin. In the electrostatics case discussed in Sect. 2.13, we had considered an axisymmetric charge distribution so that we could work out the full series expansion without too much mathematical complication. If we are interested in the expansion only up to the dipole term, then it is not complicated to carry on the analysis for a non-axisymmetric charge distribution (Exercise 2.15). Since we shall be interested in the magnetostatics case only in finding out the nature of the dipole term, we proceed without assuming axisymmetry. Let us select the $\theta = 0$ axis of the spherical coordinate system in the direction of the field point which has the position

x at a distance r from the origin. Taking the current density at the source point **x'** to be **j(x')**, we use (2.89) to conclude from (3.2) that a particular component of the vector potential is given by

$$A_\alpha(\mathbf{x}) = \frac{\mu_0}{4\pi} \frac{1}{r} \sum_{n=0}^{\infty} \int j_\alpha(\mathbf{x}') P_n(\cos\theta') \left(\frac{r'}{r}\right)^n dV'. \tag{3.23}$$

We shall focus our attention here only on the first two terms corresponding to the monopole ($n = 0$) and the dipole ($n = 1$). Since θ' is the angle between the vectors **x** and **x'**, it is obvious that

$$P_1(\cos\theta') = \cos\theta' = \frac{\mathbf{x}.\mathbf{x}'}{rr'}.$$

On making use of this, the first two terms in (3.23) become

$$A_\alpha(\mathbf{x}) \approx \frac{\mu_0}{4\pi} \left[\frac{1}{r} \int j_\alpha(\mathbf{x}') \, dV' + \frac{\mathbf{x}}{r^3} \cdot \int j_\alpha(\mathbf{x}') \mathbf{x}' \, dV' + \cdots \right]. \tag{3.24}$$

We are assuming the currents to remain confined within a finite region of space. In such a situation, if j_α is positive at some point **x'**, there has to be some return current at some other point **x'** where j_α must be negative. We thus expect

$$\int j_\alpha(\mathbf{x}') \, dV' = 0,$$

which means that the monopole term disappears, and the leading term in (3.24) is the dipole term. Approximating $A_\alpha(\mathbf{x})$ by this leading term, we can write

$$A_\alpha(\mathbf{x}) = \frac{\mu_0}{4\pi} \sum_\beta \frac{x_\beta}{r^3} \int j_\alpha x'_\beta \, dV'. \tag{3.25}$$

We can write

$$j_\alpha x'_\beta = \frac{1}{2}(j_\alpha x'_\beta + j_\beta x'_\alpha) + \frac{1}{2}(j_\alpha x'_\beta - j_\beta x'_\alpha). \tag{3.26}$$

We shall now show that the symmetric term will not contribute to the integral in (3.25), and only the anti-symmetric term will contribute. To show this, we note that

$$\nabla'.(x'_\alpha x'_\beta \mathbf{j}) = x'_\alpha x'_\beta \nabla'.\mathbf{j} + \mathbf{j}.\nabla'(x'_\alpha x'_\beta).$$

Since $\nabla'.\mathbf{j} = 0$ in a magnetostatic situation, the first term in the RHS of the above expression vanishes, and we have

$$\nabla'.(x'_\alpha x'_\beta \mathbf{j}) = j_\alpha x'_\beta + j_\beta x'_\alpha.$$

Making use of this, the integral in (3.25) arising from the symmetric term in (3.26) can be written as

$$\int (j_\alpha x'_\beta + j_\beta x'_\alpha) \, dV' = \int \nabla'.(x'_\alpha x'_\beta \mathbf{j}) \, dV' = \oint x'_\alpha x'_\beta \mathbf{j} . d\mathbf{S} \qquad (3.27)$$

by Gauss's theorem (B.13). Now, we can take the volume integral over a region much larger than the region within which the currents are confined. Since the current density at any point on the bounding surface of this volume is zero, we have from (3.27) that

$$\int (j_\alpha x'_\beta + j_\beta x'_\alpha) \, dV' = 0.$$

This concludes our proof that the symmetric term in (3.26), when substituted in the integral in (3.25), will not make a contribution. Using only the anti-symmetric term in (3.26), we write (3.25) as follows:

$$A_\alpha(\mathbf{x}) = \frac{\mu_0}{8\pi} \sum_\beta \frac{x_\beta}{r^3} \int (j_\alpha x'_\beta - j_\beta x'_\alpha) \, dV'. \qquad (3.28)$$

Given the fact that

$$\sum_\beta x_\beta (j_\alpha x'_\beta - j_\beta x'_\alpha) = [(\mathbf{x}' \times \mathbf{j}) \times \mathbf{x}]_\alpha,$$

we get from (3.28):

$$\mathbf{A}(\mathbf{x}) = \frac{\mu_0}{8\pi r^3} \int [(\mathbf{x}' \times \mathbf{j}) \times \mathbf{x}] \, dV'. \qquad (3.29)$$

It is easy to check that this is a completely general expression of the dipole field due to a localized current distribution at any arbitrary field point located far outside the region of current distribution. Our initial choice of the coordinate axis in the direction of the field point, which enabled us to write down (3.23), does not restrict the generality of the vector expression (3.29). Our discussion from (3.24) onwards does not depend on a particular orientation of the coordinate system.

Let us now write

$$\mathbf{m} = \frac{1}{2} \int (\mathbf{x}' \times \mathbf{j}) \, dV'. \qquad (3.30)$$

Then (3.29) can be written as

$$\mathbf{A}(\mathbf{x}) = \frac{\mu_0}{4\pi} \frac{\mathbf{m} \times \mathbf{x}}{r^3}. \qquad (3.31)$$

The magnetic field can be obtained by taking the curl of this. We leave it for the reader to work out Exercise 3.7 and show that the magnetic field would have exactly the same expression as the expression of the electric field (2.16) due to an electric dipole moment, with $\mu_0/4\pi$ replacing $1/4\pi\epsilon_0$ and m replacing p. This shows that $\mathbf{m}$ defined by (3.30) is the *magnetic dipole moment*, often called simply the magnetic moment. It is instructive to consider the expression of the magnetic moment for a closed circuit carrying current I. As we did in Sect. 3.1, we have to replace $\mathbf{j}\,dV'$ in (3.30) by $I\,d\mathbf{l}'$. Keeping in mind that I is constant for the circuit and can be taken outside the integral, we can write

$$\mathbf{m} = \frac{I}{2}\int (\mathbf{x}' \times d\mathbf{l}'). \tag{3.32}$$

This expression conveys the same thing as (3.7), according to which a current-carrying circuit is equivalent to a magnetic dipole of which the dipole moment per unit area is I. We therefore expect the magnetic moment of the circuit to be the current multiplied by the area. Since $(1/2)\int(\mathbf{x}' \times d\mathbf{l}')$ is the area of the circuit, clearly (3.32) implies that the equivalent magnetic moment is given by I multiplied by the area of the circuit.

On obtaining the magnetic field for a system of localized currents from (3.31), we find that the magnetic field falls as $1/r^3$ (see Exercise 3.7) as we expect in the case of a dipole field. It is useful to make some comments how various fields fall with distance when we have electric charges and currents confined within a region. The electrostatic multipole expansion discussed in Sect. 2.13 makes it clear that the leading term in the expression of the electrostatic field would fall as $1/r^2$ if we calculate the field from (2.96) by using (2.1), provided the net electric charge is non-zero. Now we find that the magnetostatic field arising from currents confined in a localized region would fall as $1/r^3$, i.e. faster than the electrostatic field. If the charges and currents do not vary with time, then we can have only electrostatic and magnetostatic fields. We shall discuss in Chap. 7 that time-varying charges and currents can give rise to electromagnetic radiation. We shall also see that the electromagnetic field associated with radiation falls as $1/r$.

An interesting question is whether a current-carrying circuit is equivalent to a magnetic dipole moment given by (3.32) in all respects. We have pointed out that it produces a magnetic field at large distances, which is exactly the same as the electric field produced by an electric dipole. Among other important properties of an electric dipole is that it experiences a torque given by (2.22) when placed in an electric field and has an energy given by (2.20) associated with it. It follows from (2.21) that there would be no net force on an electric dipole if the electric field is uniform. A current-carrying circuit placed in a uniform magnetic field indeed experiences no net force, but experiences a torque L acting on it:

$$L = \mathbf{m} \times \mathbf{B}, \tag{3.33}$$

where $\mathbf{m}$ is given by (3.32). It is instructive to show this for a rectangular current-carrying circuit by considering the Lorentz forces acting on its sides (Exercise 3.8). To prove (3.33) for a general system of localized currents, one has to do a little bit of formal mathematical manipulation, which is discussed in some of the standard textbooks (see, for example, Panofsky and Phillips [5], p. 133). Only when the magnetic dipole moment **m** is aligned with the magnetic field is there no torque. In order to orient the magnetic dipole through some angle with respect to the magnetic field, one has to do work against the torque. From such considerations, it is easy to show that the appropriate expression of the potential energy would be

$$U = -\mathbf{m} . \mathbf{B}, \tag{3.34}$$

which is similar to (2.20), just as (3.33) is similar to (2.22). On expressing the cross product in (3.33) and the dot product in (3.34) in the scalar notation (assuming θ to be the angle between **B** and **m**), the relation between these pair of equations becomes obvious. However, (3.34) should be interpreted as the potential energy in a somewhat limited sense. We should keep in mind that there are other sources of energy in a current-carrying circuit (such as the battery which supplies the current), and energy considerations for such a system happen to be more complicated than the energy considerations for a simple electric dipole. You may turn to Feynman et al. [4, Sects. 15–1 and 15–2] for a discussion of this somewhat subtle point.

Since a moving charge can be regarded like a current, we expect that a moving charge also would have an equivalent magnetic moment—especially when it moves in a closed orbital path. For a charge q moving with velocity **v**, we get the equivalent magnetic moment by simply substituting $q\mathbf{v}$ for $\mathbf{j}\,dV'$ in (3.30). This gives

$$\mathbf{m} = \frac{q}{2}\,\mathbf{x} \times \mathbf{v}, \tag{3.35}$$

there being no integration in this case. We know that the angular momentum of the charged particle is given by

$$\mathbf{L} = m\,\mathbf{x} \times \mathbf{v}. \tag{3.36}$$

On the basis of (3.35) and (3.36), we can write

$$\mathbf{m} = \Gamma \mathbf{L}, \tag{3.37}$$

where the ratio Γ of the magnetic moment to the angular momentum, called the *gyromagnetic ratio*, is given by

$$\Gamma = \frac{|\mathbf{m}|}{|\mathbf{L}|} = \frac{q}{2m}.$$

This is the expression of the gyromagnetic ratio for the simple case of a moving particle with charge q and mass m. More generally, the gyromagnetic ratio may be given by

$$\Gamma = \frac{gq}{2m}, \tag{3.38}$$

where g is a number usually of order unity.

3.6   Polarization in a Magnetic Medium

So far we have considered magnetostatics in free space. Just as we discussed electrostatics with a dielectric medium in Sect. 2.14, we now discuss how the basic magnetostatic equations are modified in the presence of a magnetic medium. In the electrostatic situation, a dielectric medium is polarized only when there is an electric field. In contrast, ferromagnetic substances display magnetic polarization even in the absence of magnetic fields. Without bothering about the question right now how the magnetic polarization is produced, let us first consider the effects of magnetic polarization.

Let $\mathbf{M}$ be the magnetic dipole density per unit volume in a magnetically polarized medium. Then the magnetic dipole of a small volume dV' at a location $\mathbf{x}'$ is $\mathbf{M}\,dV'$. We now want to determine the magnetic field produced at a field point $\mathbf{x}$ by this magnetic dipole. On the basis of (3.31), we can write down the vector potential at $\mathbf{x}$ due to this magnetic dipole:

$$d\mathbf{A}(\mathbf{x}) = \frac{\mu_0}{4\pi} \frac{\mathbf{M} \times (\mathbf{x} - \mathbf{x}')}{|\mathbf{x} - \mathbf{x}'|^3} \, dV'.$$

To get the vector potential due to the whole magnetic medium, we have to carry on a volume integral. Making use of (1.18), we write

$$\mathbf{A}(\mathbf{x}) = \frac{\mu_0}{4\pi} \int \mathbf{M} \times \nabla' \left(\frac{1}{|\mathbf{x} - \mathbf{x}'|} \right) dV'. \tag{3.39}$$

To get the correct sign in (3.39), we have to keep in mind that there is a relative minus sign between ∇ and ∇' operating on a function of $\mathbf{x} - \mathbf{x}'$. By using the vector identity (B.5), we can write (3.39) as

$$\mathbf{A}(\mathbf{x}) = \frac{\mu_0}{4\pi} \left[\int \frac{\nabla' \times \mathbf{M}}{|\mathbf{x} - \mathbf{x}'|} \, dV' - \int \nabla' \times \left(\frac{\mathbf{M}}{|\mathbf{x} - \mathbf{x}'|} \right) dV' \right]. \tag{3.40}$$

The second term in the RHS can be converted from a volume integral to a surface integral

$$\int \nabla' \times \left(\frac{\mathbf{M}}{|\mathbf{x} - \mathbf{x}'|} \right) dV' = \oint d\mathbf{S} \times \frac{\mathbf{M}}{|\mathbf{x} - \mathbf{x}'|}$$

by a vector identity similar to Gauss's theorem in vector analysis. If we take the bounding surface far away outside the magnetic material, then this term is obviously zero. We have to keep only the first term in the RHS of (3.40) so that we can write

$$\mathbf{A}(\mathbf{x}) = \frac{\mu_0}{4\pi} \int \frac{\mathbf{j}_m(\mathbf{x}')}{|\mathbf{x} - \mathbf{x}'|} \, dV', \tag{3.41}$$

where $\mathbf{j}_m$ is given by

$$\mathbf{j}_m = \nabla' \times \mathbf{M}. \tag{3.42}$$

By comparing (3.41) with (3.2), we can at once conclude that the polarization in the magnetic medium can be represented by the polarization current given by (3.42).

Now we proceed in a way very similar to the procedure followed in Sect. 2.14 where the theory of dielectric media was discussed. As there would be no need to distinguish between source and field points while developing the theory of magnetic media, we do not use primes in this discussion and take the polarization current to be given by $\nabla \times \mathbf{M}$. If we follow the convention that $\mathbf{j}$ denotes only currents flowing through conductors, then we have to take $\mathbf{j} + \nabla \times \mathbf{M}$ to be the total current that can give rise to the magnetic field $\mathbf{B}$. In that case, (1.11) gives us

$$\nabla \times \mathbf{B} = \mu_0 \left(\mathbf{j} + \nabla \times \mathbf{M} \right) \tag{3.43}$$

in the presence of magnetic media. It is often useful to separately treat the current $\mathbf{j}$ through the conductors as the source which we can control. Keeping only the term involving $\mathbf{j}$ on the RHS, (3.43) becomes

$$\nabla \times \left(\mathbf{B} - \mu_0 \mathbf{M} \right) = \mu_0 \, \mathbf{j}. \tag{3.44}$$

Let us now introduce the other magnetic field variable

$$\mathbf{H} = \frac{1}{\mu_0} (\mathbf{B} - \mu_0 \mathbf{M}). \tag{3.45}$$

Sometimes, in order to make the distinction between $\mathbf{B}$ and $\mathbf{H}$ clear, only $\mathbf{H}$ is referred to as the magnetic field, whereas $\mathbf{B}$ is referred to as the magnetic flux density or magnetic induction. We use the convention of calling $\mathbf{B}$ also the magnetic field. The symbol $\mathbf{B}$ or $\mathbf{H}$ that we use will make it clear as to which quantity we are referring to. On making use of (3.45), we can put (3.44) in the form

$$\nabla \times \mathbf{H} = \mathbf{j}. \tag{3.46}$$

In the presence of magnetic media, we often take this as the basic equation in the place of (1.11). It follows from (3.45) that

$$\mathbf{B} = \mu_0 \left(\mathbf{H} + \mathbf{M} \right). \tag{3.47}$$

The other basic Eq. (1.10) involving the magnetic field remains unchanged in the presence of magnetic media.

All the results we have so far discussed in this section hold for all kinds of magnetic materials. If the magnetization in a material is produced by an imposed magnetic field and is proportional to it, then we have some additional relations which we point out in this paragraph. In paramagnetic and diamagnetic materials, the induced magnetization **M** is proportional to the magnetic field, but that is not generally the case for ferromagnetic materials. For paramagnetic and diamagnetic materials, we can write

$$\mathbf{M} = \chi_m \, \mathbf{H}, \tag{3.48}$$

where χ_m is the *magnetic susceptibility*, which is positive for paramagnetic materials and negative for diamagnetic materials. When **M** is proportional to the magnetic field, we expect from (3.47) that **B** should be proportional to **H**, and we can write

$$\mathbf{B} = \mu \mathbf{H}. \tag{3.49}$$

On making use of (3.47) and (3.48), we find from (3.49) that

$$\mu = \mu_0 \, (1 + \chi_m). \tag{3.50}$$

We now introduce the important quantity *permeability* κ_m of a magnetic material

$$\kappa_m = \frac{\mu}{\mu_0} = 1 + \chi_m. \tag{3.51}$$

It may be noted that sometimes μ is also referred to as the permeability.

We can easily determine the boundary conditions which have to be satisfied at the interface between two magnetic media. If we allow for all kinds of magnetic materials (including ferromagnetic materials), then we can use only those relations which hold for all magnetic materials. If there is no surface current at the interface we are considering, (3.46) reduces to

$$\nabla \times \mathbf{H} = 0.$$

Proceeding exactly in the same way as we did in Sect. 2.15 for the interface between two dielectric media, we can easily show from this equation and (1.10) that

$$H_{1t} = H_{2t}, \tag{3.52}$$

$$B_{1n} = B_{2n}, \tag{3.53}$$

where the subscripts 1 and 2 refer to the two magnetic media on the two sides of the interface, whereas t and n refer respectively to the tangential and normal components. Only if the materials are paramagnetic or diamagnetic can we additionally use (3.49) to obtain a magnetostatic condition similar to the condition (2.106) in the electrostatic situation. If one of the materials is ferromagnetic, then we have to proceed with the boundary conditions (3.52) and (3.53). An illustrative example of how a magnetostatic boundary value problem with a ferromagnetic material is solved is given in Sect. 3.7.

Although **B** is not proportional to **H** in ferromagnetic materials, sometimes one can write a relation

$$\mathbf{B} = \mu(\mathbf{H})\,\mathbf{H}, \tag{3.54}$$

where $\mu(\mathbf{H})$ is a function of **H** and not a constant as in (3.49). This relation (3.54) is often useful in dealing with situations in which the material does not have a significant permanent magnetization and the hysteresis effects are small ("soft" iron being an example of such a material). The corresponding permeability $\kappa_m = \mu(\mathbf{H})/\mu_0$ for such ferromagnetic materials can be quite high ($\kappa_m \gg 1$). We found in Sect. 2.16 that the field lines of the electric displacement **D** tend to become concentrated inside a dielectric material, which is clear in Fig. 2.14. Exactly in the same manner, the field lines of the magnetic field **B** tend to become concentrated inside a material with high magnetic permeability (see Exercise 3.9). Since such materials can be useful in channelizing magnetic fields in an appropriate manner, they are used widely in electromagnetic devices.

3.7   A Boundary Value Problem in Magnetostatics: A Sphere of Magnetic Material

We have pointed out in Sect. 3.4.4 that a magnetic scalar potential Φ_m defined by (3.19) can be introduced in magnetostatics in a region where there are no currents. On substituting (3.19) into (1.10), it follows that Φ_m also satisfies Laplace's equation

$$\nabla^2 \Phi_m = 0 \tag{3.55}$$

just like the electrostatic potential. It is easy to argue that the techniques of solving electrostatic boundary value problems discussed in Sects. 2.8–2.12 can be applied to the magnetostatic situation whenever we are able to introduce the magnetic scalar potential Φ_m. As an illustration of a boundary value problem in magnetostatics, it may be instructive to consider a problem which does not have an electrostatic analogue: a magnetized ferromagnetic sphere in free space. Paramagnetic and diamagnetic substances are magnetically polarized only when placed in a magnetic field (just as a dielectric substance is polarized only in the presence of an electric field). In contrast, a ferromagnetic substance can have an internal magnetic polarization even in the absence of an imposed magnetic field. Rather, a ferromagnetic sphere will itself produce a magnetic field around it because of its internal polarization. In order to study this problem of a ferromagnetic sphere, we note that we cannot use (3.49) and have to work with the boundary conditions (3.52) and (3.53).

Let us assume that the magnetization density **M** inside the ferromagnetic sphere is constant. We take the origin at the centre of the sphere of which the radius is at $r = a$. As the general solution of Laplace's equation in spherical coordinates is given by (2.81), Φ_m either inside the sphere or outside it should be given by such an expression. Since $\mathbf{M}$ is constant in the interior, we may expect that the magnetic field arising from it will also have a constant value $\mathbf{B}_i$ in the inside. Taking the polar axis in the direction of the magnetic field (which should also be the direction of the magnetization), we know that the constant magnetic field should correspond to the scalar potential

$$\Phi_{m,r<a} = -B_i \, r \cos \theta. \tag{3.56}$$

According to (3.45), the corresponding magnetic field vector $\mathbf{H}_i$ has the constant magnitude

$$H_i = \frac{B_i}{\mu_0} - M. \tag{3.57}$$

It is clear from (3.56), that we have retained only the term involving $P_1(\cos \theta) = \cos \theta$ in the full solution (2.81) that Φ_m may have. The solution outside the sphere is also expected to have only the term involving this, i.e.

$$\Phi_{m,r>a} = \frac{C}{r^2} \cos \theta, \tag{3.58}$$

where C is a constant coefficient to be determined from the boundary conditions. Since the polarization density is zero outside the sphere, the magnetic field outside the sphere is given simply by

$$\mathbf{H} = \frac{\mathbf{B}}{\mu_0}. \tag{3.59}$$

We now have to apply the boundary conditions (3.52) and (3.53). It follows from (3.58) and (3.59) that H_θ outside the sphere is given by

$$H_{\theta,r>a} = \frac{C}{\mu_0 r^3} \sin \theta,$$

whereas it follows (3.57) that H_θ inside the sphere is

$$H_{\theta,r<a} = -\left(\frac{B_i}{\mu_0} - M \right) \sin \theta.$$

The boundary condition (3.52) applied at the surface $r = a$ then gives

$$\frac{B_i}{\mu_0} - M = -\frac{C}{\mu_0 a^3}. \tag{3.60}$$

To apply the other boundary condition (3.53), we have to first obtain B_r inside and outside the sphere by taking the negative r derivative of (3.56) and (3.58), and then match them at the surface of the sphere. This gives

$$\frac{2C}{a^3} = B_i.$$ (3.61)

From (3.60) and (3.61), we get

$$C = \frac{1}{3}\mu_0 M a^3,$$ (3.62)

$$B_i = \frac{2}{3}\mu_0 M.$$ (3.63)

Now (3.63) gives the magnetic field inside the sphere in terms of the polarization density $\mathbf{M}$, whereas the substitution from (3.62) in (3.58) would give

$$\Phi_{m,r>a} = \frac{\mu_0}{4\pi} \frac{4\pi}{3} a^3 M \frac{\cos\theta}{r^2}.$$

We have obtained a complete solution of the magnetic field inside and outside a ferromagnetic sphere with uniform magnetic polarization $\mathbf{M}$. Compared with (2.15), we note that the potential outside the sphere is essentially that due to the magnetic dipole having moment $\frac{4\pi}{3}a^3 M$. One surprising thing to note is that H_i obtained from (3.57) and (3.63) is

$$H_i = -\frac{1}{3}M$$ (3.64)

and is in the opposite direction with respect to the polarization density $\mathbf{M}$!

   After obtaining the solution for the boundary value problem with a ferromagnetic sphere, it is easy to solve the problem of a paramagnetic sphere. In this case, the sphere will have a magnetization only if an externally imposed magnetic field $\mathbf{B_0} = \mu_0 \mathbf{H_0}$ is present. This magnetic field will induce a uniform magnetization $\mathbf{M}$ inside the paramagnetic sphere. We expect that the paramagnetic sphere with magnetization $\mathbf{M}$ will produce a magnetic field given by exactly same mathematical expression as the magnetic field produced by the ferromagnetic sphere we considered above. To get the total magnetic field inside the paramagnetic sphere, we simply have to add the external magnetic field to (3.63) and (3.64). The two magnetic field vectors inside the paramagnetic sphere then are

$$B_i = B_0 + \frac{2}{3}\mu_0 M,$$ (3.65)

$$H_i = \frac{B_0}{\mu_0} - \frac{1}{3}M.$$ (3.66)

Since (3.49) or (3.51), which should not be used for a ferromagnetic substance, can now be used for the interior of a paramagnetic sphere, (3.65) and (3.66) lead us to write

$$B_0 + \frac{2}{3}\mu_0 M = \mu \left(\frac{B_0}{\mu_0} - \frac{1}{3}M \right) = \kappa_m \left(B_0 - \frac{1}{3}\mu_0 M \right). \tag{3.67}$$

From this, we get

$$M = \frac{3(\kappa_m - 1)}{(\kappa_m + 2)} \frac{B_0}{\mu_0}. \tag{3.68}$$

This expression for the magnetic polarization induced by a magnetic field inside a paramagnetic sphere is exactly the same as the expression (2.120) for the electric polarization induced by an electric field inside a dielectric sphere.

3.8   Microscopic Theory of Magnetic Materials

Readers of this book would surely know that different magnetic materials broadly display three kinds of magnetic behaviour: paramagnetism, diamagnetism and ferromagnetism. Understanding the microscopic physics which gives rise to these three kinds of magnetic phenomenology is important. The basic theory of paramagnetism proceeds along the same lines as the theory of dielectric materials briefly outlined in Sect. 2.18. However, theories of diamagnetism and ferromagnetism happen to be more complicated. This book is primarily devoted to macroscopic aspects of electromagnetic theory. Our occasional forays into the underlying microscopic physics which give rise to macroscopic behaviours of material media (such as what we discussed in Sect. 2.18 and shall discuss in Sect. 4.9) are limited to situations where the microscopic physics involves only some simple ideas about forces on charged particles or dipoles. We refrain from getting into more complicated microscopic theories. Students who take a course on advanced electromagnetic theory usually also take a course on solid state physics at this level. Standard textbooks of solid state physics have detailed discussions of the microscopic physics of magnetic materials (Kittel [6], Chaps. 15–16; Ashcroft and Mermin [7], Chaps. 31–33). We shall refrain from getting into a detailed discussion of magnetic materials in this book—not because this subject is any less important, but because students are expected to learn this subject in other courses and from other textbooks.

Let us make only a brief remark about paramagnetic materials. In a material with atoms having magnetic dipole moments, there would be a tendency for these atomic dipoles to align along an imposed magnetic field. The quantum effect that the component of the magnetic moment in the direction of the magnetic field—which would be proportional to the component of the angular moment in accordance with (3.37)—is expected to be quantized will have to be incorporated in detailed calculations. Still, the physics is very similar to the microscopic theory of dielectric materials discussed in Sect. 2.18. The average magnetic moment in the direction of the magnetic field is given by an expression similar to (2.125), implying that the average magnetic moment is directly proportional to the imposed magnetic field and inversely proportional to temperature. From the definition of magnetic susceptibility introduced through (3.48), it then follows that the magnetic susceptibility would be inversely proportional to temperature:

$$\chi_m \propto \frac{1}{T}. \tag{3.69}$$

This result is known as the *Curie law*.

3.9   Analogy Between Electric Currents and Moving Charges

A moving charge is in some ways analogous to a current. After all, currents inside conductors are produced by many moving charges. From (1.7), it follows that a charge q moving with velocity $\mathbf{v}$ makes a contribution $q\mathbf{v}$ to the current. For a static charge, the charge density can be represented by a δ function, as indicated in (2.4). If the moving charge is at a position $\mathbf{x}_q(t)$ at time t, then we can represent the charge and current densities associated with that charge in the following manner:

$$\rho(\mathbf{x}', t) = q\,\delta(\mathbf{x}' - \mathbf{x}_q(t)), \quad \mathbf{j}(\mathbf{x}', t) = q\,\mathbf{v}\,\delta(\mathbf{x}' - \mathbf{x}_q(t)). \tag{3.70}$$

It may be noted that this expression of $\mathbf{j}$ substituted in (3.30) would lead to (3.35).

We expect the moving charge to produce a magnetic field like a current. It is clear that such a magnetic field will not be static, unlike the magnetic field due to a time-independent current. We arrived at (3.2) from Maxwell's equations by throwing away the time-varying terms. We cannot expect (3.2) to be valid when things are changing with time. Still, it is tempting to substitute (3.70) in (3.2) and see what we get. Carrying on the volume integration (which is easy to do when the integrand has a δ function), we obtain

$$\mathbf{A}(\mathbf{x}, t) = \frac{\mu_0}{4\pi} \frac{q\mathbf{v}}{|\mathbf{x} - \mathbf{x}_q(t)|}. \tag{3.71}$$

It is easy to take the curl of this to obtain the magnetic field, which is

$$\mathbf{B}(\mathbf{x}, t) = \frac{\mu_0}{4\pi} q \frac{\mathbf{v} \times [\mathbf{x} - \mathbf{x}_q(t)]}{|\mathbf{x} - \mathbf{x}_q(t)|^3}, \tag{3.72}$$

in the place of (3.4). What we have done is certainly not logically consistent. Still, it is amazing that (3.72) gives the magnetic field due to a non-relativistically moving charge, for which $|\mathbf{v}| \ll c$. We shall discuss the correct general expression (more complicated!) of the magnetic field due to a moving charge in Sects. 5.10.2 and 6.4–6.5. We shall see that the general expression for the magnetic field due to a moving charge would reduce to (3.72) in the non-relativistic limit.

Exercises

3.1 Determine the magnetic field due to a line current I at a distance r from it by integrating (3.5) and show that the result is the same as what you get from Ampere's law.

3.2 (a) For a finite solenoid tightly wound with n turns per unit length and carrying a current I, show that the magnetic field on the axis is given by

$$B = \frac{\mu_0 n I}{2}(\cos \alpha_1 + \cos \alpha_2),$$

where the angles are as defined in Fig. 3.3. Note that this result is independent of the radius of the solenoid.

(b) Obtain the magnetic field in the limit when the solenoid is infinite. Show that you get the same magnetic field by a suitable application of Ampere's law and argue that the magnetic field is uniform inside an infinite solenoid, while being zero outside.

3.3 We have a uniform magnetic field $\mathbf{B} = B_0 \mathbf{e}_z$. Show that this magnetic field follows from a vector potential

$$\mathbf{A}(s) = -syB_0\mathbf{e}_x + (1 - s)xB_0\mathbf{e}_y$$

for any constant value of s. Show also that we can write

$$\mathbf{A}(s_1) - \mathbf{A}(s_2) = \nabla\psi$$

and find ψ.

3.4 Consider an axisymmetric magnetic field

$$\mathbf{B} = \nabla \times [A(r, \theta)\mathbf{e}_\phi]$$

in spherical coordinates. Show that the contours of constant $A(r, \theta)$ give the magnetic field lines lying in the (r, θ) plane.

Fig. 3.3 Refer to
Exercise 3.2

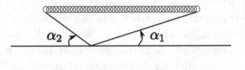

3.5 (a) Show that the magnetic flux passing through a circuit is given by the line integral

$$\Phi = \oint \mathbf{A}.\,d\mathbf{l}$$

carried out around the circuit, where $\mathbf{A}$ is the vector potential.

(b) Suppose two circuits are kept near each other. The magnetic flux linked with circuit 1 when a current I_2 passes through circuit 2 can be written as

$$\Phi_1 = M_{12}I_2,$$

where M_{12} is the *mutual inductance* between the circuits. Show that it is given by

$$M_{12} = \frac{\mu_0}{4\pi} \oint_1 \oint_2 \frac{d\mathbf{l}_1.\,d\mathbf{l}_2}{|\mathbf{x}_2 - \mathbf{x}_1|},$$

where the various symbols have the same meanings as in (3.13).

3.6 Consider an infinite solenoid of radius a tightly wound with n turns per unit length and carrying a current I. Although the magnetic field outside the solenoid is zero, show that the vector potential $\mathbf{A}$ at a distance r from the axis of the solenoid (larger than the radius a of the solenoid) has the amplitude

$$A = \frac{\mu_0}{2} \frac{nIa^2}{r}$$

with field lines of $\mathbf{A}$ making closed circles around the solenoid.

Hint. Use the result of Exercise 3.5(a).

3.7 Suppose a small magnetic dipole $\mathbf{m}$ is aligned with the polar axis. Show that A_ϕ is the only component of $\mathbf{A}$ in spherical coordinates which is non-zero and write it down in terms of $\mathbf{m}$, r and θ from the expression of the vector potential due to this dipole as given by (3.31). Work out the components of the magnetic field due to this magnetic dipole and show that it is exactly similar to the electric field due to an electric dipole as given by (2.16).

3.8 Consider a rectangular circuit of area S carrying current I. The magnetic moment corresponding to this circuit is given by

$$\mathbf{m} = SI\hat{\mathbf{n}},$$

where $\hat{\mathbf{n}}$ is a unit vector perpendicular to the plane of the circuit following the right-hand rule with respect to the direction of the current. Suppose now that this rectangular circuit is placed in a uniform magnetic field $\mathbf{B}$ in such a manner that the magnetic field is perpendicular to two opposite sides of the rectangle. Considering the Lorentz force on each of the sides, show that there is no net force on the circuit, but there is a torque given by

$$\mathbf{L} = \mathbf{m} \times \mathbf{B}.$$

3.9 We consider a spherical shell with inner radius a and outer radius b made up of a magnetic material with high permeability ($\kappa_m \gg 1$). This spherical shell is introduced in an otherwise uniform magnetic field $\mathbf{B}_0$. Determine the resulting magnetic fields everywhere and show that the magnetic field in the interior of the spherical shell in the limit $\kappa_m \gg 1$ is given by

$$\mathbf{B}_{\text{in}} = \frac{9b^3}{2\kappa_m \, (b^3 - a^3)} \mathbf{B}_0,$$

which is much smaller than the outside magnetic field $\mathbf{B}_0$ if the spherical shell is made of magnetic material with high permeability. This shows that such materials can screen a region from magnetic fields.

References

1. Jackson, J.D.: Classical Electrodynamics, 3rd edn. Wiley, New York (1999)
2. Landau, L.D., Lifshitz, E.M.: Quantum Mechanics: Volume 3 (Course of Theoretical Physics), 3rd edn. Butterworth-Heinemann, Oxford (1981)
3. Merzbacher, E.: Quantum Mechanics, 3rd edn. Wiley, New York (1998)
4. Feynman, R.P., Leighton, R.B., Sands, M.: The Feynman Lectures on Physics: Volume II, Mainly Electromagnetism and Matter. Addison-Wesley, Boston (1964)
5. Panofsky, W.K.H., Phillips, M.: Classical Electricity and Magnetism, 2nd edn. Addison-Wesley (reprinted by Dover), Boston (1962)
6. Kittel, C.: Solid State Physics, 4th edn. Wiley, New York (1971)
7. Ashcroft, N.W., Mermin, N.D.: Solid State Physics. W. B. Saunders, Philadelphia (1976)

Chapter 4
Electrodynamics and Electromagnetic Waves

4.1 Time Derivative Terms in Maxwell's Equations

In the previous two chapters, we have confined ourselves to static situations so that the time derivative terms in Maxwell's equations could be taken as zero. One of the consequences of this was that the electric and the magnetic fields became decoupled and could be treated separately. We shall now start discussing the interesting physics that comes out of the time derivative terms in Maxwell's equations which couple electric and magnetic fields together.

We have introduced Maxwell's equations in Sect. 1.2. Then, in Sects. 2.14 and 3.6, we pointed out that, in the presence of material media, it is often useful to write down the equations in such a manner that ρ and $\mathbf{j}$ do not include polarization charges and polarization currents induced in the media. If we follow this approach, then (1.1) has to be replaced by (2.100), whereas the replacement of the static part of (1.3) is given by (3.46). We write down Maxwell's equations following this approach:

$$\nabla \cdot \mathbf{D} = \rho, \tag{4.1}$$

$$\nabla \cdot \mathbf{B} = 0, \tag{4.2}$$

$$\nabla \times \mathbf{H} = \mathbf{j} + \frac{\partial \mathbf{D}}{\partial t}, \tag{4.3}$$

$$\nabla \times \mathbf{E} = -\frac{\partial \mathbf{B}}{\partial t}. \tag{4.4}$$

On using the relations $\mathbf{D} = \epsilon_0 \mathbf{E}$ and $\mathbf{B} = \mu_0 \mathbf{H}$ for vacuum, we see (4.1)–(4.4) lead to (1.1)–(1.4). It may superficially seem that (1.1)–(1.4) are Maxwell's equations for the vacuum. However, as we have stressed repeatedly (in Sects. 1.2, 2.14 and 3.6), the Eqs. (1.1)–(1.4) are completely general equations if we include polarization charges and polarization currents within ρ and $\mathbf{j}$.

© Springer Nature Singapore Pte Ltd. 2022
A. R. Choudhuri, *Advanced Electromagnetic Theory*, Lecture Notes in Physics 1009,
https://doi.org/10.1007/978-981-19-5944-8_4

As we can easily see, electric and magnetic fields will be coupled to each other when the time derivative terms are non-zero, so that these fields have to be treated together. Some of the discussions in this chapter will focus on what happens when all the terms in Maxwell's equations (4.1)–(4.4) are retained. However, just as we have made the simplification in the previous two chapters that the time derivative terms were taken to be zero, we shall introduce a different kind of simplification in the central part of this chapter (in Sects. 4.4 and 4.6–4.8): we shall take $\rho = 0$, $\mathbf{j} = 0$. We shall show that Maxwell's equations even with this simplification lead to the famous prediction of electromagnetic waves and to an elucidation of its basic properties. A detailed discussion of the theory of what happens when all the terms in Maxwell's equations are present will be first taken up in Sect. 4.10 and then developed in Chaps. 6–7.

Let us begin our discussion of the time derivative terms as given in (4.1)–(4.4) by arguing that certain requirements of self-consistency dictate that the terms need to have the forms they have.

4.1.1   The Displacement Current Term

Apart from the term $\partial \mathbf{D}/\partial t$ in (4.3)—the so-called *displacement current* term— all the other terms in Maxwell's equations were actually known before Maxwell's time. It was the great insight of Maxwell to realize that we need to include the displacement current term in order to satisfy charge conservation. Let us look at Maxwell's argument in modern vector analysis notation. Let us consider a volume V bounded by a surface S. If ρ is the charge density at a point, then the total charge inside this volume is

$$Q = \int_V \rho \, dV. \tag{4.5}$$

If electric charge is conserved, then Q can change only if some charge is moved across the bounding surface. Keeping in mind that the current density $\mathbf{j}$ gives the rate of charge flow, we can conclude that $\mathbf{j} \cdot d\mathbf{S}$ gives the amount of charge moving across the surface $d\mathbf{S}$ in unit time. Integrating this over the boundary surface gives the rate at which charge leaves the volume inside the bounding surface. We, therefore, have

$$\frac{dQ}{dt} = -\int_S \mathbf{j} \cdot d\mathbf{S}, \tag{4.6}$$

where we have introduced the minus sign to account for the fact that Q decreases when some charge leaves the volume across the bounding surface. Now, ρ at various points inside the volume is expected to change, in order to cause the change in Q. Substituting from (4.5) in (4.6) and transforming the surface integral on the RHS of (4.6) into a volume integral by the application of Gauss's theorem (B.13), we get

$$\int_V \frac{\partial \rho}{\partial t} \, dV = - \int_V (\nabla \cdot \mathbf{j}) \, dV,$$

from which

$$\int_V \left[\frac{\partial \rho}{\partial t} + \nabla \cdot \mathbf{j} \right] dV = 0.$$

Since this relation should hold for any arbitrary volume, we must have

$$\frac{\partial \rho}{\partial t} + \nabla \cdot \mathbf{j} = 0, \qquad (4.7)$$

which is the *charge conservation equation*. Let us now look at the question whether Maxwell's equations satisfy charge conservation. If the displacement current term was not there in (4.3), then its divergence would give

$$\nabla \cdot \mathbf{j} = 0,$$

keeping in mind that the divergence of a curl is zero. From (4.7), we see that this equation would imply charge conservation only when all the time derivative terms are zero. Let us now assume the displacement current term to be non-zero and again take the divergence of (4.3). This gives

$$\nabla \cdot \mathbf{j} + \frac{\partial}{\partial t} (\nabla \cdot \mathbf{D}) = 0. \qquad (4.8)$$

It is readily seen that (4.1) substituted in (4.8) would give the charge conservation equation (4.7). This concludes our demonstration that the displacement current term is essential if we want the charge conservation equation to be satisfied in a general situation allowing variations with time.

4.1.2 The Electromagnetic Induction Term

The other time derivative term in Maxwell's equations is the term $\partial \mathbf{B}/\partial t$ in (4.4). This term captures the essence of Faraday's famous 1831 discovery of the *electromagnetic induction*. When magnetic flux passing through a circuit changes, Faraday discovered that an EMF (eletromotive force) is produced in the circuit and that this EMF is proportional to the rate of change of the magnetic flux linked with the circuit. While writing down a mathematical expression of this law, we have to allow for the magnetic flux changing both due to the intrinsic temporal variation of the magnetic field and due to the motion of the circuit. The EMF is defined as the work done on a unit charge during its circulation through the circuit. From (1.5), the force on a unit charge is $\mathbf{E} + \mathbf{v} \times \mathbf{B}$ so that the EMF is clearly given by

$$\mathcal{E} = \oint (\mathbf{E} + \mathbf{v} \times \mathbf{B}) . d\mathbf{l}. \tag{4.9}$$

Since the flux linked with circuit is given by $\int \mathbf{B} . d\mathbf{S}$ integrated over a surface of which the circuit is the boundary, we can write *Faraday's law of electromagnetic induction* in the following form

$$\mathcal{E} = -k \frac{d}{dt} \int \mathbf{B} . d\mathbf{S}, \tag{4.10}$$

where k is the constant of proportionality, and the minus sign arises from the fact that the EMF is produced in a such a manner that it tries to resist further changes in the magnetic flux—a result often known as *Lenz's law*. Equating (4.9) and (4.10), we are now going to argue that k is going to be equal to 1 in SI units for the sake of consistency.

We first have to determine how to calculate the time derivative of the flux $\int \mathbf{B} . d\mathbf{S}$ associated with a moving circuit, which appears in (4.10). Suppose at time t_i we consider an element of area $d\mathbf{S}_i$ of the surface of which the circuit is the boundary. If $\mathbf{B}_i(t_i)$ is the magnetic field there, then the flux linked with the circuit at time t_i is $\int \mathbf{B}_i(t_i) . d\mathbf{S}_i$. Defining $d\mathbf{S}_f$ and $\mathbf{B}_f(t_f)$ in a similar fashion with respect to the position of the circuit at time $t_f = t_i + \Delta t$, the flux linked with the circuit at that time should be $\int \mathbf{B}_f(t_i + \Delta t) . d\mathbf{S}_f$. We obviously have

$$\frac{d}{dt} \int \mathbf{B} . d\mathbf{S} = \lim_{\Delta t \to 0} \frac{\int \mathbf{B}_f(t_i + \Delta t). d\mathbf{S}_f - \int \mathbf{B}_i(t_i). d\mathbf{S}_i}{\Delta t}.$$

Substituting the Taylor expansion

$$\mathbf{B}_f(t_i + \Delta t) = \mathbf{B}_f(t_i) + \frac{\partial \mathbf{B}_f}{\partial t} \Delta t,$$

we get

$$\frac{d}{dt} \int \mathbf{B} . d\mathbf{S} = \int \frac{\partial \mathbf{B}}{\partial t} . d\mathbf{S} + \lim_{\Delta t \to 0} \frac{\int \mathbf{B}_f(t_i). d\mathbf{S}_f - \int \mathbf{B}_i(t_i). d\mathbf{S}_i}{\Delta t}. \tag{4.11}$$

We now have to evaluate the second term in the RHS.

Figure 4.1 shows the positions of the circuit at times t_i and t_f. Along with the side surface which would be generated by the motion of the circuit, we consider the bottom and the top surfaces of which the positions of the circuit at t_i and t_f are the boundaries. These surfaces together make up a closed cylinder, as seen in Fig. 4.1. Since $\nabla.\mathbf{B}$ is zero, the sum of the magnetic fluxes passing through all the surfaces of this cylinder at time t_i must be zero. Note that $\int \mathbf{B}_f(t_i).d\mathbf{S}_f$ and $- \int \mathbf{B}_i(t_i).d\mathbf{S}_i$ are the outward magnetic fluxes through the top surface and the bottom surface both at time t_i. To determine the magnetic flux through the side surface, we consider an element $d\mathbf{l}$ of the circuit and point out that the strip of the cylinder indicated in Fig. 4.1

Fig. 4.1 A sketch showing
the displacement of a circuit.
See the text for discussion

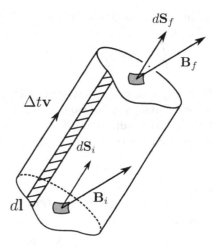

has the vector area $-\Delta t\, \mathbf{v} \times d\mathbf{l}$, keeping in mind that the side surface is generated by the displacement $\Delta t\, \mathbf{v}$ of the circuit element $d\mathbf{l}$ moving with velocity $\mathbf{v}$. It is clear that the total magnetic flux through the side surface is given by

$$-\Delta t \oint \mathbf{B} \cdot (\mathbf{v} \times d\mathbf{l}) = \Delta t \oint (\mathbf{v} \times \mathbf{B}) \cdot d\mathbf{l},$$

where the line integral has to be carried out over the full circuit. Now, the sum of the magnetic fluxes through all the surfaces of the cylinder at time t_i is

$$\int \mathbf{B}_f(t_i) \cdot d\mathbf{S}_f - \int \mathbf{B}_i(t_i) \cdot d\mathbf{S}_i + \Delta t \oint (\mathbf{v} \times \mathbf{B}) \cdot d\mathbf{l} = 0,$$

from which

$$\lim_{\Delta t \to 0} \frac{\int \mathbf{B}_f(t_i) \cdot d\mathbf{S}_f - \int \mathbf{B}_i(t_i) \cdot d\mathbf{S}_i}{\Delta t} = - \oint (\mathbf{v} \times \mathbf{B}) \cdot d\mathbf{l}.$$

Substituting this in (4.11), we finally get

$$\frac{d}{dt} \int \mathbf{B} \cdot d\mathbf{S} = \int \frac{\partial \mathbf{B}}{\partial t} \cdot d\mathbf{S} - \oint (\mathbf{v} \times \mathbf{B}) \cdot d\mathbf{l}. \qquad (4.12)$$

It should be clear that the first term in the RHS is due to the intrinsic variation of the magnetic field with time, whereas the second term arises because of the change in magnetic flux due to the motion of the circuit.

Substituting (4.9) and (4.12) into (4.10), we get

$$\oint \mathbf{E} \cdot d\mathbf{l} + \oint (\mathbf{v} \times \mathbf{B}) \cdot d\mathbf{l} = -k \int_S \frac{\partial \mathbf{B}}{\partial t} \cdot d\mathbf{S} + k \oint (\mathbf{v} \times \mathbf{B}) \cdot d\mathbf{l}. \tag{4.13}$$

Since the velocity $\mathbf{v}$ with which an element of the circuit moves can be arbitrary, the terms involving $\mathbf{v}$ in the two sides of (4.13) have to balance each other, as they cannot be balanced by other terms. This requirement leads to the very important conclusion that the constant of proportionality appearing in (4.10) has to have the value

$$k = 1 \tag{4.14}$$

in SI units which we are using. If the terms involving $\mathbf{v}$ in (4.13) balance each other, then we are left with

$$\int (\vec{\nabla} \times \mathbf{E}) \cdot d\mathbf{S} = - \int_S \frac{\partial \mathbf{B}}{\partial t} \cdot d\mathbf{S},$$

in which the term on the LHS has been transformed by using Stokes's theorem (B.15). This is merely the integral form of (4.4). This completes our demonstration that Faraday's law of electromagnetic induction (4.10) requires for the sake of consistency that k has to be 1 and (4.4) has to have this particular form.

One often comes across the statement that (4.4) is the mathematical expression of Faraday's law of electromagnetic induction. Whether this statement is justified depends on exactly what one means by Faraday's law of electromagnetic induction. If (4.10) with the specification $k = 1$ is taken to be the statement of Faraday's law of electromagnetic induction, then this law definitely incorporates more physical effects than what is given by (4.4). To make this point clear, let us consider the example of a relative motion between a magnet and a circuit. According to (4.10), we expect an EMF to be induced in the circuit depending merely on the relative motion. Let us now consider the frame S_C in which the circuit is at rest and the frame S_M in which the magnet is at rest. In the frame S_C, the magnetic field changes with time due to the motion of the magnet, and the EMF, which is given by the $\oint \mathbf{E} \cdot d\mathbf{l}$ term in the RHS of (4.9) in this case, can indeed be calculated from (4.4). On the other hand, in the frame S_M, the magnetic field does not change with time, and the EMF obviously cannot be calculated from (4.4). Rather, the EMF has to be calculated from the second term in the RHS of (4.9) arising from the force $q \mathbf{v} \times \mathbf{B}$ acting on the charged particles inside the circuit which would be seen as moving in this frame. Our discussion in this subsection implies that the EMF calculated in the two frames out of these different considerations would be the same. But it is conceptually unsatisfactory that we have to give completely different kinds of arguments to explain the same phenomenon seen from two frames of reference. Einstein began his famous 1905 paper on special relativity by pointing out this asymmetry in the explanation of the same phenomenon seen from two frames of reference (Lorentz et al. [1], p. 37). Only with the help of relativistic transformation laws of electromagnetic fields to be discussed in Sect. 5.10, it is possible to unify the explanations given with respect to the two frames (see especially Sect. 5.10.3). In other words, (4.4) alone does not contain all the physics behind Faraday's law of electromagnetic induction. However, (4.4) in conjunction with the transformation laws of electromagnetic fields turn out to be equivalent to Faraday's law of electromagnetic induction as given by (4.10) with $k = 1$.

If we take (4.10) with $k = 1$ as the fundamental law (often called the 'flux rule') instead of (4.4), it may seem that we eliminate the above-mentioned conceptual difficulty even without invoking special relativity. However, this approach also has its limitations because we have to clarify very carefully as to what we mean by a 'circuit' through which the magnetic flux $\int \mathbf{B} . d\mathbf{S}$ has to be considered. Feynman, Leighton and Sands ([2], Sect. 17-2) discuss some 'exceptions' to the 'flux rule' which arise from the confusion regarding what can be taken as a 'circuit'. Even within the non-relativistic framework, it may be safer to work with (4.4) and invoke (4.9) to cover the case of a moving circuit. For a moving circuit, it should follow from our discussion in this subsection that the 'flux rule' would give the same EMF as one would get by using (4.9) which follows from the forces experienced by the charged particles inside the circuit (see Exercise 4.3).

4.2 Energy of Electromagnetic Fields

Electromagnetic fields can make charged particles move by exerting forces on them, thereby imparting some kinetic energy to them. From the point of view of the principle of conservation of energy, the most obvious interpretation is that the electromagnetic field must be having some energy associated with it, of which a part is imparted to the charged particles. The rate at which the electromagnetic field transfers the energy to the charged particles must be given by the rate at which the electromagnetic field does work on the charged particles. From the rate of work done, we may try to determine the energy stored in the electromagnetic field.

Let us assume that in a region there are n particles per unit volume with charge q moving with the same velocity $\mathbf{v}$. Since the force on each charged particle is given by (1.5), the rate of work done by the electromagnetic field per unit volume is

$$nq \, (\mathbf{E} + \mathbf{v} \times \mathbf{B}) . \mathbf{v} = \mathbf{j} . \mathbf{E},$$

where we have made use of (1.7). Note the well-known fact the Lorentz force is perpendicular to the direction of motion and does not do any work. By integrating $\mathbf{j} . \mathbf{E}$ over a certain volume, we get the rate at which work is done on the charged particles within that volume, which should be equal to rate of change of the mechanical energy $\mathcal{E}_{\text{mech}}$ of these particles, that is,

$$\frac{\partial \mathcal{E}_{\text{mech}}}{\partial t} = \int \mathbf{j} . \mathbf{E} \, dV. \tag{4.15}$$

Substituting for $\mathbf{j}$ from (4.3), we get

$$\frac{\partial \mathcal{E}_{\text{mech}}}{\partial t} = \int (\nabla \times \mathbf{H}). \mathbf{E} \, dV - \int \frac{\partial \mathbf{D}}{\partial t}. \mathbf{E} \, dV.$$

We now use the vector identity (B.7) to transform the first term on the RHS, whereas we make use of the relation $\mathbf{D} = \epsilon \mathbf{E}$ given by (2.102) in the second term. This gives

$$\frac{\partial \mathcal{E}_{\text{mech}}}{\partial t} = \int \nabla . (\mathbf{H} \times \mathbf{E}) \, dV + \int \mathbf{H} . (\nabla \times \mathbf{E}) \, dV - \int \epsilon \frac{\partial \mathbf{E}}{\partial t}. \mathbf{E} \, dV.$$

Using (4.4) to replace $\nabla \times \mathbf{E}$ in the second term of the RHS, we get

$$\frac{\partial \mathcal{E}_{\text{mech}}}{\partial t} = -\int \nabla .(\mathbf{E} \times \mathbf{H}) \, dV - \int \mathbf{H} . \frac{\partial \mathbf{B}}{\partial t} \, dV - \int \frac{\partial}{\partial t} \left(\frac{1}{2} \epsilon E^2 \right) dV. \quad (4.16)$$

It is obvious that we can write $\epsilon E^2 = \mathbf{E}.\mathbf{D}$. By making use of $\mathbf{B} = \mu \mathbf{H}$ as given by (3.49), it is easy to show that

$$\mathbf{H} . \frac{\partial \mathbf{B}}{\partial t} = \frac{\partial}{\partial t} \left(\frac{1}{2} \mathbf{H} . \mathbf{B} \right).$$

We can thus write (4.16) in the form

$$\frac{\partial \mathcal{E}_{\text{mech}}}{\partial t} = -\oint (\mathbf{E} \times \mathbf{H}). d\mathbf{S} - \int \frac{\partial}{\partial t} \left(\frac{1}{2} \mathbf{H} . \mathbf{B} + \frac{1}{2} \mathbf{E} . \mathbf{D} \right) dV, \quad (4.17)$$

where the first term in the LHS of (4.16) has been transformed by Gauss's theorem (B.13) into a surface integral over the surface bounding the volume we are considering. We now write the Eq. (4.17) in the following manner:

$$\frac{\partial \mathcal{E}_{\text{mech}}}{\partial t} = -\oint \mathbf{N} . d\mathbf{S} - \int \frac{\partial U}{\partial t} \, dV, \quad (4.18)$$

where

$$\mathbf{N} = \mathbf{E} \times \mathbf{H} \quad (4.19)$$

is called the *Poynting vector*, and U is given by

$$U = \frac{1}{2} \mathbf{E} . \mathbf{D} + \frac{1}{2} \mathbf{H} . \mathbf{B}. \quad (4.20)$$

Now our job is to give a physical interpretation of (4.18). We have already noted in Sect. 2.17 that $(1/2)\mathbf{E}.\mathbf{D}$ can be interpreted as the energy density of an electrostatic field. This tempts us to conclude that U given by (4.20) may be interpreted as the total energy density of the electromagnetic field. We now justify that this is indeed the case. In the case where U is the energy density of the electromagnetic field, the total energy $\mathcal{E}_{em}$ of the electromagnetic field within the volume under consideration is

$$\mathcal{E}_{em} = \int U \, dV. \qquad (4.21)$$

We now put (4.18) in the form

$$\frac{\partial}{\partial t}(\mathcal{E}_{mech} + \mathcal{E}_{em}) = - \oint \mathbf{N} . d\mathbf{S}. \qquad (4.22)$$

Let us first interpret this equation in the situation when $\mathbf{N}$ is zero. In this situation, clearly $\mathcal{E}_{mech} + \mathcal{E}_{em}$ is conserved. If we expect the charged particles to get their mechanical energy from the electromagnetic field, then we would anticipate that the sum of the energies of the charged particles and the electromagnetic field to be conserved. On this basis, (4.22) leads us to interpret $\mathcal{E}_{em}$ given by (4.21) as the energy of the electromagnetic field inside the volume under consideration. This means that U as given by (4.20) must indeed be the energy density. It is now not difficult to interpret the Poynting vector $\mathbf{N}$. When this is non-zero, the RHS of (4.22) corresponds to an energy loss across the bounding surface, the Poynting vector $\mathbf{N}$ giving the electromagnetic energy flux.

By making use of (4.15) and (4.21), we can write (4.22) in the form

$$\int \left(\frac{\partial U}{\partial t} + \mathbf{j} . \mathbf{E} \right) dV = - \int (\nabla . \mathbf{N}) \, dV,$$

where the surface integral in the RHS of (4.22) has been converted into a volume integral by Gauss's theorem (B.13). Since this equation has to be true inside any arbitrary volume, we must have

$$\frac{\partial U}{\partial t} + \nabla . \mathbf{N} = - \mathbf{j} . \mathbf{E}. \qquad (4.23)$$

If the RHS is zero, then this equation has the same form as (4.7). Just as (4.7) gives the conservation of charge, (4.23) with the RHS zero gives the conservation of the electromagnetic energy, for which U is the density, and $\mathbf{N}$ is the flux. The RHS corresponds to the energy of the electromagnetic field supplied to the mechanical energy of the charged particles on which the electromagnetic field does work. When this happens, the pure electromagnetic energy is no longer conserved.

To summarize, $\mathbf{N}$ given by (4.19) is the expression for energy flux (i.e. the rate of energy flow) within an electromagnetic field. We shall repeatedly use (4.19) to calculate the energy flux carried by electromagnetic radiation under various circumstances. However, keep in mind that our discussion in this section merely makes it plausible that U given by (4.20) can be interpreted as energy density and $\mathbf{N}$ given by (4.19) can be interpreted as energy flux. Our discussion cannot be said to 'prove' these assertions. The reader may wish to look at a discussion of this point in

Feynman, Leighton and Sands ([2], Sects. 27-4 and 27-5), where some counterintuitive examples of energy flow which result from the expression (4.19) can be found.

4.3 Momentum of Electromagnetic Fields

An electromagnetic field must have momentum associated with it, along with energy. When charged particles on which the electromagnetic field acts gain momentum, it has to come from the momentum of the electromagnetic field. In a manner similar to the manner in which we obtained an expression of the energy density (by considering the energy supply rate to charged particles), we can also determine the momentum density of the electromagnetic field by considering the rate of momentum supplied to the charged particles. Since the charged particles give rise to the charge density ρ and the current density $\mathbf{j}$, the force exerted by the electromagnetic field on the charges and currents should be the collective force on the particles and must be equal to the rate of change of momentum of the charged particles.

Let us begin with the expression (1.6) giving the force per unit volume exerted by the electromagnetic field. Making a substitution for ρ from (4.1) and a substitution for $\mathbf{j}$ from (4.3), we get

$$\mathbf{F}_v = (\nabla \cdot \mathbf{D})\,\mathbf{E} + \left(\nabla \times \mathbf{H} - \frac{\partial \mathbf{D}}{\partial t} \right) \times \mathbf{B}. \qquad (4.24)$$

We can write the last term in the RHS in the following manner:

$$\frac{\partial \mathbf{D}}{\partial t} \times \mathbf{B} = \frac{\partial}{\partial t}(\mathbf{D} \times \mathbf{B}) - \mathbf{D} \times \frac{\partial \mathbf{B}}{\partial t} = \frac{\partial}{\partial t}(\mathbf{D} \times \mathbf{B}) + \mathbf{D} \times (\nabla \times \mathbf{E})$$

on making use of (4.4). Substituting this for the last term in (4.24), we arrive at

$$\mathbf{F}_v = (\nabla \cdot \mathbf{D})\,\mathbf{E} + (\nabla \cdot \mathbf{B})\,\mathbf{H} - \mathbf{B} \times (\nabla \times \mathbf{H}) - \mathbf{D} \times (\nabla \times \mathbf{E}) - \frac{\partial}{\partial t}(\mathbf{D} \times \mathbf{B}),$$
$$(4.25)$$

in which we have introduced the term $(\nabla \cdot \mathbf{B})\mathbf{H}$, which is zero by virtue of (4.2). Note that we began with the expression (1.6) for $\mathbf{F}_v$, in which electric and magnetic field vectors seemed to appear somewhat differently. By making use of all four of the Eqs. (4.1)–(4.4), we now have arrived at the expression (4.25), in which electric and magnetic field vectors appear in a very symmetrical manner.

We now have to interpret (4.25), for which purpose we proceed as in the previous section. The force $\mathbf{F}_v$ acting on the charges and currents imparts momentum to the particles making up the charges and currents. If $\mathbf{P}_{mech}$ is the momentum of the charged particles inside a certain volume, then we must have

$$\frac{\partial \mathbf{P}_{mech}}{\partial t} = \int \mathbf{F}_v \, dV, \qquad (4.26)$$

since we have to integrate $\mathbf{F}_v$ over this volume to determine the rate of change of momentum within this volume. Now, we tentatively identify $\mathbf{D} \times \mathbf{B}$ as the momentum density of the electromagnetic field. We shall later see that such an identification makes various things come out consistently. Then the momentum of the electromagnetic field inside the volume under consideration is

$$\mathbf{P}_{\text{em}} = \int (\mathbf{D} \times \mathbf{B}) \, dV = \epsilon \mu \int (\mathbf{E} \times \mathbf{H}) \, dV \tag{4.27}$$

on making use of (2.102) and (3.49). We now integrate (4.25) over the volume under consideration and use (4.26) and (4.27). This gives

$$\frac{\partial}{\partial t} (\mathbf{P}_{\text{mech}} + \mathbf{P}_{\text{em}}) = \int [(\nabla \cdot \mathbf{D}) \mathbf{E} - \mathbf{D} \times (\nabla \times \mathbf{E}) + (\nabla \cdot \mathbf{B}) \mathbf{H} - \mathbf{B} \times (\nabla \times \mathbf{H})] \, dV. \tag{4.28}$$

To proceed further, we consider the first two terms inside the volume integral in the RHS of (4.28). Indicating the three directions in three-dimensional space by 1, 2 and 3, the 1 component of these two terms gives

$$[(\nabla \cdot \mathbf{D})\mathbf{E} - \mathbf{D} \times (\nabla \times \mathbf{E})]_1 = E_1 \left(\frac{\partial D_1}{\partial x_1} + \frac{\partial D_2}{\partial x_2} + \frac{\partial D_3}{\partial x_3} \right) - D_2 \left(\frac{\partial E_2}{\partial x_1} - \frac{\partial E_1}{\partial x_2} \right) + D_3 \left(\frac{\partial E_1}{\partial x_3} - \frac{\partial E_3}{\partial x_1} \right).$$

It is easy to check that this can be written in the following form:

$$[(\nabla \cdot \mathbf{D})\mathbf{E} - \mathbf{D} \times (\nabla \times \mathbf{E})]_1 = \frac{\partial}{\partial x_1} (E_1 D_1) + \frac{\partial}{\partial x_2} (E_1 D_2) + \frac{\partial}{\partial x_3} (E_1 D_3) - \frac{1}{2} \frac{\partial}{\partial x_1} (\mathbf{E} . \mathbf{D}).$$

This means that the α component can be written as follows:

$$[(\nabla \cdot \mathbf{D}) \mathbf{E} - \mathbf{D} \times (\nabla \times \mathbf{E})]_\alpha = \sum_\beta \frac{\partial}{\partial x_\beta} \left(E_\alpha D_\beta - \frac{1}{2} \delta_{\alpha\beta} \mathbf{E} . \mathbf{D} \right), \tag{4.29}$$

where the summation sign implies that we have to sum over the three possible values 1, 2 and 3 of β. One can easily show that the last two terms inside the volume integral in the RHS of (4.28) also can be put in a similar form involving the corresponding magnetic quantities. We now define what is called the *Maxwell stress tensor* in the following way:

$$T_{\alpha\beta} = E_\alpha D_\beta + H_\alpha B_\beta - \frac{1}{2} \delta_{\alpha\beta} (\mathbf{E} . \mathbf{D} + \mathbf{H} . \mathbf{B}). \tag{4.30}$$

Then, on the basis of (4.29), we can put (4.28) in the form

$$\frac{\partial}{\partial t} (\mathbf{P}_{\text{mech}} + \mathbf{P}_{\text{em}})_\alpha = \int \sum_\beta \frac{\partial T_{\alpha\beta}}{\partial x_\beta} \, dV. \tag{4.31}$$

If we consider $T_{\alpha 1}$, $T_{\alpha 2}$ and $T_{\alpha 3}$ to be somewhat like the three components of a vector, then

$$\sum_{\beta} \frac{\partial T_{\alpha \beta}}{\partial x_{\beta}}$$

is clearly like the divergence of that. We can transform the volume integral in the RHS of (4.31) into a surface integral by Gauss's theorem (B.13) to obtain

$$\frac{\partial}{\partial t}(\mathbf{P}_{\text{mech}} + \mathbf{P}_{\text{em}})_{\alpha} = \oint \sum_{\beta} T_{\alpha \beta} \, dS_{\beta}. \tag{4.32}$$

Now our job is to give a physical interpretation of (4.32). If the Maxwell stress tensor over the bounding surface is zero, then clearly $\mathbf{P}_{\text{mech}} + \mathbf{P}_{\text{em}}$ will be conserved. This justifies our identification of $\mathbf{D} \times \mathbf{B}$ as the momentum density of the electromagnetic field in (4.27). The rate of change of the total momentum inside a volume must be equal to the force acting on the volume, which is given by the surface integral in the RHS of (4.32). This means that $\sum_{\beta} T_{\alpha \beta} dS_{\beta}$ must be the α component of the force acting across the surface element dS_{β} of the bounding surface. To elucidate the nature of this force, let us consider one particularly illuminating example. Suppose we have only an electric field in a region (i.e. the magnetic field is zero). We select the coordinate system in such a manner that the z axis is in the direction of the local electric field. It is easy to see that $T_{\alpha \beta}$ given by (4.30) in this situation becomes

$$T_{\alpha \beta} = \begin{pmatrix} -\frac{1}{2} E_z D_z & 0 & o \\ 0 & -\frac{1}{2} E_z D_z & 0 \\ 0 & 0 & \frac{1}{2} E_z D_z \end{pmatrix}. \tag{4.33}$$

Let us now determine the force exerted across a surface element dS_z perpendicular to the local electric field. It is obvious that the α component of this force is

$$T_{\alpha z} \, dS_z.$$

If $T_{\alpha \beta}$ is given by (4.33), then only the component $\alpha = z$ is non-zero, which would give a force

$$\frac{1}{2} E_z D_z \, dS_z$$

along the z direction. Since this is positive, it is of the nature of a pull on a volume across an element of the bounding surface. We can say that the electric field exerts a tension $\frac{1}{2} E_z D_z$ along field lines. Next, we consider a surface element dS_x parallel to the local electric field. It is easy to check that the force across this surface element is given by

$$-\frac{1}{2} E_z D_z \, dS_x$$

in the x direction. The minus sign implies that the force would be of the nature of a push on a volume across this element of the bounding surface. We can say that the electric field exerts a sideways pressure $\frac{1}{2} E_z D_z$. Similar considerations would clearly hold for the magnetic field if it is non-zero. The magnetic field also would exert a tension $\frac{1}{2} H_z B_z$ along field lines and a sideways pressure $\frac{1}{2} H_z B_z$.

While discussing the momentum of the electromagnetic fields, apart from concluding that the momentum density is given by $\mathbf{D} \times \mathbf{B}$ in accordance with (4.27), we have obtained another very important result. We have established that electric and magnetic fields exert tension along field lines and apply sideways pressure. The general expression of the force exerted by an electromagnetic field across a surface can be obtained from the expression of the Maxwell stress tensor given by (4.30).

It is clear from our discussion in Sect. 4.2 that the electric field around a charged particle has some energy (which is equivalent to mass by special relativity) associated with it. If the charged particle is moving, then there will be a magnetic field around it, as discussed in Sect. 3.9, and it would follow from the results of this section that the electromagnetic field around the charged particle would have a momentum density at a point in space, giving rise to a total momentum. Readers are asked to work this out in Exercise 4.5. Since an effective mass and an effective momentum will be associated with a moving charged particle in this manner, we are faced with the intriguing question whether the mass of a particle like the electron could have arisen in this manner. In fact, there were efforts to build a model of the electron along these lines in the early decades of the twentieth century. As a reader who works out Exercise 4.5 will find, the effective mass arising from energy considerations and the effective mass arising from momentum considerations have a factor $4/3$ between them. This factor creates very serious difficulties if we want to build an electromagnetic model of the electron mass consistent with special relativity. Ultimately, efforts to build a model of the electron in this manner were abandoned. Readers interested in knowing more about this subject are strongly encouraged to read Chap. 28 of Feynman, Leighton and Sands [2].

4.4 Electromagnetic Wave in an Infinite Medium

We have shown that Maxwell's equations (4.1)–(4.4) lead to some conclusions of great importance, such as the demonstration that electromagnetic fields have energy and momentum associated with them. However, so far in this chapter, we have not tried to solve these equations. In the previous two chapters, we have discussed various solutions for the static situation, when the absence of the time derivative terms simplifies the equations considerably. We now want to present the first simple solutions of Maxwell's equations with the time derivative terms. While keeping the time derivative terms, we can simplify the equations by considering the situation in which $\rho = 0$ and $\mathbf{j} = 0$. We shall now see that, even with these simplifications, we are able to derive a famous result—the existence of electromagnetic waves. Although we are able to study many characteristics of propagating electromagnetic waves with the simplifications $\rho = 0$ and $\mathbf{j} = 0$, we cannot address the question of how these waves are emitted when these simplifications are made. It is necessary to include the charge and current densities in Maxwell's equations when we want to study the emission of electromagnetic waves. This subject of emission of electromagnetic waves will be taken up in Chaps. 6–7.

We shall now carry on an analysis based on Maxwell's equations (4.1)–(4.4) with $\rho = 0$ and $\mathbf{j} = 0$. To proceed with this analysis, we need to relate $\mathbf{D}$ with $\mathbf{E}$ and $\mathbf{H}$ with $\mathbf{B}$. The simplest relationships we can use are $\mathbf{D} = \epsilon\mathbf{E}$ and $\mathbf{B} = \mu\mathbf{H}$, as given in (2.102) and (3.49), respectively. We shall keep using these relations throughout this chapter. However, we remind the reader that these relations assume an isotropic nature of the medium. As mentioned briefly in Sect. 2.14, $\mathbf{D}$ and $\mathbf{E}$ in an anisotropic medium have to be related through a tensor. One important problem connected with electromagnetic waves is to study the propagation of electromagnetic waves through crystals, which are typically anisotropic. In such a study, it is necessary to use a tensorial relation between $\mathbf{D}$ and $\mathbf{E}$, which makes the mathematical analysis much more complicated. We shall not discuss this topic in this book (see Exercise 4.8), and refer the interested reader to Born and Wolf ([3], Chap. XIV).

Let us first consider an infinite medium with constant ϵ and μ. On putting $\rho = 0$, $\mathbf{j} = 0$ and using $\mathbf{D} = \epsilon\mathbf{E}$, $\mathbf{B} = \mu\mathbf{H}$, Maxwell's equations (4.1)–(4.4) lead to:

$$\nabla \cdot \mathbf{E} = 0, \tag{4.34}$$

$$\nabla \cdot \mathbf{B} = 0, \tag{4.35}$$

$$\nabla \times \mathbf{B} = \epsilon\mu \frac{\partial \mathbf{E}}{\partial t}, \tag{4.36}$$

$$\nabla \times \mathbf{E} = -\frac{\partial \mathbf{B}}{\partial t}. \tag{4.37}$$

Note that we have multiplied (4.3) by μ to arrive at (4.36). We now take the curl of (4.37), which gives

$$\nabla \times (\nabla \times \mathbf{E}) = -\frac{\partial}{\partial t}(\nabla \times \mathbf{B}).$$

We make use of the vector identity (B.12) in the LHS and substitute for $\nabla \times \mathbf{B}$ from (4.36) in the RHS. This gives

$$\nabla(\nabla \cdot \mathbf{E}) - \nabla^2\mathbf{E} = -\frac{\partial}{\partial t}\left(\epsilon\mu \frac{\partial \mathbf{E}}{\partial t}\right).$$

Since the first term on the LHS would disappear because of (4.34), we are led to

$$\nabla^2\mathbf{E} = \epsilon\mu \frac{\partial^2 \mathbf{E}}{\partial t^2}. \tag{4.38}$$

This is the wave equation, one of the well-known equations of mathematical physics, demonstrating that Maxwell's equations predict electromagnetic waves. We can write (4.38) in the form

$$\frac{\partial^2 \mathbf{E}}{\partial t^2} = u_{em}^2 \nabla^2 \mathbf{E}, \qquad (4.39)$$

where the speed of propagation u_{em} of the wave is given by

$$u_{em} = \frac{1}{\sqrt{\epsilon \mu}}. \qquad (4.40)$$

From (2.103) and (3.51), we can write $\epsilon = \epsilon_0 \kappa$ and $\mu = \mu_0 \kappa_m$ so that

$$u_{em} = \frac{c}{\sqrt{\kappa \kappa_m}}, \qquad (4.41)$$

where

$$c = \frac{1}{\sqrt{\epsilon_0 \mu_0}}. \qquad (4.42)$$

Since $\kappa = 1$ and $\kappa_m = 1$ for free space or vacuum, it is obvious from (4.41) that the speed of light in free space or vacuum is c given by (4.42). We have pointed out in Sect. 3.2 that in the SI system we choose μ_0 in such a manner that the unit ampere (A) of current comes out to be a convenient unit for everyday life. Once the unit of current is fixed from magnetic measurements for the given value of μ_0, the unit coulomb (C) of charge also gets fixed, as mentioned in Sect. 2.1, and it is in principle possible to determine ϵ_0 from (2.7) by making purely electrical measurements. Using the value of ϵ_0 obtained from such measurements and combining it with μ_0 given by (3.11), one can calculate the speed of light in vacuum by using (4.42). In other words, no experiment involving light needs to be done in order to obtain the value of the speed of light. When Maxwell derived the theory of electromagnetic waves, he calculated its speed in this manner purely on the basis of electrical and magnetic measurements. It may be noted that Maxwell's derivation was based on Gaussian units, which will be discussed in Appendix A. The speed of light obtained by Maxwell in this manner was found to be very close to the speed of light known at that time, suggesting that light is an electromagnetic wave. The present practice in introducing SI units, however, is to proceed in a somewhat different manner. The speed of light c is now taken to be more fundamental than the permittivity of free space ϵ_0. Once we have introduced μ_0 through (3.11) and we have an accurate experimentally determined value of c, we can use (4.42) to obtain the value of ϵ_0. The best value of the speed of light in vacuum available to us is

$$c = 3.00 \times 10^8 \text{ m s}^{-1}, \qquad (4.43)$$

which leads to the value of ϵ_0 given by (2.8). One can, of course, check this value of ϵ_0 independently through electrical measurements to ensure that different parts of the theory are consistent.

Let us discuss some basic physical characteristics of electromagnetic waves by considering plane electromagnetic waves (i.e. waves which have plane wave fronts) propagating in a particular direction, which we can take as the z direction without any loss of generality. A simple solution of (4.39) is given by

$$\mathbf{E} = \mathbf{E}_0 \, e^{i(kz - \omega t)}, \tag{4.44}$$

with the relation

$$\omega = k \, u_{\mathrm{em}}, \tag{4.45}$$

which one can easily find on substituting (4.44) into (4.39). Since the wave equation (4.39) is a linear equation, a linear superposition of solutions with different values of k will also be a solution of (4.39). The most general type of plain wave propagating in the z direction is, therefore, given by the following superposition of the solutions of type (4.44):

$$\mathbf{E} = \int \mathbf{E}_0(k) \, e^{i(kz - k \, u_{\mathrm{em}} t)} dk, \tag{4.46}$$

where we have made use of (4.45).

We now derive some properties of electromagnetic waves by focussing our attention on a component of the wave given by (4.44). The time derivative terms in (4.36) and (4.37) suggest that there has to be a magnetic field associated with the electric field given by (4.44). We leave it as an exercise for the reader to show from (4.34)–(4.37) that the magnetic field also would satisfy the same type of wave equation as (4.38) satisfied by the electric field:

$$\nabla^2 \mathbf{B} = \epsilon \mu \, \frac{\partial^2 \mathbf{B}}{\partial t^2}, \tag{4.47}$$

and we can write down a solution of the magnetic field

$$\mathbf{B} = \mathbf{B}_0 \, e^{i(kz - \omega t)}. \tag{4.48}$$

It is clear that $\partial/\partial t$ operating on (4.44) or (4.48) would yield $-i\omega$, whereas ∇ operating on them would yield $i\mathbf{k}$, where $\mathbf{k} = k\hat{\mathbf{e}}_z$ is the wavenumber vector. We then conclude from (4.34) and (4.35) that

$$\mathbf{k} \cdot \mathbf{E} = 0, \quad \mathbf{k} \cdot \mathbf{B} = 0, \tag{4.49}$$

implying that both $\mathbf{E}$ and $\mathbf{B}$ would lie in the plane perpendicular to the propagation direction. In other words, an electromagnetic wave is a transverse wave. On substituting (4.44) and (4.48) in (4.37), we get

$$i \, \mathbf{k} \times \mathbf{E} = i \, \omega \mathbf{B},$$

from which

$$\mathbf{B} = \frac{\mathbf{k}}{\omega} \times \mathbf{E}. \tag{4.50}$$

This shows that the magnetic field $\mathbf{B}$ of the electromagnetic wave associated with the electric field $\mathbf{E}$ has to be perpendicular to it. It follows from (4.49) and (4.50) that $\mathbf{E}$, $\mathbf{B}$ and $\mathbf{k}$ make up a triad of orthogonal vectors. Both $\mathbf{E}$ and $\mathbf{B}$, which are perpendicular to each other, have to lie in the plane perpendicular to the propagation direction. It may be noted that (4.49) and (4.50) hold for electromagnetic waves with propagation vector $\mathbf{k}$ in any direction (i.e. it is not necessary to assume $\mathbf{k}$ to be in the z direction as we have done in order to obtain these relations).

From (4.45) and (4.50), it follows that the amplitudes of the magnetic field and the electric field are related by

$$B_0 = \frac{E_0}{u_{\mathrm{em}}}. \tag{4.51}$$

In vacuum where u_{em} equals c according to (4.41), we must have

$$E_0 = c \, B_0. \tag{4.52}$$

We sometimes have to consider the force exerted by an electromagnetic wave on a charged particle, based on (1.5). The ratio of the magnetic force to the electric force is obviously of order

$$\frac{|\text{Magnetic force}|}{|\text{Electric force}|} \approx \frac{|v| \, |B|}{|E|} \approx \frac{|v|}{c} \tag{4.53}$$

on making use of (4.52). Under normal circumstances, the velocity v induced in the charged particle by the electromagnetic wave would be completely negligible compared to c. According to (4.53), we can neglect the magnetic force when we consider the effect of an electromagnetic wave on a charged particle.

4.4.1 Polarization of Electromagnetic Waves

We have been considering electromagnetic waves propagating in the z direction. Certainly the electric field amplitude $\mathbf{E}_0$ appearing in (4.44) has to lie in the xy plane. Since it is possible for $\mathbf{E}_0$ to be in any direction in the xy plane, we readily recognize the possibility of two polarization states. For an electromagnetic wave with wavenumber k propagating in the z direction, the amplitude $\mathbf{E}_0$ of the electric field appearing in (4.44) is given quite generally by an expression of the kind

$$\mathbf{E}_0 = a_x e^{i\delta_x} \hat{\mathbf{e}}_x + a_y e^{i\delta_y} \hat{\mathbf{e}}_y, \tag{4.54}$$

where a_x, δ_x, a_y and δ_y are assumed to be real. The magnetic field corresponding to this electric field can easily be obtained by using (4.50). Clearly $\delta_y - \delta_x$ gives the phase difference between the oscillating electric fields in the y and x directions. In a completely unpolarized light beam, this phase difference $\delta_y - \delta_x$ will keep changing randomly with time. On the other hand, if this phase difference does not change with time, then it is easy to see that we get elliptically polarized light. The physical value of the electric field would be given by the real part of the complex expression. We write down the electric field by equating it to the real part of the expression we get by substituting (4.54) into (4.44) and by considering a specific value of z, say $z = 0$:

$$\mathbf{E} = a_x \cos(\omega t - \delta_x) \hat{\mathbf{e}}_x + a_y \cos(\omega t - \delta_y) \hat{\mathbf{e}}_y. \tag{4.55}$$

We leave it for the reader to argue that the tip of the electric field vector given by (4.55) will trace out an ellipse in time. If $a_y = a_x$, then a phase difference $\delta_y - \delta_x = \pi/2$ will give circular polarization. Putting $\delta_y = \delta_x + \pi/2$ in (4.55), we get

$$\mathbf{E} = a_x \cos(\omega t - \delta_x) \hat{\mathbf{e}}_x + a_x \sin(\omega t - \delta_x) \hat{\mathbf{e}}_y,$$

which corresponds to circularly polarized light.

4.4.2 Energy Density and Energy Flux of Electromagnetic Waves

It is rather instructive to consider the energy density and the energy flux associated with electromagnetic waves, which we can calculate from (4.20) and (4.19). Since the electric and magnetic fields are varying harmonically with time, the values of energy density or energy flux at a point will keep changing with time so that we need to consider their average values. Now, both the energy density and the energy flux are quadratic in field variables. For two quantities $\mathbf{E}$ and $\mathbf{D}$ harmonically varying in the same way, we consider the quadratic quantity

$$\mathbf{E} \cdot \mathbf{D} = \epsilon E^2 = \epsilon E_0^2 \cos^2(kz - \omega t)$$

where we have substituted the real part of (4.44). It is clear that the time-averaged value of this is given by

$$\overline{\mathbf{E} \cdot \mathbf{D}} = \frac{1}{2} \epsilon E_0^2.$$

It follows from (4.20) that the average energy density of the electromagnetic wave would be given by

$$\overline{U} = \frac{1}{4}\epsilon E_0^2 + \frac{1}{4}\frac{B_0^2}{\mu} \tag{4.56}$$

on making use of the fact that $\mathbf{H} = \mathbf{B}/\mu$. In the magnetic energy density term, we make substitutions from (4.40) and (4.51), which gives

$$\frac{1}{4}\frac{B_0^2}{\mu} = \frac{1}{4}\frac{E_0^2}{u_{em}^2\mu} = \frac{1}{4}\epsilon E_0^2. \tag{4.57}$$

Thus, we arrive at the very important conclusion that the magnetic energy density associated with an electromagnetic wave is exactly equal to the electric energy density. It may seem at first a little bit counter-intuitive that, although the effect of the magnetic field on a charged particle, according to (4.53), is negligible compared to the effect of the electric field, the energy densities associated with the fields are equal! Taking note of this equality (4.57), the energy density of the electromagnetic wave, as given by (4.56), becomes

$$\overline{U} = \frac{1}{2}\epsilon E_0^2. \tag{4.58}$$

We now turn to the Poynting vector given by (4.19). Since the electric and magnetic fields of an electromagnetic wave both vary harmonically in phase with each other, it is easy to see that the time-averaged value of the Poynting vector would be given by

$$\overline{\mathbf{N}} = \frac{1}{2}\mathbf{E}_0 \times \mathbf{H}_0 = \frac{1}{2}\mathbf{E}_0 \times \frac{\mathbf{B}_0}{\mu}.$$

Making use of (4.50), we get

$$\overline{\mathbf{N}} = \frac{1}{2}\frac{1}{\mu\omega}\mathbf{E}_0 \times (\mathbf{k} \times \mathbf{E}_0) = \frac{1}{2}\frac{1}{\mu\omega}E_0^2\mathbf{k}$$

keeping in mind that $\mathbf{E}_0$ is perpendicular to $\mathbf{k}$. For an electromagnetic wave propagating in the z direction, we can write $\mathbf{k} = k\hat{\mathbf{e}}_z$ and use (4.45) to get

$$\overline{\mathbf{N}} = \frac{1}{2}\frac{1}{\mu u_{em}^2}E_0^2\mathbf{u}_{em},$$

where $\mathbf{u}_{em} = u_{em}\hat{\mathbf{e}}_z$ is the velocity with which the wave is propagating. Making use of (4.40), we finally arrive at

$$\overline{\mathbf{N}} = \frac{1}{2}\epsilon E_0^2\mathbf{u}_{em} = \overline{U}\mathbf{u}_{em} \tag{4.59}$$

after using (4.58). This important equation indicates that the average energy flux in an electromagnetic wave is given by average energy density multiplied by the propagation velocity, which we expect on physical grounds.

4.5 Electromagnetic Waves Inside Conductors

While discussing electrostatics and magnetostatics in the previous two chapters, we have extensively dealt with conductors—as repositories of electric charge in electrostatics and as conduits of electric current in magnetostatics. Now we consider the propagation of electromagnetic waves inside conductors. We have pointed out in Sect. 2.6 that the electric field is taken to be zero inside a conductor in electrostatics. This is because any unbalanced electric field inside a conductor would cause a movement of charges until the electric field is neutralized. Although we have to take the electric field to be zero inside a conductor in a static situation, while dealing with situations involving time evolution we have to allow for the possibility of electric fields inside a conductor giving rise to currents. Since currents are caused by the movements of charges subject to the force law (1.5), we expect the current to be given by an expression of the kind

$$\mathbf{j} = \sigma(\mathbf{E} + \mathbf{v} \times \mathbf{B}), \tag{4.60}$$

where σ is the electrical conductivity.

If $\mathbf{E}$ and $\mathbf{B}$ refer to the electric and magnetic fields of an electromagnetic wave, we saw in (4.53) that the magnetic force is negligible compared to the electric force. In this situation, we can neglect the magnetic force term in (4.60) and write

$$\mathbf{j} = \sigma\mathbf{E}. \tag{4.61}$$

This is often called *Ohm's law for the continuum*. We assume that readers are familiar with Ohm's law for conductors, which is discussed extensively in elementary textbooks on electromagnetic theory. While the magnetic force term $\mathbf{v} \times \mathbf{B}$ in (4.60) is negligible when $\mathbf{B}$ refers to the magnetic field of the electromagnetic wave, we may point out that in the presence of an external magnetic field, this term can give rise to an interesting effect known as the *Hall effect*, which we shall not discuss in this book.

In the discussion of the previous section pertaining to non-conducting media, we had taken ρ and $\mathbf{j}$ equal to zero. While considering conductors, we now have to keep the $\mathbf{j}$ term in (4.3) and take $\mathbf{j}$ to be given by (4.61). Multiplying (4.3) by μ, we arrive at

$$\nabla \times \mathbf{B} = \mu\sigma\mathbf{E} + \epsilon\mu\frac{\partial \mathbf{E}}{\partial t}. \tag{4.62}$$

This equation now replaces (4.36). We have to use it along with (4.34), (4.35) and (4.37), which remain unchanged. On taking the curl of (4.37) and substituting for $\nabla \times \mathbf{B}$ from (4.62), we get

$$\nabla \times (\nabla \times \mathbf{E}) = -\frac{\partial}{\partial t}\left(\mu\sigma\mathbf{E} + \epsilon\mu\frac{\partial \mathbf{E}}{\partial t}\right).$$

Using the vector identity (B.12) in the LHS of this equation and noting (4.34), we arrive at

$$\nabla^2\mathbf{E} = \mu\sigma\frac{\partial \mathbf{E}}{\partial t} + \epsilon\mu\frac{\partial^2 \mathbf{E}}{\partial t^2}. \tag{4.63}$$

The electric field of an electromagnetic wave propagating inside a conductor would satisfy this equation.

It is easily seen that (4.63) would reduce to (4.38) in the previous section if we put $\sigma = 0$. Since the electrical conductivity σ would be zero for a non-conductor, the derivation of Sect. 4.4 tacitly assumed the medium to be non-conducting when we took $\mathbf{j} = 0$. If the medium is electrically conducting, then we cannot have $\mathbf{j} = 0$ in the presence of an electric field. We solved (4.38) by trying out a solution of the type (4.44). Let us again try a solution of the type (4.44) for our more complicated Eq. (4.63). In our discussion of Sect. 4.4, both ω and k turned out to be real. We shall now see that it is not possible to make both ω and k real when we try out a solution of type (4.44) for (4.63). Let us consider the situation of an electromagnetic wave harmonically varying with time impinging on a conductor. Then we have to take the time variation of the form $e^{-i\omega t}$ with a real ω. In this situation, we cannot demand k to be real. On substituting (4.44) into (4.63), we get

$$k^2 = i\mu\sigma\omega + \epsilon\mu\omega^2 = i\mu\sigma\omega\left(1 - \frac{i\epsilon\omega}{\sigma}\right).$$

For a good conductor, we may expect σ to be very high such that $\sigma \gg \epsilon\omega$. For example, $\epsilon_0\omega$ for yellow light is of order 3×10^4 in SI units, whereas the conductivity σ for a metal like copper is about 5.8×10^7 in SI units. In such a situation, we approximately have

$$k^2 = i\mu\sigma\omega,$$

from which we get

$$k = \pm\frac{(1+i)}{\sqrt{2}}\sqrt{\mu\sigma\omega}. \tag{4.64}$$

Let us now consider a semi-infinite conductor extending over all positive values of z, with its surface being at $z = 0$. Suppose an electromagnetic wave propagating from the negative z direction with its electric field given by (4.44) impinges on this conductor. As this electromagnetic wave propagates inside the conductor, the electric field inside the conductor will be given by (4.44) with k substituted by the expression given in (4.64). We expect the wave to be attenuated as it propagates inside the conductor. For this to happen, we have to take the positive sign in (4.64). With such a choice, the electric field inside the conductor is given by

$$\mathbf{E} = \mathbf{E}_0 \, e^{-\sqrt{\frac{\mu\sigma\omega}{2}} \, z} e^{i(\sqrt{\frac{\mu\sigma\omega}{2}} \, z - \omega t)}. \tag{4.65}$$

We now define the *skin depth*

$$\delta = \sqrt{\frac{2}{\mu\sigma\omega}}. \tag{4.66}$$

It is obvious from (4.65) that the electromagnetic wave gets attenuated within a surface layer of thickness of the order of the skin depth. For radio waves having the wavelength of a few cm, the skin depth turns out to be of order 10^{-3} cm, i.e. a very small fraction of the wavelength. We see from (4.66) that the skin depth becomes thinner when ω is higher. This suggests that a high-frequency electromagnetic wave cannot penetrate much inside a conductor. However, our simple theory has to be modified if the frequency is sufficiently high.

We end our discussion about electromagnetic waves inside conductors by pointing out the limitation of the simple theory we have worked out. The expression (4.61) between the electric field $\mathbf{E}$ and the current density $\mathbf{j}$ caused by it tacitly assumes that a charged particle in the conductor gets accelerated for time of order τ between two collisions, τ being typically of the order of 10^{-14} s for an electrically conducting metal. If the frequency ω of the electromagnetic wave is such that $1/\omega \gg \tau$ (which is true for radio waves and microwaves, but not for visible light), then many collisions can take place during the period of the wave so that statistical considerations lead to (4.61). In other words, (4.61) is really a relation valid for low-frequency waves, and the expression for skin depth we have obtained also would hold specifically for such waves. For electron gases in metals, we approach the opposite limit of $1/\omega \ll \tau$ for ultraviolet radiation and electromagnetic waves with still higher frequencies. The theory has to be modified for high-frequency electromagnetic waves, as indicated in Exercise 4.12. In fact, as we shall point out in Sect. 8.3.1, metals can become transparent to high-frequency ultraviolet radiation. A good discussion of the optics of metals can be found in Born and Wolf ([3], Chap. XIII).

4.6   Reflection and Refraction of Electromagnetic Waves at an Interface

In the previous two sections, we have discussed the propagation of electromagnetic waves in a uniform medium—Sect. 4.4 dealing with propagation in a non-conducting medium and Sect. 4.5 dealing with propagation in a conducting medium. In this section and the next two sections, we shall discuss what happens if there is a surface separating two different media on two sides. This section is devoted to a discussion

Fig. 4.2 A beam of an electromagnetic wave falling at an interface between two media at an incidence angle θ, giving rise to a refracted beam at refraction angle θ' and a reflected beam at reflection angle θ''

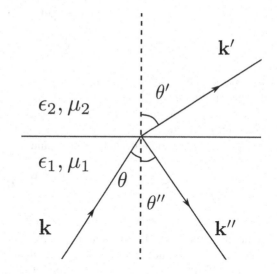

of what happens when an electromagnetic wave propagates towards an interface between two non-conducting media. We have made some comments in Sect. 4.5 on the interface between a non-conducting and a conducting medium, which will be further taken up in Sects. 4.7–4.8.

Let the values of ϵ and μ inside the two media on two sides of an interface be ϵ_1, μ_1 and ϵ_2, μ_2, respectively. The speed of the electromagnetic wave in these two media would be given by

$$u_1 = \frac{1}{\sqrt{\epsilon_1 \mu_1}}, \quad u_2 = \frac{1}{\sqrt{\epsilon_2 \mu_2}} \tag{4.67}$$

in accordance with (4.40). We now consider an incident beam of an electromagnetic wave moving through medium 1 towards the interface, making an angle θ with respect to the normal to the interface, as indicated in Fig. 4.2. If $\mathbf{k}$ is the wavenumber vector of this beam, which is in the propagation direction, the electric field associated with this electromagnetic wave can be written as

$$\mathbf{E} = \mathbf{E}_0 \, e^{i(\mathbf{k} \cdot \mathbf{x} - \omega t)}, \tag{4.68}$$

which is a trivial generalization of (4.44). When this incident beam falls on the interface, we expect that it will give rise to a refracted beam and a reflected beam. As shown in Fig. 4.2, these beams have wavenumber vectors $\mathbf{k}'$ and $\mathbf{k}''$ associated with them, making angles θ' and θ'' with the normal to the interface. Note that we do not a priori assume that the reflection angle is equal to the incidence angle. The electric fields associated with the refracted and reflected waves can be written as

$$\mathbf{E}' = \mathbf{E}_0' \, e^{i(\mathbf{k}'.\mathbf{x}-\omega t)}, \tag{4.69}$$

$$\mathbf{E}'' = \mathbf{E}_0'' \, e^{i(\mathbf{k}''.\mathbf{x}-\omega t)}. \tag{4.70}$$

We have assumed the same frequency ω for all the waves. Since the refracted and the reflected waves are generated from the incident wave, we certainly expect this to be the case. The magnetic fields associated with these waves can be obtained from (4.50). Making use of (3.49), the magnetic field vectors $\mathbf{H}$ associated with the incident, refracted and reflected waves should be

$$\mathbf{H} = \frac{\mathbf{k} \times \mathbf{E}}{\mu_1 \omega}, \ \mathbf{H}' = \frac{\mathbf{k}' \times \mathbf{E}'}{\mu_2 \omega}, \ \mathbf{H}'' = \frac{\mathbf{k}'' \times \mathbf{E}''}{\mu_1 \omega}. \tag{4.71}$$

It is easy to check that the boundary conditions (2.107) and (3.52) for tangential components of $\mathbf{E}$ and $\mathbf{H}$ across an interface, which we had earlier derived for the static situation, should hold in the present situation also. We can now apply these boundary conditions for the incident, refracted and reflected waves at some point on the interface. Now, the boundary conditions satisfied at one point, say $\mathbf{x}_I$, of the interface will ensure that they are satisfied at any other point, say $\mathbf{x}_{II}$, of the interface as well, only if the phase difference between these two arbitrary points $\mathbf{x}_I$ and $\mathbf{x}_{II}$ is the same for the incident, refracted and reflected waves. From (4.68)–(4.70), it is easy to see that this condition demands

$$\mathbf{k} \cdot (\mathbf{x}_I - \mathbf{x}_{II}) = \mathbf{k}' \cdot (\mathbf{x}_I - \mathbf{x}_{II}) = \mathbf{k}'' \cdot (\mathbf{x}_I - \mathbf{x}_{II}).$$

A look at Fig. 4.2 should readily convince the reader that this is equivalent to

$$k \sin \theta = k' \sin \theta' = k'' \sin \theta'', \tag{4.72}$$

where k, k' and k'' are the magnitudes of the three wavenumber vectors. It now follows from (4.45) that

$$u_1 = \frac{\omega}{k} = \frac{\omega}{k''}, \tag{4.73}$$

$$u_2 = \frac{\omega}{k'}. \tag{4.74}$$

We see from (4.73) that k'' has to be equal to k. On the basis of (4.72), this implies that

$$\theta'' = \theta. \tag{4.75}$$

This is the well-known *law of reflection* that the reflection angle should be equal to the incidence angle. From (4.73) and (4.74), we can obtain

$$u_1 k = u_2 k'.$$

It then follows from (4.72) that

$$\frac{\sin \theta}{\sin \theta'} = \frac{k'}{k} = \frac{u_1}{u_2}. \tag{4.76}$$

Since u_1/u_2 must be constant for two given media, it follows from (4.76) that $\sin \theta / \sin \theta'$ has to be a constant. This is the *law of refraction*, often called *Snell's law*. We have also found that this constant, called the refractive index, depends on the ratio of the speeds of light in the two media.

We have shown above that the laws of reflection and refraction follow easily from basic considerations. We now want to show how the amplitudes of the refracted and reflected beams can be found by using the boundary conditions. For this calculation, we have to treat the two polarizations of the electromagnetic wave separately. We have seen in (4.49) that the electric field has to be perpendicular to the propagation direction. We consider the two following cases: (i) the electric fields $\mathbf{E}$, $\mathbf{E}'$ and $\mathbf{E}''$ associated with all the three beams are perpendicular to the plane of propagation, which is the plane of Fig. 4.2; and (ii) all these electric fields lie in the plane of propagation, i.e. the plane of Fig. 4.2.

For case (i), the boundary condition (2.107) leads to

$$E_0 + E_0'' = E_0'. \tag{4.77}$$

In this case, the magnetic fields $\mathbf{H}$ for all the three beams have to lie in the plane of propagation. It is easy to check that the tangential components of $\mathbf{H}$ for the three beams are $H_0 \cos \theta$, $H_0' \cos \theta'$ and $-H_0'' \cos \theta''$. Then the boundary condition (3.52) implies

$$H_0 \cos \theta - H_0'' \cos \theta'' = H_0' \cos \theta'. \tag{4.78}$$

Since $\mathbf{k}$, $\mathbf{E}$ and $\mathbf{H}$ make up a triad of orthogonal vectors, (4.71) suggests that their amplitudes should be connected by the following relations:

$$H_0 = \frac{kE_0}{\mu_1 \omega}, \quad H_0' = \frac{k'E_0'}{\mu_2 \omega}, \quad H_0'' = \frac{k''E_0''}{\mu_1 \omega}. \tag{4.79}$$

Now, in many situations of an electromagnetic wave propagating from one medium to another, the change in the speed of the wave, as given by (4.40), is caused by a change in ϵ rather than a change in μ. If that is the case, then we can write $\mu_1 = \mu_2 = \mu$. Further, on making use of (4.73)–(4.74), we are led from (4.79) to

$$H_0 = \frac{E_0}{\mu u_1}, \quad H_0' = \frac{E_0'}{\mu u_2}, \quad H_0'' = \frac{E_0''}{\mu u_1}. \tag{4.80}$$

Substituting from (4.80) into (4.78), we get

$$E_0 \cos \theta - E_0'' \cos \theta = \frac{u_1}{u_2} E_0' \cos \theta', \quad (4.81)$$

where we have made use of (4.75). From (4.77) and (4.81), we get

$$\frac{E_0'}{E_0} = \frac{2 \cos \theta}{\cos \theta + (u_1/u_2) \cos \theta'}, \quad (4.82)$$

$$\frac{E_0''}{E_0} = \frac{\cos \theta - (u_1/u_2) \cos \theta'}{\cos \theta + (u_1/u_2) \cos \theta'}. \quad (4.83)$$

These give the amplitudes of the reflected and refracted waves.

We now consider case (ii) in which the electric fields of all the three beams lie in the plane of polarization, i.e. the plane of Fig. 4.2. Then the magnetic field **H** for all these waves must be perpendicular to this plane. Then the boundary condition (3.52) readily gives us

$$H_0 + H_0'' = H_0'. \quad (4.84)$$

The tangential components of **E** for the three beams lying in the plane of propagation are $E_0 \cos \theta$, $E_0' \cos \theta'$ and $-E_0'' \cos \theta''$. From the boundary condition (2.107), we get

$$E_0 \cos \theta - E_0'' \cos \theta'' = E_0' \cos \theta'.$$

On making use of (4.80) in conjunction with (4.75), we get from this

$$u_1 (H_0 \cos \theta - H_0'' \cos \theta) = u_2 H_0' \cos \theta'. \quad (4.85)$$

From (4.84) and (4.85), we arrive at

$$\frac{H_0'}{H_0} = \frac{2 \cos \theta}{\cos \theta + (u_2/u_1) \cos \theta'}, \quad (4.86)$$

$$\frac{H_0''}{H_0} = \frac{\cos \theta - (u_2/u_1) \cos \theta'}{\cos \theta + (u_2/u_1) \cos \theta'}. \quad (4.87)$$

These formulae (4.82), (4.83), (4.86) and (4.87) are known as *Fresnel's formulae*. Several decades before Maxwell showed that light is an electromagnetic wave, Fresnel arrived at these formulae by assuming that light is an elastic wave in ether filling all space!

One can draw several obvious conclusions from Fresnel's formulae. First of all, if the two media on the two sides of the interface are the same, then $u_2 = u_1$, and we expect the incident beam to propagate undisturbed. It follows from (4.76) that $\theta' = \theta$ in this case. Then we easily see from (4.83) and (4.87) that there will be no reflected beam for both of the polarizations. It also follows from (4.82) and (4.86) that $E_0' = E_0$ and $H_0' = H_0$ for the two polarizations, respectively. This means that

Fig. 4.3 Sketch of a
situation in which the
reflected and refracted rays
make a right angle between
them

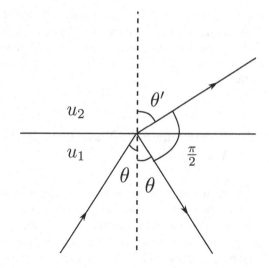

the incident beam propagates in a straight line with unchanged amplitude and without
any reflection. When there is a real interface between two distinct media, certainly
reflection and refraction do take place. We expect the energy to be conserved in this
case. In other words, the sum of the energy fluxes in the refracted and reflected beams
should be equal to the energy flux in the incident beam. We leave it as an exercise
for the reader to calculate the various energy fluxes by using (4.59) and to show this
for both the polarizations (Exercise 4.11).

A particularly interesting case arises when the reflected beam and the refracted
beam make an angle of $\pi/2$ as shown in Fig. 4.3. It is obvious that θ, θ' and this
angle $\pi/2$ together should be equal to π. This means that

$$\theta + \theta' = \frac{\pi}{2},$$

from which it follows that

$$\cos\theta = \sin\theta', \quad \cos\theta' = \sin\theta.$$

When we substitute these in (4.87) and keep (4.76) in mind, we find that

$$H_0'' = 0.$$

This means that in this case there would be no reflected beam for the polarization
case (ii) corresponding to the electric fields lying in the plane of propagation. If an
unpolarized beam falls on a surface in such a way that the reflected beam would
be perpendicular to the refracted beam, then we would find the reflected beam to
be linearly polarized, since the polarization involving the electric field lying in the
plane of propagation will be absent in the reflected beam, and only the polarization

involving the electric field perpendicular to the plane will be present in this beam. This fact that a light beam can be polarized in this manner is called *Brewster's law* and was known from optics experiments for many years before the advent of the electromagnetic theory of light. This theory at last gave a very elegant explanation of this law.

4.7 Electromagnetic Wave Propagation Through Waveguides

After discussing in Sect. 4.6 the propagation of an electromagnetic wave across an interface between two non-conducting media, we now consider an interface between a non-conducting and a conducting medium. We have shown in Sect. 4.5 that an electromagnetic wave cannot penetrate into the interior of a conducting medium beyond a distance of the order of the skin depth given by (4.66), provided the current inside the conducting medium can be written in the form (4.61). In such a situation, we have to consider the propagation of the electromagnetic wave only inside the non-conducting medium and treat the surface of the conducting medium with a suitable boundary condition. Since the skin depth given by (4.66) will be very thin for a conductor with high electrical conductivity σ, we can idealize the problem by assuming that the skin depth is infinitesimally small. Then we assume that the electric field inside the conductor just below the surface is zero and we can use the boundary condition (2.107) to conclude that the electric field inside the non-conducting medium has to satisfy the boundary condition that its tangential component has to be zero at the surface of the conductor. We shall see below that we can use (3.53) to obtain a suitable boundary condition for the magnetic field on the conducting surface. We may point out that we cannot use the boundary conditions (2.105) or (3.52) in this situation, since they were obtained from $\nabla . \mathbf{D} = 0$ and $\nabla \times \mathbf{H} = 0$ based on the assumption that there was no surface charge or surface current. This is not true for the surface of a conductor on which there can be surface charges or surface currents.

We now study the propagation of an electromagnetic wave inside a non-conducting medium bounded by conducting surfaces within which the wave cannot penetrate. This problem of an electromagnetic wave propagating through a region bounded by conducting surfaces has a great practical application. Suppose we want to transfer some energy from one region to another through some electromagnetic means. We can certainly transfer the energy through currents flowing through metal wires. That is how energy reaches a modern home from a power-generating station. However, an alternating current flowing through a wire is accompanied by energy loss through the emission of electromagnetic radiation. When we discuss the radiation from an oscillating electric dipole in Sect. 7.4, we shall see that the power radiated goes as the fourth power of the frequency ω. For alternating currents having frequency in the range 50–100 Hz, the energy loss due to radiation is tolerable. However, sending the energy through metal wires is certainly not a viable option if the frequency is

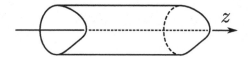

Fig. 4.4 Sketch of a waveguide with a cross section which is invariant along its axis in the z direction

considerably higher. If energy has to be transferred at frequencies of the order of frequencies of radio waves or microwaves, then the best strategy is to transfer it in the form of electromagnetic waves. If a source emits electromagnetic radiation in all directions, then it is easy to show that the intensity will fall off with the radial distance r as r^{-2} (Exercise 4.9). The electromagnetic wave has to be suitably channelized so that the energy from the source is delivered in the place where it is needed. We can accomplish this by sending the electromagnetic wave through a waveguide—a tube-like conduit bounded by conducting surfaces, as shown in Fig. 4.4. Let us take the z axis along the length of the waveguide. The cross section of the waveguide in the xy plane (i.e. in the plane perpendicular to its length) is a two-dimensional region of non-conducting medium bounded by conducting surfaces. We shall assume a simple kind of waveguide of which the cross section is invariant in the z direction along its length.

We shall be interested in an electromagnetic wave propagating in the z direction along the axis of the waveguide. Since we expect different quantities associated with the wave to vary as $e^{i(kz-\omega t)}$, we can assume the electric and the magnetic fields inside the waveguide to have the forms

$$\mathbf{E} = \mathbf{E}_0(x, y)\, e^{i(kz-\omega t)}, \tag{4.88}$$

$$\mathbf{B} = \mathbf{B}_0(x, y)\, e^{i(kz-\omega t)}. \tag{4.89}$$

In Sect. 4.4, we had considered a plane wave in infinite space of the form (4.44) and (4.48), from which the Eqs. (4.34) and (4.35) led to the conclusion that the electric and the magnetic fields are perpendicular to the propagation direction z. Since we are allowing variations in x and y in (4.88) and (4.89), the application of (4.34) and (4.35) will not lead to the same conclusion here. Therefore, we have to allow for the possibility that $\mathbf{E}$ and $\mathbf{B}$ have non-zero components in the propagation direction z. It is not the case that our important conclusion about the transverse nature of electromagnetic waves derived in Sect. 4.4 becomes invalidated now. Rather, we do not have a plane wave with a wavefront perpendicular to the z direction in the present situation. We can think of simple plane electromagnetic waves being repeatedly reflected from the walls of the waveguide leading to an overall propagation in the z direction with the fields being given by (4.88) and (4.89).

We certainly expect Eqs. (4.34)–(4.37) to hold in the non-conducting medium inside the waveguide. On substituting (4.88) and (4.89) into (4.37), the x and y components give us

$$\frac{\partial E_{0,z}}{\partial y} - ikE_{0,y} = i\omega B_{0,x}, \tag{4.90}$$

$$ikE_{0,x} - \frac{\partial E_{0,z}}{\partial x} = i\omega B_{0,y}. \tag{4.91}$$

Similarly, x and y components of (4.36) on substituting (4.88) and (4.89) become

$$\frac{\partial B_{0,z}}{\partial y} - ikB_{0,y} = -\frac{i\omega}{c^2} E_{0,x}, \tag{4.92}$$

$$ikB_{0,x} - \frac{\partial B_{0,z}}{\partial x} = -\frac{i\omega}{c^2} E_{0,y}, \tag{4.93}$$

where we have made use of (4.42) assuming that $\kappa = 1$ and $\kappa_m = 1$ for the non-conducting medium inside the waveguide. Eliminating $B_{0,y}$ between (4.91) and (4.92), we get

$$\frac{\partial B_{0,z}}{\partial y} - \frac{k}{\omega} \left(ikE_{0,x} - \frac{\partial E_{0,z}}{\partial x} \right) = -\frac{i\omega}{c^2} E_{0,x},$$

from which it follows that

$$E_{0,x} = \frac{i}{\left(\frac{\omega^2}{c^2} - k^2 \right)} \left(k\frac{\partial E_{0,z}}{\partial x} + \omega\frac{\partial B_{0,z}}{\partial y} \right). \tag{4.94}$$

We thus see that if we know $E_{0,z}$ and $B_{0,z}$, then we can calculate $E_{0,x}$ from these. We leave it as an exercise for the reader to show that one can derive equations similar to (4.94) for $E_{0,y}$, $B_{0,x}$ and $B_{0,y}$ as well, showing that all these quantities can be calculated from $E_{0,z}$ and $B_{0,z}$. Thus, the problem of finding the electromagnetic fields inside the waveguide is reduced to finding solutions for $E_{0,z}$ and $B_{0,z}$. All other field components can be obtained from them.

In order to find $E_{0,z}$ and $B_{0,z}$, the first step is to note that the electromagnetic fields inside the waveguide satisfy the wave equations (4.38) and (4.47), which follow from (4.34)–(4.37) holding inside the waveguide. Note that $E_{0,z}$ and $B_{0,z}$ are the z components of the electromagnetic fields given by (4.88) and (4.89), which can be substituted in the wave equations (4.38) and (4.47) to obtain

$$\left[\frac{\partial^2}{\partial x^2} + \frac{\partial^2}{\partial y^2} + \left(\frac{\omega^2}{c^2} - k^2 \right) \right] \left[\begin{matrix} E_{z,0}(x, y) \\ B_{z,0}(x, y) \end{matrix} \right] = 0, \tag{4.95}$$

where we have written $1/c^2$ for $\epsilon\mu$. Our job now is to solve (4.95), which clearly should be solved over the cross section of the waveguide. We have to determine the boundary conditions to use at the conducting surface bounding the waveguide. Let us consider a small region of the boundary and choose the coordinate system in such a manner that the x axis is perpendicular to this small boundary region. Since E_z

is a tangential component at this boundary, and we assume E_z to be zero inside the conductor, the boundary condition (2.107) suggests that we must have

$$E_z = 0 \tag{4.96}$$

on the boundary, which is the appropriate boundary condition for E_z. Obviously, the other tangential component E_y also should be zero at the boundary. To find the boundary for B_z, we make use of the boundary condition (3.53), which suggests in this case that

$$B_x = 0$$

on the small region of the boundary we are considering. Now we consider (4.93) and make use of the fact that E_y and B_x are zero on the boundary. This leads to the conclusion

$$\frac{\partial B_z}{\partial x} = 0.$$

For any arbitrary region of the conducting boundary, we must have

$$\frac{\partial B_z}{\partial n} = 0, \tag{4.97}$$

where n refers to the direction normal to the boundary. The waveguide problem has now become the mathematical problem of solving (4.95) over the cross section of the waveguide with E_z and B_z satisfying the boundary conditions (4.96) and (4.97), respectively, on the surface of the waveguide. Once we have found E_z and B_z, we can find all the other components of the electromagnetic field inside the waveguide from (4.94) and similar equations for other components.

Apart from assuming the waveguide to have a uniform cross section, so far we have not assumed anything about the nature of the cross section. The theory of waveguides we have developed so far should be applicable to waveguides with any arbitrary cross section. To illustrate the nature of the solutions, we now consider a waveguide with a rectangular cross section. In Cartesian coordinates, let the inside of the waveguide correspond to the region

$$0 < x < \alpha, \;\; 0 < y < \beta \tag{4.98}$$

with conducting boundaries around. We have already pointed out that, unlike a plane electromagnetic wave in infinite space, the electromagnetic wave inside the waveguide does not need to have the electromagnetic fields transverse to the propagation direction. However, we can consider modes inside the waveguide in which *either* the electric field *or* the magnetic field is perpendicular to the axis of the waveguide (which is the propagation direction). We now consider these modes.

Let us first discuss the transverse electric or TE mode inside the waveguide. For this mode,

$$E_z = 0$$

everywhere. We have to obtain B_z which would be a solution of (4.95) satisfying the boundary condition (4.97) on all the surfaces of the waveguide. It is easy to check that

$$B_z = B^0 \cos\frac{m\pi x}{\alpha} \cos\frac{n\pi y}{\beta} e^{i(kz-\omega t)} \quad (4.99)$$

satisfies the boundary condition on all the boundary surfaces, provided m and n are integers. Looking at (4.89), we conclude that we have

$$B_{z,0}(x, y) = B^0 \cos\frac{m\pi x}{\alpha} \cos\frac{n\pi y}{\beta}$$

in this case, and this will be a solution of (4.95) if

$$\omega^2 = c^2 \left[k^2 + \left(\frac{m\pi}{\alpha}\right)^2 + \left(\frac{n\pi}{\beta}\right)^2 \right]. \quad (4.100)$$

Thus, the TE mode is given by (4.99) in which k can have continuous values, whereas ω is given by (4.100) for specified k, m and n. It should be obvious that either m or n has to be non-zero for the TE mode to exist. For the case $\alpha > \beta$, we have the lowest possible value of ω for the combination $m = 1$, $n = 0$ with vanishingly small k. This lowest possible value of ω is given by

$$\omega_c = c \frac{\pi}{\alpha}. \quad (4.101)$$

If ω is slightly larger than ω_c, then (4.100) can be satisfied with a small k. However, if ω is less than ω_c, then it is not possible to have a wave solution of the form (4.99) satisfying (4.100). In other words, ω_c given by (4.101) is a cut-off frequency: an electromagnetic wave with a frequency $\omega < \omega_c$ will not be able to propagate through the waveguide in the form of a TE mode. The waveguide can be used for transmitting TE wave modes having only frequencies higher than ω_c given by (4.101).

We now turn to the transverse magnetic or TM mode. For this mode inside the waveguide, we have

$$B_z = 0$$

everywhere. We need to find E_z which would satisfy (4.95) and would obey the boundary condition (4.96) on all the boundary surfaces. It is not difficult check that the appropriate solution is

$$E_z = E^0 \sin\frac{m\pi x}{\alpha} \sin\frac{n\pi y}{\beta} e^{i(kz-\omega t)} \quad (4.102)$$

where m and n are integers satisfying the same Eq. (4.100) as above, as we can find by substituting (4.102) in (4.95). For this mode to exist, both m and n have to be non-zero. We easily conclude that the TM mode also has a cut-off frequency, which can be obtained in this case by taking $m = n = 1$ with $k = 0$, that is,

$$\omega_c = c \left(\frac{\pi^2}{\alpha^2} + \frac{\pi^2}{\beta^2} \right)^{1/2}. \tag{4.103}$$

If an electromagnetic wave has frequency less than this, it cannot propagate as a TM mode through the waveguide.

It would appear from our discussion that a wave mode with both $E_z = 0$ and $B_z = 0$ should not propagate through the waveguide. It is indeed true that such a transverse electric and magnetic (TEM) mode cannot propagate through a waveguide which has a region of free space of any shape surrounded on all the sides by conducting surfaces. However, if one introduces another conductor inside this free space, then the propagation of a TEM mode becomes possible. Work out Exercise 4.14 and then think carefully how we get around some of the results of this section which seem to imply that there should not be a TEM mode.

The kind of waveguide which we have discussed is particularly convenient for transferring energy from one region to another through radio waves or microwaves. We remind the reader that the expression (4.66) for the skin depth holds for such waves but not for electromagnetic waves with higher frequencies, as we pointed out in Sect. 4.5. Visible light can be channelized through a somewhat different kind of waveguide—the *optical fibre*—which has revolutionized all types of communication networks around us within the last few years. While we shall not discuss optical fibres in detail in this book, we make a few comments.

What we have discussed in this section is the simplest kind of waveguide within which we have free space surrounded by conducting walls. If the electromagnetic wave cannot penetrate into the walls, then it is channelized along the free space. On the other hand, inside an optical fibre, we have a core region made of some material which has a higher refractive index than the surrounding region. This implies that electromagnetic waves propagating along the core region and falling on the interface of the core region with the surrounding region with a sufficiently high angle of incidence may undergo total internal reflection, due to the lower refractive index of the surrounding region. This implies that the electromagnetic wave is transmitted through the core region of the fibre. We certainly would have (4.90)–(4.93) holding in the core region of the fibre, with c replaced by the speed of light within the material of the core. However, the boundary conditions which hold at the interface between the core and the surrounding region are different from (4.96) and (4.97). In fact, one has to allow for the electromagnetic wave in the surrounding region also. This makes the mathematical analysis more complicated, although the basic principles are straightforward. The interested readers are referred to Ghatak and Thyagarajan ([4], Chap. 11) for a discussion of the mathematical theory of electromagnetic wave transmission through optical fibres. The widespread use of optical fibres has been possible due to the fact that they can be made from flexible materials, allowing light to be propagated through curved paths along the axis of the optical fibre cable.

4.8   Rectangular Cavity Resonator

We now discuss a problem which played a tremendously important role in the history of physics. One of the most pressing problems of physics in the closing years of the nineteenth century was to provide a theoretical explanation of the spectrum of blackbody radiation. Let us consider a rectangular cavity occupying the region of space

$$0 < x < \alpha, \ \ 0 < y < \beta, \ \ 0 < z < \gamma \tag{4.104}$$

with conducting walls on all the sides and kept at a fixed temperature. The radiation inside this cavity, which is in equilibrium with matter, will surely have the blackbody spectrum. To calculate this spectrum theoretically, we have to proceed in two steps. We first need to apply electromagnetic theory to determine the possible wave modes inside the cavity. Then we need to apply statistical mechanics to determine the average energy that would be associated with each mode. Here, we restrict ourselves to a discussion of only the first electromagnetic part of this problem: finding possible electromagnetic wave modes inside the cavity.

Let us try to determine the electric field of the radiation trapped inside the cavity. We first determine the boundary conditions which have to be satisfied on the conducting surfaces at $z = 0$ and $z = \gamma$ perpendicular to the z axis. Since E_x and E_y are transverse to this surface, we must have

$$E_x = 0, \ \ E_y = 0 \tag{4.105}$$

at these surfaces, exactly like the boundary condition (4.96) for the waveguide. We also have (4.34) satisfied inside the cavity very close to these boundaries. Given (4.105), we easily see that (4.34) would imply

$$\frac{\partial E_z}{\partial z} = 0 \tag{4.106}$$

at these surfaces perpendicular to the z axis we are considering. Since readers will be able to easily determine similar boundary conditions on the surfaces perpendicular to x and y axes, we do not bother to write them down explicitly. We have already seen in Sect. 4.4 that the electric field in the absence of sources (ρ or $\mathbf{j}$) has to satisfy (4.38). Hence, our job now is to obtain an electric field which is a solution of (4.38) and satisfies all the necessary boundary conditions.

We consider E_z to be given by the following expression:

$$E_z = E_{z,0} \sin\frac{l\pi x}{\alpha} \sin\frac{m\pi y}{\beta} \cos\frac{n\pi z}{\alpha} e^{-i\omega t}, \tag{4.107}$$

where l, m and n are integers. This expression obviously satisfies the boundary conditions that this component, which corresponds to a transverse electric field at the surfaces perpendicular to the x and y axes, has to go to zero at these surfaces. The expression (4.107) also satisfies the boundary condition (4.106) on the surfaces perpendicular to the z axis. It is clear that boundary conditions on all the surfaces surrounding the rectangular cavity are satisfied. We write down similar expressions for the other components of the electric field

$$E_x = E_{x,0} \cos\frac{l\pi x}{\alpha} \sin\frac{m\pi y}{\beta} \sin\frac{n\pi z}{\alpha} e^{-i\omega t}, \tag{4.108}$$

$$E_y = E_{y,0} \sin\frac{l\pi x}{\alpha} \cos\frac{m\pi y}{\beta} \sin\frac{n\pi z}{\alpha} e^{-i\omega t}. \tag{4.109}$$

It is easy to check that these other components also satisfy the appropriate boundary conditions on all the surrounding surfaces. The only other requirement for the electric given by (4.107)–(4.109) is that is that it has to be a solution of (4.38). On substituting this expression of the electric field in (4.38), we find that (4.38) will indeed be satisfied if the frequency ω is given by the relation

$$\frac{\omega^2}{c^2} = \pi^2 \left(\frac{l^2}{\alpha^2} + \frac{m^2}{\beta^2} + \frac{n^2}{\gamma^2} \right). \tag{4.110}$$

Since the electric field given by (4.107)–(4.109) along with the relation (4.110) satisfies (4.38) and all the boundary conditions, it must correspond to a possible wave mode inside the rectangular cavity. All the different modes inside the cavity are given by different integral values of l, m and n. The frequencies of the different modes are given by (4.110). Thus, one can determine the distribution of possible frequencies from (4.110), keeping in mind that we have to consider only the positive integral values of l, m and n (it is easy to check that negative values of these will not give independent modes).

We also note that (4.34) leads to the requirement

$$\frac{l}{\alpha} E_{x,0} + \frac{m}{\beta} E_{y,0} + \frac{n}{\gamma} E_{z,0} = 0, \tag{4.111}$$

which we get by substituting (4.107)–(4.109) in (4.34). In other words, the amplitudes of the three components of the electric field associated with a mode have to be connected by the relation (4.111).

4.9   Theory of Optical Dispersion

We have already derived Snell's law of refraction (4.76) in Sect. 4.6. Comparing
(4.76) with the way this law is written down in terms of refractive indices of the
media in standard treatments of optics, it is obvious that the refractive index of a
medium must be given by

$$n = \frac{c}{u_{em}}, \tag{4.112}$$

where u_{em} is the velocity of the electromagnetic wave in the medium. The definition
of the refractive index through (4.112) ensures that it is 1 for a vacuum and higher
than 1 for a material medium (since we expect $u_{em} < c$ in a medium). Using (4.41),
we conclude from that (4.112) that

$$n = \sqrt{\kappa \kappa_m}. \tag{4.113}$$

For many media through which we may be interested in studying the propagation of
electromagnetic waves, we can take $\kappa_m = 1$. Using (2.103), we get from (4.113) that

$$n = \sqrt{1 + \chi}. \tag{4.114}$$

If the electric field **E** of the electromagnetic wave polarizes the medium giving rise
to the polarization density **P** (which we expect to be proportional to **E**), then we can
use (2.101) to calculate the susceptibility χ and substitute it in (4.114) to obtain the
refractive index.

The variation of the refractive index with the frequency of the electromagnetic
wave is referred to as *dispersion*. For many optical media (though not for all), it is
experimentally found that the refractive index increases with increasing frequency.
This is called *normal dispersion*. We shall now show that one can give a theory of
this based on a rather simple model of how the polarization density **P** arises from the
electric field **E** of the electromagnetic wave. Since we shall find in the theory that
the refractive index may turn out to be complex, let us first discuss the significance
of this before proceeding with the theory.

Let us write

$$n = n_r + i n_i, \tag{4.115}$$

where n_r and n_i are real. For an electromagnetic wave of frequency ω, it follows
from (4.45) and (4.112) that the wavenumber k is given by

$$k = n \frac{\omega}{c} = (n_r + i n_i) \frac{\omega}{c} \tag{4.116}$$

on using (4.115). We have already discussed a case of complex k in Sect. 4.5. Substituting (4.116) in (4.44), we conclude that the electric field **E** of an electromagnetic
wave propagating in the z direction is given by

$$\mathbf{E} = \mathbf{E}_0 \, e^{-\frac{\omega}{c} n_i z} \, e^{i \frac{\omega}{c} (n_r z - ct)}. \tag{4.117}$$

This shows that a positive n_i implies absorption in the medium leading to attenuation of the wave. In fact, we can argue that n_i has to be positive in order to be physically meaningful. It is also clear from (4.117) that the electromagnetic wave inside the medium propagates with velocity c/n_r.

We now discuss a simple microscopic model of how the medium may become polarized due to an electric field $\mathbf{E}$. Since electrons are much lighter than the atomic nuclei, an electric field produces a much larger acceleration in the electrons, which are clearly much more mobile than the atomic nuclei. In the present discussion, we shall consider only the motions of electrons. According to (4.53), the magnetic force exerted on an electron by an electromagnetic wave will be negligible compared to the electric force, provided the velocities induced in the electrons are non-relativistic. We, therefore, consider only the electric force exerted on electrons while developing our theory. Let us make the extremely simplifying assumption that an electron is bound inside an atom or a molecule by a spring-like restoring force. In other words, when the electron is displaced by an amount $\mathbf{x}$ from its mean position, there is a restoring force $-m_e\omega_0^2\mathbf{x}$ acting on it. Clearly, ω_0 is like the normal frequency of the electron because the electron, when displaced from its mean position and left to itself without any other force acting on it, will oscillate with this frequency. The equation of motion of the electron is given by

$$m_e \frac{d^2\mathbf{x}}{dt^2} = -e\mathbf{E} - m_e\omega_0^2\mathbf{x} - m_e\gamma\frac{d\mathbf{x}}{dt}, \tag{4.118}$$

where $-e$ is the charge of the electron having mass m_e. Note that we have included a possible damping term opposing the motion of the electron. Since the electric field of a monochromatic electromagnetic wave varies sinusoidally at a point, and we can write $\mathbf{E} = \mathbf{E}_0 e^{-i\omega t}$, (4.118) leads to

$$\frac{d^2\mathbf{x}}{dt^2} + \gamma\frac{d\mathbf{x}}{dt} + \omega_0^2\mathbf{x} = -\frac{e}{m_e}\mathbf{E}_0\,e^{-i\omega t}. \tag{4.119}$$

Trying out a solution of the form

$$\mathbf{x} = \mathbf{x}_0\,e^{-i\omega t} \tag{4.120}$$

and substituting in (4.119), we get

$$(-\omega^2 - i\gamma\omega + \omega_0^2)\,\mathbf{x}_0 = -\frac{e}{m_e}\mathbf{E}_0.$$

Obtaining $\mathbf{x}_0$ from this and substituting in (4.120), we finally have

$$\mathbf{x} = \frac{e/m_e}{(\omega^2 - \omega_0^2) + i\gamma\omega} \, \mathbf{E}_0 e^{-i\omega t}. \tag{4.121}$$

This gives the displacement of an electron bound inside an atom or molecule due to the electric field of the electromagnetic wave.

Now that we have obtained the displacement of an electron, we need to determine the polarization that would arise due to electrons displaced in this manner. We assume that the atom or the molecule did not have an electric dipole moment when the electron was in its undisturbed position, i.e. when $\mathbf{x} = 0$. Then a displacement of a charge $-e$ by $\mathbf{x}$ would give to an electric dipole moment $-e\mathbf{x}$. If there are N_e such electrons per unit volume, then the polarization density is obviously

$$\mathbf{P} = -N_e e\mathbf{x} = -\frac{N_e e^2/m_e}{(\omega^2 - \omega_0^2) + i\gamma\omega} \, \mathbf{E}_0 e^{-i\omega t}$$

on using (4.121). We then conclude from (2.101) that

$$\chi = -\frac{N_e e^2}{\epsilon_0 m_e} \frac{1}{(\omega^2 - \omega_0^2) + i\gamma\omega}. \tag{4.122}$$

From (2.103), we than have

$$\kappa = 1 - \frac{N_e e^2}{\epsilon_0 m_e} \frac{1}{(\omega^2 - \omega_0^2) + i\gamma\omega}. \tag{4.123}$$

While deriving this equation, we have assumed the macroscopic electric field $\mathbf{E}$ to be the effective field acting on the electrons. If we take the effective electric field to be given by (2.129), then our result gets slightly modified (see Exercise 4.16). We also point out that we have derived the expression (4.123) for the dielectric constant by treating the dynamics of the electrons classically. A more complete treatment requires the application of quantum mechanics. Still, (4.123) is a very important and comprehensive result which has connections with several topics discussed in different portions of this book. We point out two special cases which follow from (4.123).

• In a static situation with a constant electric field, we have to put the time derivative terms in (4.118) to zero, giving a simpler expression of displacement

$$\mathbf{x} = -\frac{e}{m_e \omega_0^2} \, \mathbf{E}$$

in the place of (4.121). This would lead to the atomic polarization given by (2.126) in our discussion the electrostatic theory of dielectric media in Sect. 2.18. It is easy to check that we get the following expression of the dielectric constant:

Fig. 4.5 Plots of $n_r - 1$ and n_i as functions of frequency ω. These plots are based on (4.125) and (4.126), suitably normalized through division by $N_e e^2/2\epsilon_0 m_e \omega_0^2$, with $\gamma = 0.1\omega_0$

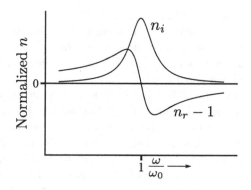

$$\kappa = 1 + \frac{N_e e^2}{\epsilon_0 m_e \omega_0^2}. \tag{4.124}$$

- When we consider free electrons inside conductors (such as metals), we have to set $\omega_0 = 0$. This corresponds to the situation considered in Exercise 4.12. There are two interesting limits. If $\omega \ll \gamma$, then we are led to the case of the electromagnetic waves inside conductors discussed in Sect. 4.5. The other limit $\omega \gg \gamma$ corresponds to the collisionless electron plasma which will be discussed in Sect. 8.3.

Additionally, the theory of scattering to be discussed in Sect. 7.8.3 also has connections with the discussion in this section. We shall make some comments about this connection when we discuss the theory of scattering.

The refracting index that we would get on substituting (4.122) in (4.114) will clearly have an imaginary part if γ is non-zero, leading to the attenuation of the electromagnetic wave. It is clear from (4.122) that the susceptibility χ is proportional to the electron density N_e. For low-density materials like gases, we often have $\chi \ll 1$. Then it becomes particularly easy to separate out the real and imaginary parts of the refractive index, as (4.114) would imply $n \approx 1 + \frac{1}{2}\chi$. On the basis of (4.115) and (4.122), we finally write down the expressions for the real and imaginary parts of the refractive index:

$$n_r = 1 - \frac{N_e e^2}{2\epsilon_0 m_e} \frac{(\omega^2 - \omega_0^2)}{(\omega^2 - \omega_0^2)^2 + \gamma^2 \omega^2}, \tag{4.125}$$

$$n_i = \frac{N_e e^2}{2\epsilon_0 m_e} \frac{\gamma\omega}{(\omega^2 - \omega_0^2)^2 + \gamma^2 \omega^2}. \tag{4.126}$$

Figure 4.5 shows how $n_r - 1$ and n_i varies with the frequency ω of the electromagnetic wave. Outside a small range of ω around $\omega = \omega_0$ (the range becomes smaller for smaller γ), we find that n_r increases with increasing ω, as we expect for normal dispersion. In the small range around $\omega = \omega_0$, however, we have n_r decreasing with ω. This is called *anomalous dispersion*. We see in Fig. 4.5 that n_i has significantly non-zero values in the same range of frequencies where we see anomalous dispersion. As we have seen in (4.117), non-zero values of n_i imply the absorption of the electromagnetic wave in the medium. We thus conclude that significant absorption of the electromagnetic wave will take place at the same frequencies where we have anomalous dispersion. This is found in experimental studies. In fact, if we have a plot of either n_r or n_i as a function of ω, we can readily guess what the plot of the other should be like. One can derive some elegant mathematical relations known as *Kramers–Kronig relations*, by using which one can quantitatively obtain either $n_i(\omega)$ from $n_r(\omega)$ or $n_r(\omega)$ from $n_i(\omega)$. The derivation of the Kramers–Kronig relations, which we shall not discuss in this book, can be found in other books (Panofsky and Phillips [5], Sect. 22-9; Jackson [6], Sect. 7.10).

Finally, we would like to point out that (4.126) provides the classical theory of absorption lines. If γ is small, then (4.126) suggests that the absorption of the electromagnetic wave takes place only in a narrow range of frequencies around ω_0. We, of course, know that the absorption at a certain frequency ω_0 means that photons of frequency $\hbar\omega_0$ are absorbed by atoms or molecules, leading to a transition to a higher energy level. We need quantum mechanics for a proper study of this process. Within the framework of classical physics, it is remarkable that a simple model based on the assumption that electrons are bound inside atoms or molecules with the normal frequency ω_0 leads to the conclusion that there would be an absorption line at this frequency ω_0.

4.10 Inhomogeneous Wave Equation

In the first three sections of this chapter, we have discussed some of the important implications of Maxwell's equations. After that, Sects. 4.4–4.8 have been devoted to situations where we have $\rho = 0$ and $\mathbf{j} = 0$, that is, where no source of electromagnetic fields is present in the region of interest, although in Sect. 4.5 we have allowed currents to be induced inside a conductor by the presence of electric fields. We have seen that electromagnetic waves can exist in regions of space where there are no sources of electromagnetic fields. We have been able to arrive at many important conclusions regarding electromagnetic waves without including the source terms in our equations. However, one question we could not address so far is how electromagnetic waves are produced. To answer this question, we have to analyse Maxwell's equations with the source terms present, which means that we cannot put any terms in these equations equal to zero. A full discussion of this topic will be taken up in Chap. 6. We shall now merely give some indications how we need to proceed when all the terms are present in Maxwell's equations. When we discuss the relation between special relativity and electrodynamics in the next chapter, some of the results we are going to present in this section will be very crucial. After that, a further discussion of how to handle Maxwell's equations when all the terms are present will be taken up in Chap. 6.

Since the handling of Maxwell's equations with all terms present is a challenging subject, we shall proceed with the simplifying assumption that ρ and $\mathbf{j}$ include all charges and currents (i.e. charges and currents induced in the surrounding media are not treated separately as in Sects. 2.14 and 3.6). With this assumption, we can use the versions (1.1)–(1.4) of Maxwell's equations. This means that either we are considering a medium like vacuum or air in which charges and currents are not induced, or ρ and $\mathbf{j}$ have to include the charges and currents induced in the medium. Several portions of this book have been devoted to topics involving material media such as the macroscopic theory of dielectric materials (Sects. 2.14–2.17) and magnetic materials (Sects. 3.6–3.8), as well as the dynamical theory of electromagnetic fields (Sects. 4.1–4.3) and electromagnetic waves (Sects. 4.4, 4.6). The approach we had followed in these discussions involving material media was to use the version of Maxwell's equations (4.1)–(4.4) containing $\mathbf{D}$ and $\mathbf{H}$ which had to be related to $\mathbf{E}$ and $\mathbf{B}$, respectively. From now onwards, however, the rest of the book will be based only on the original version (1.1)–(1.4) of Maxwell's equations.

Making use of (4.42), we can put (1.3) in the form

$$\nabla \times \mathbf{B} = \mu_0 \mathbf{j} + \frac{1}{c^2} \frac{\partial \mathbf{E}}{\partial t}. \tag{4.127}$$

The other Eqs. (1.1), (1.2) and (1.4) will be taken as they are. Clearly, (1.2) suggests that we can write the magnetic field in the form

$$\mathbf{B} = \nabla \times \mathbf{A}. \tag{4.128}$$

Since (1.2) is one of the starting equations in magnetostatics, it is clear that we can use the vector potential $\mathbf{A}$ introduced through (4.128) even in a static situation, and some of the uses of the vector potential $\mathbf{A}$ have been discussed in Chap. 3. Substituting for $\mathbf{B}$ from (4.128) in (1.4), we get

$$\nabla \times \left(\mathbf{E} + \frac{\partial \mathbf{A}}{\partial t} \right) = 0. \tag{4.129}$$

This equation will be satisfied if we introduce a scalar potential Φ in the following way:

$$\mathbf{E} + \frac{\partial \mathbf{A}}{\partial t} = -\nabla \Phi.$$

This means that if we know the scalar potential Φ and the vector potential $\mathbf{A}$, then we can obtain the electric field as follows:

$$\mathbf{E} = -\nabla \Phi - \frac{\partial \mathbf{A}}{\partial t}. \tag{4.130}$$

When discussing electrostatics, we could use (1.9) rather than (1.4), which enables us to write down $\mathbf{E}$ only in terms of the scalar potential as in (2.1). We can view (4.130) as a generalization of (2.1) in a dynamic situation.

If we express $\mathbf{B}$ and $\mathbf{E}$ in terms of $\mathbf{A}$ and Φ as in (4.128) and (4.130), then (1.2) and (1.4) are automatically satisfied. Now our job is to determine how we can obtain

A and Φ for a given distribution of ρ and **j**. For this purpose, we turn to the other Eqs. (1.1) and (1.3)—or rather to (1.1) and (4.127). Let us substitute for **E** and **B** from (4.128) and (4.130) into (1.1) and (4.127). From (1.1) we get

$$-\nabla^2\Phi - \frac{\partial}{\partial t}\nabla.\mathbf{A} = \frac{\rho}{\epsilon_0}, \tag{4.131}$$

whereas (4.127) gives

$$\nabla(\nabla.\mathbf{A}) - \nabla^2\mathbf{A} = \mu_0\mathbf{j} - \frac{1}{c^2}\frac{\partial}{\partial t}\left(\nabla\Phi + \frac{\partial\mathbf{A}}{\partial t}\right) \tag{4.132}$$

on making use of the vector identity (B.12).

To proceed further, we note that there is some arbitrariness in defining Φ and **A**. Suppose we have a scalar potential Φ and a vector potential **A** from which we can get the electromagnetic fields by using (4.128) and (4.130). Now let us write

$$\Phi = \Phi' - \frac{\partial\psi}{\partial t}, \tag{4.133}$$

$$\mathbf{A} = \mathbf{A}' + \nabla\psi, \tag{4.134}$$

where ψ is an arbitrary scalar function of t and spatial coordinates. On substituting (4.133) and (4.134) into (4.128) and (4.130), we readily find

$$\mathbf{B} = \nabla \times \mathbf{A}',$$

$$\mathbf{E} = -\nabla\Phi' - \frac{\partial\mathbf{A}'}{\partial t}.$$

This means that Φ' and $\mathbf{A}'$ provide scalar and vector potentials which are as good as Φ and **A** for calculating the electromagnetic fields. For any given scalar potential Φ and vector potential **A**, we can consider many different possible scalar functions ψ, and for each of these scalar functions ψ we obtain admissible scalar and vector potentials Φ' and $\mathbf{A}'$ by using (4.133) and (4.134). Thus, starting from any given Φ and **A**, we can get many different sets of Φ' and $\mathbf{A}'$ which are equally capable of giving the electromagnetic fields. This arbitrariness in defining Φ and **A** is known as *gauge freedom*.

One can use some additional criterion to choose a convenient combination of Φ and **A** out of what seems like endless possibilities. This is called *fixing the gauge*. We now fix the gauge by requiring that Φ and **A** satisfy the relation

$$\nabla.\mathbf{A} + \frac{1}{c^2}\frac{\partial\Phi}{\partial t} = 0. \tag{4.135}$$

The gauge fixed in this way is called the *Lorentz gauge*. One may raise the question whether it is possible to introduce the Lorentz gauge in all situations. To demonstrate this, let us assume that we initially have Φ' and $\mathbf{A}'$ which do not satisfy (4.135). Let us write

$$\nabla.\mathbf{A}' + \frac{1}{c^2}\frac{\partial \Phi'}{\partial t} = \chi, \tag{4.136}$$

where χ clearly has to be a scalar function of t and the spatial variables. Let us now obtain a scalar function ψ by solving the equation

$$\left(\nabla^2 - \frac{1}{c^2}\frac{\partial^2}{\partial t^2}\right)\psi = -\chi. \tag{4.137}$$

From this ψ and the initial Φ', $\mathbf{A}'$, we now introduce new Φ and $\mathbf{A}$ as given by (4.133) and (4.134). Substituting

$$\Phi' = \Phi + \frac{\partial \psi}{\partial t}, \quad \mathbf{A}' = \mathbf{A} - \nabla\psi$$

in (4.136), we find that the new Φ and $\mathbf{A}$ would satisfy (4.135) if ψ satisfies (4.137). This shows that it is always possible to introduce Φ and $\mathbf{A}$ which satisfy the Lorentz gauge condition (4.135), provided we can solve (4.137) to obtain ψ. The method of obtaining a general solution of (4.137) will be discussed in Sect. 6.2.

We now have come to last step of checking what happens to (4.131) and (4.132) when Φ and $\mathbf{A}$ are chosen to be in the Lorentz gauge satisfying (4.135). Replacing $\nabla.\mathbf{A}$ in (4.131) by $-(1/c^2)(\partial\Phi/\partial t)$ in accordance with (4.135), we get

$$\left(\nabla^2 - \frac{1}{c^2}\frac{\partial^2}{\partial t^2}\right)\Phi = -\frac{\rho}{\epsilon_0}. \tag{4.138}$$

We also note that the relation (4.135) makes the first term in the LHS of (4.132) equal to the second term in the RHS. On cancelling these terms, (4.132) gives

$$\left(\nabla^2 - \frac{1}{c^2}\frac{\partial^2}{\partial t^2}\right)\mathbf{A} = -\mu_0\mathbf{j}. \tag{4.139}$$

These equations are called *inhomogeneous wave equations* for the scalar and vector potentials. We have derived these equations starting from Maxwell's equations (1.1)–(1.4) with c given by (4.42). Suppose we have the charge and the current distributions $\rho(\mathbf{x}, t)$ and $\mathbf{j}(\mathbf{x}, t)$ given to us. If we can obtain Φ and $\mathbf{A}$ by solving (4.138) and (4.139), and then get the electromagnetic fields by using (4.128) and (4.130), the electromagnetic fields obtained in this way will clearly satisfy Maxwell's equations (1.1)–(1.4). Our goal is to solve Maxwell's equations to find electromagnetic fields for any given charge and the current distribution $\rho(\mathbf{x}, t)$ and $\mathbf{j}(\mathbf{x}, t)$ when no term is zero in the equations. This problem is now reduced to the problem of solving the

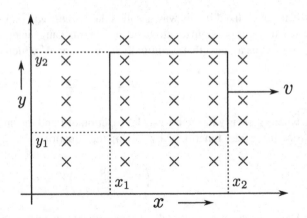

Fig. 4.6 A rectangular circuit in the xy plane with a perpendicular magnetic field $B(x)\mathbf{e}_z$. Refer to Exercise 4.3

inhomogeneous wave equations (4.138) and (4.139) for the given $\rho(\mathbf{x}, t)$ and $\mathbf{j}(\mathbf{x}, t)$. A full discussion of this problem will be taken up in Chap. 6.

One very important result we have derived in this chapter is that there can be electromagnetic waves propagating in vacuum with speed c, the speed of light. Since special relativity is based on the premise that the speed of light has to be equal to c in all inertial frames of reference, we expect that there would be a deep relation between electrodynamics and special relativity. This relation is explored in the next chapter. So far, we have avoided a discussion of the important question as to which are the frames of reference in which Maxwell's equations hold. This question will be addressed when we study the relationship between electrodynamics and special relativity. Only after discussing these topics in the next chapter shall we come back in Chap. 6 to a discussion of how the inhomogeneous wave equations (4.138) and (4.139) can be solved. As we shall see, the solution of these equations will lead to an understanding of how electromagnetic waves are produced.

Exercises

4.1 Consider a spherically symmetric distribution of charge density $\rho(r)$ around a central point, with an outward current density $j(r)\,\mathbf{e}_r$ which is also symmetric. Calculate the displacement current at the distance r from the central point and show that its sum with the current $j(r)$ is equal to zero so that no magnetic field arises according to (4.3).

4.2 Suppose some charge represented by the charge density $\rho(\mathbf{x}, t = 0)$ is introduced in a region inside a dielectric medium of uniform dielectric constant κ. Any electric field arising from this charge gives rise to a current density $\mathbf{j} = \sigma \mathbf{E}$, where σ is the electrical conductivity of the medium. Using the charge conservation equation (4.7), show that the charge evolves in time in the following manner:

$$\rho(\mathbf{x}, t) = \rho(\mathbf{x}, t = 0) \exp\left(-\frac{\sigma t}{\epsilon_0 \kappa}\right).$$

4.3 Consider a rectangular circuit in the xy plane with its arms parallel to the x and y axes (as shown in Fig. 4.6). There is a magnetic field $B(x)\mathbf{e}_z$ perpendicular to this plane which varies with x but not with y. We now consider a translational motion of this circuit with speed v in the x direction.

(a) Show that the identical expression for the EMF produced in the circuit

$$\mathcal{E} = [B(x_2) - B(x_2)](y_2 - y_1)v$$

follows from the two considerations of (i) calculating the EMF from the flux rule that it is equal to the rate of change of magnetic flux; and (b) calculating the EMF from the from the Lorentz force $q\,\mathbf{v} \times \mathbf{B}$ acting on any charged particle inside the circuit.

(b) If the resistance of the circuit is R, then the rate of energy dissipation in the circuit is $\mathcal{E}^2/R$. Show that this is equal to the rate of work done on the circuit by the Lorentz force given by (3.9).

4.4 Suppose we have a distribution of circuits, with I_k being the current flowing through the kth circuit with which a magnetic flux Φ_k is linked. Starting from the expression

$$\frac{1}{2}\mathbf{H}.\mathbf{B}$$

for magnetic energy density, show that the total energy of the system of circuits can be written as

$$W = \frac{1}{2}\sum_k I_k \Phi_k.$$

Hint. You have to use (3.1) and the result of Exercise 3.5(a) along with the vector identity (B.7).

4.5 (a) Consider a conducting sphere of radius a kept in free space with charge Q distributed uniformly over its surface. By integrating the energy density of the electric field around this sphere, show that the total energy is

$$U_{\text{elec}} = \frac{Q^2}{8\pi\epsilon_0 a} = m_{\text{elec}}c^2,$$

where m_{elec} is the mass equivalent to this energy given by

$$m_{\text{elec}} = \frac{Q^2}{8\pi\epsilon_0 c^2 a}.$$

(b) Now suppose the sphere is set into motion with non-relativistic speed $\mathbf{v}$ and the magnetic field around the sphere is given by (3.72). Write down the momentum density of the electromagnetic field at a point in space outside the sphere and integrate this momentum density over all space around the sphere to show that the total momentum is

$$\mathbf{P} = \frac{4}{3} m_{\mathrm{elec}} \mathbf{v}.$$

4.6 A charged sphere is kept in a uniform magnetic field. Show that the momentum density arising from this magnetic field and the electric field of the charged sphere is generally non-zero in a local region and argue that the total momentum of the electromagnetic field is zero. Can you think of any physical interpretation of this non-zero local momentum density?

4.7 (a) Two electric charges, each equal to $+q$, are placed at points $(a, 0, 0)$ and $(-a, 0, 0)$ in Cartesian coordinates. Calculate the Maxwell stresses at a point in the mid-plane $x = 0$. Write down the force exerted across an element of area of this mid-plane and then integrate it to show that the total force exerted across the mid-plane is equal to the force between the two charges.

   (b) Do the similar calculation for two electric charges $+q$ and $-q$ placed at points $(a, 0, 0)$ and $(-a, 0, 0)$

4.8 In an anisotropic crystal, $\mathbf{E}$ and $\mathbf{D}$ are generally not in the same direction, but are related by

$$D_\alpha = \sum_\beta \epsilon_0 \kappa_{\alpha\beta} E_\beta.$$

A plane electromagnetic wave varying in space and time as $\exp[i(\mathbf{k}.\mathbf{x} - \omega t)]$ is propagating through the crystal. Assuming $\mathbf{B} = \mu_0 \mathbf{H}$, show that $\mathbf{E}$, $\mathbf{D}$, $\mathbf{k}$ and the Poynting vector are in the same plane, with $\mathbf{H}$ perpendicular to this plane. Among the four coplanar vectors, which pairs will be mutually orthogonal? Make a sketch of all these vectors at a point.

4.9 Suppose we have a source of a spherically symmetric wave at the origin of spherical coordinates. This wave is associated with some wave variable ψ which satisfies the wave equation with wave speed c. Starting from the three-dimensional wave equation, show that we can get the following one-dimensional equation in the spherically symmetric situation

$$\frac{\partial^2}{\partial r^2}(r\psi) - \frac{1}{c^2}\frac{\partial^2}{\partial t^2}(r\psi) = 0,$$

of which the solution is of the form

$$\psi(r, t) = \frac{f(r - ct)}{r}.$$

If the energy flux due to this wave at a point is proportional to $|\psi|^2$, show that the same amount of the energy flux passes through successive spherical surfaces (centred around the source) of increasing r.

4.10 (a) Suppose a source is giving out energy isotropically in the form of light at the rate of 100 W. Estimate the amplitude of the electric field associated with the light at a distance of 5 m from the source of light.

(b) Suppose that the light source is surrounded by a metallic mesh of radius 5 cm which is kept at a potential of 10 V. Estimate the electrostatic field produced by this mesh at the distance of 5 m. Beyond what distance will the electric field associated with the light be stronger than this electrostatic field?

4.11 Consider a plane electromagnetic wave incident on an interface between two dielectric media. Calculate the reflectance R (the ratio of the components of the Poynting vector normal to the surface for the reflected and incident waves) and transmittance T (defined similarly) for both the polarizations and show that

$$R + T = 1$$

for both the cases.

4.12 We may like to develop a theory of dispersion inside conductors with free electrons. Then we have to put $\omega_0 = 0$ in (4.118) for electrons which are not bound inside atoms. The damping of motion is caused by the collisions suffered by the electrons, and we can take $\gamma = 1/\tau$, where τ is the typical time between two successive collisions which an electron may undergo. Show that the velocity of the electrons would be given by

$$\mathbf{v} = -i \frac{e\omega}{m_e} \frac{1}{\omega^2 + i\omega/\tau} \mathbf{E},$$

if the electric field varies in time as $e^{-i\omega t}$. In the limit $1/\omega \gg \tau$, argue that we would get the result (4.61) with σ given by

$$\sigma = \frac{N_e e^2 \tau}{m_e},$$

where N_e is the number density of electrons. In the opposite limit $1/\omega \ll \tau$, determine the polarization density and show that the dielectric constant is given by

$$\kappa = 1 - \frac{\omega_p^2}{\omega^2},$$

where $\omega_p = \sqrt{N_e e^2/\epsilon_0 m_e}$.

4.13 For a waveguide with a circular cross section (radius $r = a$), obtain the solution for the axisymmetric TM model and determine the cut-off frequency. (You need to be familiar with Bessel's equation and the Bessel functions to solve this problem.)

4.14 Suppose you have free space between two conducting cylindrical surfaces at $r = a$ and $r = b$ ($b > a$). Show that the electromagnetic fields

$$\mathbf{E}(r, \theta, z, t) = A \frac{e^{i(kz-\omega t)}}{r} \mathbf{e}_r,$$

$$\mathbf{B}(r, \theta, z, t) = \frac{k}{\omega} A \frac{e^{i(kz-\omega t)}}{r} \mathbf{e}_\theta.$$

obey Maxwell's equations for free space (4.34)–(4.37) in the region between the conducting cylinders with $\omega = ck$ and also satisfy all the appropriate boundary conditions which should hold at the surfaces of these conducting cylinders. Therefore, argue that this corresponds to a transverse electric and magnetic (TEM) mode of an electromagnetic wave propagating through this waveguide between the two coaxial cylindrical surfaces. Can this mode exist if the inner cylinder of radius a is removed?

4.15 For the rectangular cavity resonator, use Maxwell's equations to determine the magnetic field corresponding to the electric field (4.107)–(4.109) inside the cavity. Show that the magnetic field satisfies all the appropriate boundary conditions.

4.16 If the effective electric field acting on an electron inside a material medium is given by (2.129) rather than the macroscopic electric field, show from the theory of optical dispersion discussed in Sect. 4.9 that the refractive index n is given by

$$3\frac{n^2 - 1}{n^2 + 2} = \frac{N_e e^2}{\epsilon_0 m_e} \frac{1}{(\omega_0^2 - \omega^2) - i\gamma\omega},$$

where all the symbols have the same meanings as in Sect. 4.9

References

1. Lorentz, H.A., Einstein, A., Minkowski, H., Weyl, H.: The Principle of Relativity. Methuen and Co. (Reprinted by Dover) (1923)
2. Feynman, R.P., Leighton, R.B., Sands, M.: The Feynman Lectures on Physics: Volume II, Mainly Electromagnetism and Matter. Addison-Wesley (1964)
3. Born, M., Wolf, E.: Principles of Optics, 6th edn. Pergamon Press (1980)
4. Ghatak, A., Thyagarajan, K.: Optical Electronics. Cambridge University Press (1989)
5. Panofsky, W.K.H., Phillips, M.: Classical Electricity and Magnetism, 2nd edn. Addison–Wesley (reprinted by Dover) (1962)
6. Jackson, J.D.: Classical Electrodynamics, 3rd edn. John Wiley (1999)

Chapter 5
Relativity and Electrodynamics

5.1 Lorentz Transformation

Many important laws of physics are of the nature of dynamical laws, that is, laws which involve spatial variables and dependence on time. Since Maxwell's equations involve derivatives with respect to space and time, they are clearly dynamical laws. When stating a dynamical law of physics, one has to specify the frame of reference with respect to which the spatial coordinates and time are to be measured. So far in this book, we have avoided a discussion of frames of reference. Now we cannot postpone this discussion any further, in view of the important result derived in Sect. 4.4 that light propagates in free space with speed c. Presumably, this result holds in the frame of reference in which Maxwell's equations hold and not necessarily in all frames. This brings us to the question as to which are the frames in which Maxwell's equations hold and light propagates with speed c isotropically in all directions.

The question of the frame of reference has been a tricky question in Newtonian mechanics. The frames of reference in which Newton's laws of motion hold are called *inertial frames*. What decides whether a frame is inertial or not is a difficult question, and we shall not get into a discussion of that. However, one issue is easy to settle. If one frame of reference is inertial, then all frames moving with uniform speed with respect to that frame also will be inertial. Till the end of the nineteenth century, physicists assumed that measurements in different frames were connected by the Galilean transformation. It is easy to show that laws of mechanics would hold in all inertial frames if they are connected by the Galilean transformation. We now present a discussion of the Galilean transformation and show the problems we get into when applying this transformation to electromagnetism.

Suppose a frame K' is moving with constant speed v in the x direction with respect to a frame K. If (x, y, z, t) are the coordinates of an event as measured in frame K and (x', y', z', t') are the coordinates of the same event as measured in frame K', then according to the Galilean transformation

$$x = x' + vt', \tag{5.1}$$

© Springer Nature Singapore Pte Ltd. 2022

A. R. Choudhuri, *Advanced Electromagnetic Theory*, Lecture Notes in Physics 1009,
https://doi.org/10.1007/978-981-19-5944-8_5

$$y = y', \tag{5.2}$$

$$z = z', \tag{5.3}$$

$$t = t'. \tag{5.4}$$

We have assumed that the origins of the two frames ($x = 0$ and $x' = 0$) coincide at time $t = t' = 0$. The important point to note is that (5.4) implies time to be absolute and independent of observers. Differentiating with respect to t or t' would be the same thing by virtue of (5.4). On differentiating (5.1) with respect to t or t', we get

$$u_x = u'_x + v, \tag{5.5}$$

which relates the x component of velocity $u_x = dx/dt$ and $u'_x = dx'/dt'$ as measured in frames K and K'. Another differentiation of (5.5) with respect to t or t' would show that the acceleration would be the same in both the frames. Hence, if the force is also the same in both the frames, then Newton's second law of motion should hold in frame K' if it holds in K. In other words, if the Galilean transformation was the correct transformation, the laws of mechanics could hold in all inertial frames, and everything would be consistent as long as laws of mechanics were concerned.

We now come to the question if laws of electromagnetism are consistent with the Galilean transformation as well. One may naively expect that Maxwell's equations should also hold in all inertial frames connected by the Galilean transformation. A little reflection shows that it is not possible for this to be true if the speed of light is c in a frame in which Maxwell's equations hold. If (5.5) holds, then the speed of light cannot be c in both the inertial frames moving with respect to each other with a non-zero velocity. This means that Maxwell's equations cannot hold in different frames connected by the Galilean transformation. It is possible for the speed of light to be c in all directions only in one frame. This implies that Maxwell's equations can hold only in this frame and not in other inertial frames. On the other hand, if Maxwell's equations are to hold in all inertial frames, then the transformation law among them has to be different from the Galilean transformation. Thus, physicists in the closing decades of the nineteenth century realized that the following two statements could not be simultaneously correct:

Statement I. Maxwell's equations of electrodynamics hold in all inertial frames of reference exactly like Newton's laws of motion.

Statement II. All inertial frames of reference are connected to each other through the Galilean transformation.

Only one of these statements could be correct. The question was which one.

Physics is an experimental science, and questions like this are to be ultimately settled by experiments. But that does not stop physicists from making intuitive guesses. If two friends tell us opposite things, we often tend to trust the friend whom we have known for a longer time. That was the situation with the majority of physicists towards the end of the nineteenth century. By that time, Statement II had been assumed to be true for about two centuries (since the publication of Newton's *Principia Mathematica* in 1689). On the other hand, Maxwell's equations were novel entities at that time. The majority of physicists of that era took Statement II to be the correct statement. It was suggested that light needed a medium—christened *luminiferous ether*—for its transmission. Only in one frame of reference, this medium would be at rest. Although Newtonian mechanics suggested that all frames of references were completely equivalent, most of the physicists at the end of the nineteenth century thought that there was one special inertial frame in which ether was at rest. It was believed that this is the frame in which Maxwell's equations hold and light has the speed c in all directions. In other frames, light was expected to have a speed different from c—at least in some directions.

Even if the Earth happened to be at rest with respect to ether at some time, this certainly could not be the case year-round. As the Earth moved round the Sun in the year, there must be times when the Earth would be moving with respect to ether. One important question was whether it is experimentally possible to determine the motion of the Earth with respect to ether. The most important experiment to attempt this was the Michelson–Morley experiment (see, for example, Kleppner and Kolenkow [7], Sect. 11.2). The surprising and unexpected result of this experiment, reported in 1890, was that it was not possible to detect any motion of the Earth through ether. Since the experiment was based on interfering light beams, the negative result of the experiment implied that the speed of light was always c with respect to the Earth. Physicists were virtually forced to accept Statement I, which suggested that the speed of light has to be c in all inertial frames. This clearly meant that Statement II could not be correct. The transformation laws between inertial frames have to be such that the speed of light can be c in all these frames.

Lorentz in 1904 established the mathematical result that Maxwell's equations can be true in all inertial frames if they are connected to each other by what we now call the Lorentz transformation:

$$x = \gamma (x' + vt'), \tag{5.6}$$

$$y = y', \tag{5.7}$$

$$z = z', \tag{5.8}$$

$$t = \gamma \left(t' + \frac{v}{c^2} x' \right), \tag{5.9}$$

where

$$\gamma = \frac{1}{\sqrt{1 - v^2/c^2}}. \tag{5.10}$$

Note that $x' = 0, t' = 0$ corresponds to $x = 0, t = 0$, implying that time is set to zero in both the frames at the instant when the origins of the two frames coincide. The revolutionary implication of (5.9) was that time measured by different observers can be different. This was an idea which flatly contradicted the idea that time is absolute and the same for all observers—an idea which no physicist questioned till that time. While the Lorentz transformation (5.6–5.9) would mathematically allow Maxwell's equations to hold in all inertial frames, the physical basis of this transformation remained obscure till Einstein formulated special relativity in 1905 and analysed the concept of spacetime carefully. According to Einstein, we should think of a frame of reference in the following way. The frame should consist of a hypothetical three-dimensional grid constructed by ticking off the three coordinate axes at regular intervals with the help of a metre scale, which has to be *kept at rest in this frame* when making spatial measurements. Additionally, identical clocks should be placed at all the grid points to measure time locally at those points. All these clocks should be synchronized in the following way. Suppose a light signal starts from a grid point at time t_1 measured by the local clock placed there. This light signal reaches the origin and is immediately sent back to reach the grid point from where it started at time t_2. If the clock at the origin recorded the time of the signal reaching there to be $(t_1 + t_2)/2$, then we would say that the clock at the grid point and the clock at the origin are synchronized. In a frame of reference K' moving with some velocity with respect to the frame of reference K we considered, the clocks have to be synchronized with each other in exactly the same way. Additionally, the coordinate axes of this moving frame K' have to ticked with the same metre scale (with which the axes of the first frame were ticked) after bringing this scale *at rest* into this frame. If frames of reference are conceived in this manner, then there is no a priori reason why the time for an event measured by two observers should be the same.

It is easy to show from (5.6–5.10) that the reverse transformation is given by

$$x' = \gamma (x - vt), \qquad (5.11)$$

$$y' = y, \qquad (5.12)$$

$$z' = z, \qquad (5.13)$$

$$t' = \gamma \left(t - \frac{v}{c^2} x \right). \qquad (5.14)$$

The main aim of this chapter is to study the relationship between special relativity and electrodynamics. We shall assume that readers are familiar with the basics of special relativity, which are discussed in many elementary textbooks such as Kleppner and Kolenkow ([7], Chaps. 11 and 12). We shall not discuss how the Lorentz transformation (5.6–5.10) can be derived. We expect readers to know that the constancy of the speed of light c in different inertial systems leads to this transformation between frames. See Kleppner and Kolenkow ([7], Sect. 11.5) for a discussion of how we can arrive at the Lorentz transformation. We also expect the readers to be familiar with some of the famous consequences of special relativity, such as length contraction, time dilation and the twin paradox. We do not discuss these topics in this book. We recommend that readers not very familiar with these basic consequences of special relativity study these topics from some other book like Kleppner and Kolenkow ([7],

Chaps. 11 and 12) before proceeding further with this chapter. The next few sections of this chapter will be devoted to the discussion of some concepts and techniques which will be needed for an understanding of the relationship between relativity and electrodynamics. After that, we shall begin our discussion of relativistic electrodynamics from Sect. 5.9.

5.2 Transformation of Velocity Between Frames

Suppose a particle is at a point (x, y, z) at time t, as seen from a frame K. This can be thought of as an event with spacetime coordinates (x, y, z, t). If the particle reaches a point $(x + dx, y + dy, z + dz)$ at a later time $t + dt$, this can be viewed as another event with spacetime coordinates $(x + dx, y + dy, z + dz, t + dt)$. From another frame K', the spacetime coordinates of these events would appear as (x', y', z', t') and $(x' + dx', y' + dy', z' + dz', t' + dt')$. It is obvious that $(x + dx, y + dy, z + dz, t + dt)$ and $(x' + dx', y' + dy', z' + dz', t' + dt')$, which are the spacetime coordinates of the same event as seen from two frames, should satisfy the transformation law (5.6–5.9). Since this transformation law is linear, it is easy to check that the increments (dx, dy, dz, dt) and (dx', dy', dz', dt') also must be related by the same transformation law. In other words,

$$dx = \gamma (dx' + v\, dt'), \tag{5.15}$$

$$dy = dy', \tag{5.16}$$

$$dz = dz', \tag{5.17}$$

$$dt = \gamma \left(dt' + \frac{v}{c^2} dx' \right). \tag{5.18}$$

On dividing (5.15) by (5.18), we have

$$\frac{dx}{dt} = \frac{dx' + v\, dt'}{dt' + \frac{v}{c^2} dx'}. \tag{5.19}$$

It is apparent that the components of the velocity of the particle in the frame K are given by

$$u_x = \frac{dx}{dt}, \quad u_y = \frac{dy}{dt}, \quad u_z = \frac{dz}{dt}, \tag{5.20}$$

whereas these components of velocity as seen from the frame K' are

$$u'_x = \frac{dx'}{dt'}, \quad u'_y = \frac{dy'}{dt'}, \quad u'_z = \frac{dz'}{dt'}. \tag{5.21}$$

It follows from (5.19) that

$$u_x = \frac{u'_x + v}{1 + \frac{vu'_x}{c^2}}. \tag{5.22}$$

The first thing to check from (5.22) is that on substituting $u'_x = c$, we get $u_x = c$. In other words, if the speed of light is c in one frame, then it is also c in another frame connected to the first frame by Lorentz transformation—one of the tenets of special relativity.

To obtain the transformation law of another component of velocity, say the y component, we have to divide (5.16) by (5.18). On making use of (5.20) and (5.21), we easily get

$$u_y = \frac{u'_y}{\gamma \left(1 + \frac{vu'_x}{c^2}\right)}. \tag{5.23}$$

We can write down an analogous expression for the transformation of the z component of velocity.

Let us now consider a particle moving in the xy plane in the frame K. It is easy to show that $u_z = 0$ would imply $u'_z = 0$. Hence, the particle will be seen to move in the $x'y'$ plane in the frame K'. If θ is the angle which the velocity vector of the particle makes with the x axis in the frame K, then we have

$$\tan \theta = \frac{u_y}{u_x} = \frac{u'_y}{\gamma (u'_x + v)} \tag{5.24}$$

on using (5.22) and (5.23). If θ' is the angle which the velocity vector of the particle makes with the x' axis in the frame K', then obviously $u'_x = u' \cos \theta'$, $u'_y = u' \sin \theta'$ so that (5.24) gives

$$\tan \theta = \frac{u' \sin \theta'}{\gamma (u' \cos \theta' + v)}. \tag{5.25}$$

The important Eq. (5.25) relates the angles θ and θ' made by the moving particle with the x direction in the two frames of reference.

Suppose we have a source of light at rest in the frame K'. We consider a ray of light emitted by this source making an angle θ' with the x' direction (which is the same as the x direction). We now want to determine the angle θ with respect to the x direction which an observer in the frame K will find the ray of light to make. Substituting $u' = c$ for a light ray in (5.25), we get

$$\tan \theta = \frac{\sin \theta'}{\gamma (\cos \theta' + v/c)}. \tag{5.26}$$

This is the relativistic formula for the *aberration of light*.

It is interesting to consider a ray of light emitted by the source in a direction perpendicular to its direction of motion in its frame. In this situation, $\theta' = \pi/2$ so that we get from (5.26)

$$\tan \theta = \frac{c}{\gamma v}.$$

If the source of light (i.e. the frame K') is moving relativistically with respect to us (i.e the frame K), then we can take $v \approx c$ so that

$$\tan \theta \approx \frac{1}{\gamma}. \tag{5.27}$$

It follows from (5.10) that γ is much larger than 1 when v is close to c. This means that θ as given by (5.27) would be a very small angle. Let us consider the significance of this result. If a source of light is moving relativistically with respect to us, then, even if it emits a ray of light in a direction perpendicular to its motion, it would appear to us that the ray of light is making a small angle with the direction of motion. Even if the relativistically moving source emits light isotropically in its own frame of reference, it would appear to us that the emitted light is making a cone with a small angle around the direction of motion. This interesting effect is called the *relativistic beaming effect*.

5.3 Proper Time

It can be easily shown from (5.15–5.18) that

$$c^2\,dt^2 - dx^2 - dy^2 - dz^2 = c^2\,dt'^2 - dx'^2 - dy'^2 - dz'^2.$$

Such a relation should be true between any two inertial frames connected by the Lorentz transformation from which we observe two nearby spacetime events. In other words, if (dt, dx, dy, dz) is the spacetime separation between the nearby events, the quantity

$$c^2\,dt^2 - dx^2 - dy^2 - dz^2 = c^2\,d\tau^2 \tag{5.28}$$

should have the same value in all inertial frames. Such quantities, of which the values remain unchanged under any Lorentz transformation, are called *Lorentz scalars*. Let us now consider the physical significance of the quantity τ introduced in (5.28). As we have already pointed out, the two nearby events we are considering may correspond to a particle being at (x, y, z) at time t and being at $(x + dx, y + dy, z + dz)$ at $t + dt$. We now consider a frame of reference moving with a velocity equal to the instantaneous velocity of the particle. In that frame, the particle will be at rest between the two events, and its spatial coordinates will not change. If $dt_{particle}$ is the time elapsed in that frame between the two events, then from (5.28), we write

$$c^2 \, dt^2_{\text{particle}} = c^2 \, d\tau^2.$$

This implies that τ is nothing but the time as measured by a clock attached to the moving particle. This variable τ is called the *proper time*. It follows from (5.28) that

$$d\tau^2 = dt^2 \left(1 - \frac{dx^2 + dy^2 + dz^2}{c^2 \, dt^2} \right) = dt^2 \left(1 - \frac{v^2}{c^2} \right), \qquad (5.29)$$

where **v** is clearly the velocity of the frame attached to the moving particle. Making use of (5.10), we get from (5.29) the following relation

$$d\tau = \frac{dt}{\gamma}. \qquad (5.30)$$

It may be noted that the frame attached to the particle may not be an inertial frame if the particle is moving non-uniformly with respect to an inertial frame. But we can always introduce a frame moving with a velocity **v** equal to the velocity of the particle at one instant. That will be an inertial frame. Thus, we can think of the non-uniformly moving particle to be at rest in successive inertial frames having slightly different velocities from each other, even when the frame attached to the particle is not an inertial frame.

If (dt, dx, dy, dz) is the spacetime separation between two arbitrary events, then there is no reason why $d\tau^2$ introduced in (5.28) will always be positive. However, only when $d\tau^2$ is positive, is it possible for the two nearby spacetime events to correspond to positions of a moving particle at slightly different times, and only in such a situation, is it meaningful to interpret τ as the proper time measured by the clock attached to the moving particle. When $d\tau^2$ is negative, then such an interpretation is not useful. However, if $d\tau^2$ is negative for two nearby events, one can argue from (5.28) that it is possible to introduce a frame of reference in which the time difference dt between the events is zero, and the two events would appear simultaneous. On the other hand, if $d\tau^2$ for two nearby events is positive, we see from (5.28) that it is not possible to have a frame of reference in which these two events are simultaneous.

5.4   A Brief Note on Vectors and Tensors

The main aim of this chapter is to discuss how the basic principles of electrodynamics can be written in a relativistic form and to show that Maxwell's equations hold in all inertial frames connected by Lorentz transformation. For this purpose, we need to introduce the concept of Lorentz vectors and tensors. Let us begin with a discussion of vectors and tensors.

In elementary textbooks on physics, a vector is usually defined as a quantity which has both magnitude and direction—in contrast to a scalar which has only magnitude. Now we shall be concerned with a more sophisticated definition of vectors. A vector

Fig. 5.1 A sketch indicating how the components of a two-dimensional vector **A** change when the coordinate system is rotated through an angle α

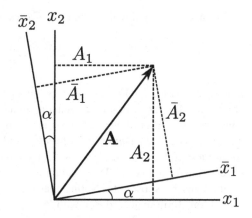

has a number components in a coordinate system (the number of components being equal to the dimension of the space). These components transform in a certain manner when we go to another coordinate system. Let (A_1, A_2) be the components of the vector **A** in two-dimensional space with respect to the coordinate system (x_1, x_2). We now consider another coordinate system $(\overline{x}_1, \overline{x}_2)$ which has coordinate axes inclined at an angle α with respect to the unprimed coordinate system, as shown in Fig. 5.1. It is easy to show that the components $(\overline{A}_1, \overline{A}_2)$ of the vector **A** in this new coordinate system are related to the components (A_1, A_2) in the earlier coordinate system by the relation

$$\overline{A}_1 = A_1 \cos \alpha + A_2 \sin \alpha, \quad \overline{A}_2 = -A_1 \sin \alpha + A_2 \cos \alpha. \tag{5.31}$$

Since the position vector of a point measured with respect to the origin will also transform this way, we have

$$\overline{x}_1 = x_1 \cos \alpha + x_2 \sin \alpha, \quad \overline{x}_2 = -x_1 \sin \alpha + x_2 \cos \alpha. \tag{5.32}$$

On the basis of (5.31) and (5.32), we can write

$$\overline{A}_i = \sum_k A_k \frac{\partial \overline{x}_i}{\partial x_k}. \tag{5.33}$$

From (5.32), we can easily arrive at the following reverse relations

$$x_1 = \overline{x}_1 \cos \alpha - \overline{x}_2 \sin \alpha, \quad x_2 = \overline{x}_1 \sin \alpha + \overline{x}_2 \cos \alpha. \tag{5.34}$$

On the basis of (5.34), along with (5.31), we can also write

$$\overline{A}_i = \sum_k A_k \frac{\partial x_k}{\partial \overline{x}_i}. \tag{5.35}$$

As long as we deal with rectangular Cartesian coordinates, it happens that

$$\frac{\partial \overline{x}_i}{\partial x_k} = \frac{\partial x_k}{\partial \overline{x}_i},$$

and we use either (5.33) or (5.35) as the transformation law for the components of a vector. However, in a more general situation, the components of a vector may transform *either* according to (5.33) *or* according to (5.35). A vector which transforms like (5.33) is called a *contravariant vector*, and from now onwards, we shall use the convention of denoting its components by superscripts. On the other hand, a vector which transforms like (5.35) is called a *covariant vector*, and we shall denote its components by subscripts. Noting that the component dx_i of a length increment should transform in accordance with (5.33) and should be treated as a contravariant vector, we write (5.33) and (5.35) in our new notation in the following manner:

$$\overline{A}^i = \sum_k A^k \frac{\partial \overline{x}^i}{\partial x^k}, \tag{5.36}$$

$$\overline{A}_i = \sum_k A_k \frac{\partial x^k}{\partial \overline{x}^i}. \tag{5.37}$$

Just as dx^i transforms as a contravariant vector, it is easy to see that that the gradient of a scalar field V, of which the components are $\partial V/\partial x^i$, would transform in accordance with (5.37) and should be regarded as a covariant vector. Note that i in x^i or $\overline{x}^i$ appearing at the bottom of a derivative has to be treated like a subscript. We now introduce the well-known *summation convention* by which if an index is repeated twice in a term, once as a subscript and once as a superscript, then it automatically implies summation over the possible values of that index and it is not necessary to put the summation sign explicitly. Using this summation convention, the transformation laws (5.36) and (5.37) for a contravariant vector A^i and a covariant vector A_i would be

$$\overline{A}^i = A^k \frac{\partial \overline{x}^i}{\partial x^k}, \tag{5.38}$$

$$\overline{A}_i = A_k \frac{\partial x^k}{\partial \overline{x}^i}. \tag{5.39}$$

A component of a vector is associated with only one coordinate axis. In the case of a general tensor, a component can be associated with several coordinate axes. So a component will generally have several indices. The transformation law of a general tensor is the following:

$$\overline{T}^{ab..d}_{l..n} = T^{\alpha\beta..\delta}_{\lambda..\nu} \frac{\partial \overline{x}^a}{\partial x^\alpha} \frac{\partial \overline{x}^b}{\partial x^\beta} \cdots \frac{\partial \overline{x}^d}{\partial x^\delta} \frac{\partial x^\lambda}{\partial \overline{x}^l} \cdots \frac{\partial x^\nu}{\partial \overline{x}^n}. \tag{5.40}$$

Note that some indices are put as superscripts and some as subscripts depending on whether the corresponding parts of the transformation are like contravariant vectors or covariant vectors.

From the transformation law (5.40) of tensors, it is very easy to show that the product $A_i B_k$ of two vectors A_i and B_k should transform like a tensor with two covariant components i and k. This can be generalized to the result that the product of the two tensors gives a tensor of higher rank. One very important operation is the *contraction* of tensors. Suppose we write $n = d$ in the tensor $\overline{T}^{ab..d}_{l..n}$. This, by the summation convention, implies that we are summing over all possible values of $n = d$. It follows from (5.40) that

$$\overline{T}^{ab..d}_{l..d} = T^{\alpha\beta..\delta}_{\lambda..\nu} \frac{\partial \overline{x}^a}{\partial x^\alpha} \frac{\partial \overline{x}^b}{\partial x^\beta} \cdot \cdot \frac{\partial \overline{x}^d}{\partial x^\delta} \frac{\partial x^\lambda}{\partial \overline{x}^l} \cdot \cdot \frac{\partial x^\nu}{\partial \overline{x}^d}.$$

Using the fact that

$$\frac{\partial \overline{x}^d}{\partial x^\delta} \frac{\partial x^\nu}{\partial \overline{x}^d} = \delta^\nu_\delta,$$

we easily get

$$\overline{T}^{ab..d}_{l..d} = T^{\alpha\beta..\delta}_{\lambda..\delta} \frac{\partial \overline{x}^a}{\partial x^\alpha} \frac{\partial \overline{x}^b}{\partial x^\beta} \cdot \cdot \frac{\partial x^\lambda}{\partial \overline{x}^l} \cdot \cdot \tag{5.41}$$

on making use of the obvious result $T^{\alpha\beta..\delta}_{\lambda..\nu} \delta^\nu_\delta = T^{\alpha\beta..\delta}_{\lambda..\delta}$ following from the properties of the Kronecker δ. We leave it as an exercise for the reader to verify that δ^i_k transforms like a tensor, but not δ_{ik} or δ^{ik} (the Kronecker δ in any form is assumed to have the value 1 if $i = k$ and 0 otherwise). It is clear from (5.41) that $T^{\alpha\beta..\delta}_{\lambda..\delta}$ transforms like a tensor with one contravariant index and one covariant index less compared to $T^{\alpha\beta..\delta}_{\lambda..\nu}$. Thus, the operation of contraction reduces the rank of the tensor (by reducing one contravariant rank and one covariant rank).

We now introduce the very important *metric tensor*. The distance between two nearby points in space in which we have a coordinate system is given by

$$ds^2 = g_{ik} \, dx^i dx^k, \tag{5.42}$$

where g_{ij} is the metric tensor, which is a tensor of covariant rank 2. Since dx^i is a contravariant vector, $g_{ik}dx^l ds^m$ is clearly a tensor which, after two contractions, would give a scalar, implying that ds^2 is a scalar. Needless to add, in a discussion of tensors, a scalar is regarded as a quantity which remains unchanged during a coordinate transformation. From a contravariant vector A^i, it is possible to construct the corresponding covariant vector in the following way:

$$A_i = g_{ik} \, A^k. \tag{5.43}$$

This is often called the *lowering of an index*. Suppose dx^i and dx^k have corresponding covariant vectors dx_i and dx_k obtained according to (5.43). It should be possible to write the metric in the form

$$ds^2 = g^{ik} dx_i \, dx_k. \tag{5.44}$$

By requiring that ds^2 given by (5.42) and (5.44) should be equal, the reader is asked to show that

$$g^{ik} g_{kl} = \delta^i_l. \tag{5.45}$$

After introducing the contravariant form g^{ik} of the metric tensor, we can use it to raise an index and obtain a contravariant vector from a covariant vector in the following way:

$$A^i = g^{ik} A_k. \tag{5.46}$$

Starting from a contravariant vector A^i, if we once lower the index by (5.43) and then raise it again by (5.46), it is easy to use (5.45) to check that we get back the original vector A^i. It may be noticed that we are repeatedly leaving various simple steps to the readers so that they can carry out these steps themselves and become conversant with tensor operations.

5.5 Lorentz Four-Vectors

After acquiring the basic knowledge about the properties of tensors as presented in the previous section, the readers should now be ready to learn about Lorentz vectors and tensors.

We can write down the spacetime coordinates of an event in the form of column vector with four rows:

$$x^\mu = \begin{pmatrix} ct \\ x \\ y \\ z \end{pmatrix}. \tag{5.47}$$

We shall be using the convention $x^0 = ct, x^1 = x, x^2 = y, x^3 = z$. We now introduce the following 4×4 matrix

$$\Lambda^\mu_\nu = \begin{pmatrix} \gamma & -\beta\gamma & 0 & 0 \\ -\beta\gamma & \gamma & 0 & 0 \\ 0 & 0 & 1 & 0 \\ 0 & 0 & 0 & 1 \end{pmatrix}, \tag{5.48}$$

where

$$\beta = \frac{v}{c}. \tag{5.49}$$

It is easy to check that the matrix equation

$$x'^{\mu} = \Lambda^{\mu}_{\nu} x^{\nu} \tag{5.50}$$

would give the Lorentz transformation laws (5.11–5.14). Clearly, (5.50) is a compact way of writing these laws. It follows easily from (5.50) that

$$\Lambda^{\mu}_{\nu} = \frac{\partial x'^{\mu}}{\partial x^{\nu}}. \tag{5.51}$$

As we shall see below, there are some quantities with one time-like and three space-like components which transform exactly like x^{μ} as given by (5.50). In other words, such a quantity A^{μ} would transform from one inertial frame to another following the transformation law

$$A'^{\mu} = \Lambda^{\mu}_{\nu} A^{\nu}. \tag{5.52}$$

Given the relation (5.51), it is obvious that this transformation law is the same as the transformation law of contravariant vectors given by (5.38). Quantities like A^{μ} which transform from one inertial frame to another in accordance with the transformation law (5.52) are called *Lorentz four-vectors*.

Let us now consider what, in the present situation, would be like the metric tensor introduced through (5.42). We have seen in Sect. 5.3 that $c^2 d\tau^2$ as given by (5.28) does not change under Lorentz transformation, that is, it is a Lorentz scalar. Writing it as $-ds^2$, we have

$$ds^2 = -c^2 dt^2 + dx^2 + dy^2 + dz^2. \tag{5.53}$$

Clearly ds can be interpreted as the spacetime distance between two events. Introducing the tensor $\eta_{\mu\nu}$ defined by

$$\eta_{\mu\nu} = \begin{pmatrix} -1 & 0 & 0 & 0 \\ 0 & 1 & 0 & 0 \\ 0 & 0 & 1 & 0 \\ 0 & 0 & 0 & 1 \end{pmatrix}, \tag{5.54}$$

we can write (5.53) in the form

$$ds^2 = \eta_{\mu\nu} dx^{\mu} dx^{\nu}. \tag{5.55}$$

A comparison with (5.42) makes it clear that we can interpret $\eta_{\mu\nu}$ as the metric tensor when discussing special relativity. By interchanging the indices μ and ν, one can easily argue that the metric tensor $\eta_{\mu\nu}$ has to be symmetric. For Lorentz four-vectors, (5.43) leads to

$$A_{\mu} = \eta_{\mu\nu} A^{\nu},$$

showing how we can obtain a covariant Lorentz four-vector. From (5.54), we conclude that the components of the covariant four-vector A_μ associated with the contravariant four-vector (often written as a row matrix in contrast to the components of a contravariant vector written as a column matrix) are

$$A_\mu = (A_0 = -A^0, A_1 = A^1, A_2 = A^2, A_3 = A^3). \tag{5.56}$$

Let us now consider an inertial frame K' indicated by primed coordinates. Since the metric tensor given by (5.54) will be the same for all frames, the same spacetime interval ds will be given in that frame by

$$ds^2 = \eta_{\sigma\tau} dx'^{\sigma} dx'^{\tau}. \tag{5.57}$$

Note that we can use any symbol for the indices which are summed up. We are choosing these indices in such a way that we shall be able to arrive at an elegant result. Keeping in mind that we must have $dx'^{\sigma} = \Lambda^\sigma_\mu dx^\mu$ and $dx'^{\tau} = \Lambda^\tau_\nu dx^\nu$ in accordance with (5.50), we get from (5.57) that

$$ds^2 = \eta_{\sigma\tau} \Lambda^\sigma_\mu \Lambda^\tau_\nu dx^\mu dx^\nu.$$

Comparing this with (5.55), we conclude that

$$\eta_{\sigma\tau} \Lambda^\sigma_\mu \Lambda^\tau_\nu = \eta_{\mu\nu}. \tag{5.58}$$

It is easy to check that this is equivalent to a matrix equation

$$\overline{\Lambda} \eta \Lambda = \eta, \tag{5.59}$$

where $\overline{\Lambda}$ is the transpose matrix of Λ.

We have pointed out that Λ corresponding to a Lorentz transformation in the x direction is given by (5.48). Let us now consider all possible kinds of transformations between inertial frames. Apart from Lorentz transformations in all the three spatial directions, we need to include rotations between frames because the transformation due to a rotation also can be represented by a transformation law like (5.52). Certainly (5.59) holds for all these possible transformations and puts some restrictions on the nature of the transformation matrices. Let us determine how many independent parameters the general transformation matrix Λ can have. Since Λ and η are 4×4 matrices, they each have 16 matrix elements. As the metric tensor is symmetric, it is easy to see that (5.59) introduces 10 restrictions on the values of the 16 matrix elements of Λ. It may be noted that when Λ represents pure Lorentz transformation, it is symmetric as in (5.48), whereas, when Λ represents pure rotation, it is antisymmetric as can be seen from (5.32). In a general situation, Λ does not have any specific symmetry properties. If 16 matrix elements have to satisfy 10 constraints, then we clearly can have six independent parameters. Out of these, three correspond to Lorentz transformations in three spatial directions, whereas the other three correspond to rotations around the three coordinate axes. Readers familiar with group theory should be able to verify that all the transformation matrices Λ together make up a mathematical group. This group of all transformations among inertial frames having six independent parameters is called the *Lorentz group*.

5.6 Doppler Effect of Light

When a source emitting light or sound is moving with respect to an observer, the observer finds the frequency of the light or sound to be different from the frequency with which it was emitted by the source. We assume that all the readers are familiar with this well-known phenomenon of the *Doppler effect*. Many elementary textbooks of physics give a derivation of the Doppler effect for sound. One may be inclined to adopt the results for sound for the case of light, although there are clear conceptual differences. Sound propagates with a certain speed, say c_s, through a medium like air. For sound, we have to handle the case of the source moving with respect to the medium and the case of the observer moving with respect to the medium somewhat differently. In the case of light, however, we have seen that no medium is required. Light would move with speed c with respect to both the source and the observer, provided the frames of both are inertial frames. Also, if the relative speed between the source and the observer is relativistic, we need to take that also into account. We present a derivation of the Doppler effect of light, which will provide the first demonstration of the power of the four-vector machinery.

Let us consider a monochromatic light wave emitted by a source with frequency ω and wavenumber $\mathbf{k}$ in the rest frame K of the source. The phase of this wave at position $\mathbf{x}$ at time t is given by $\mathbf{k}.\mathbf{x} - \omega t$. We expect this phase to be the same for all observers. If ω and $\mathbf{k}$ made up a four-vector

$$
k^{\mu} = \begin{pmatrix} \omega/c \\ k_x \\ k_y \\ k_z \end{pmatrix},
\tag{5.60}
$$

then the quantity

$$
\eta_{\mu\nu} k^{\mu} x^{\nu} = -\omega t + \mathbf{k}.\mathbf{x},
$$

as follows from (5.47), (5.54) and (5.60), should be a scalar. Now, this quantity is the phase, and we know it to be a scalar. Hence, k^{μ}, as defined by (5.60), has to be a true four-vector to ensure that the phase of the wave comes out as a scalar. We can also construct another scalar

$$
\eta_{\mu\nu} k^{\mu} k^{\nu} = -\frac{\omega^2}{c^2} + k^2.
$$

Fig. 5.2 A sketch showing
the configuration used for
discussing the Doppler effect
of light. The observer O is
moving with velocity **v** with
respect to the source S, the x
axis being taken in the
direction of **v**. The light
signal emitted by the source
S has to make an θ with the x
axis to reach the observer O

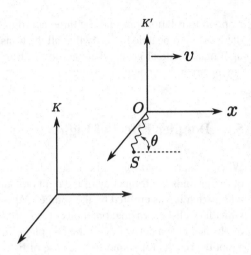

Since we have $\omega = kc$ for light wave propagating through a vacuum, this above
quantity is zero in all inertial frames and is certainly a scalar.

To determine the four-vector k'^{μ} in the frame K' of an observer moving with
speed v in the x direction, we have to use the fact that k^{μ} as given by (5.60), being
a four-vector, should transform in accordance with the transformation law (5.52). It
is easy to check that

$$\frac{\omega'}{c} = \gamma\left(\frac{\omega}{c} - \beta k_x\right) = \gamma\left(\frac{\omega}{c} - \frac{v}{c}k\cos\theta\right). \tag{5.61}$$

To understand the significance of the angle θ, refer to Fig. 5.2. The observer O is
moving with velocity **v** with respect to the source S assumed at rest in the frame
K. The x axis is taken in the direction of **v** as indicated in Fig. 5.2. The light signal
emitted by the source S has to make an θ with the x axis to reach the observer O.
Using the fact $k = \omega/c$, it follows from (5.61) that

$$\omega' = \gamma\omega\left(1 - \frac{v}{c}\cos\theta\right). \tag{5.62}$$

This is the relativistic Doppler shift formula for light. If $v\cos\theta$ is positive, then we
have the observer O moving away from the source S of light, which should be obvious
from Fig. 5.2. In this case, ω' turns out to be less than ω, implying that the frequency of
light measured by the observer is less than the frequency at which the light is emitted
in the frame of the source. On the other hand, if $v\cos\theta$ is negative, corresponding
to the case of the observer moving towards the source, then the observed frequency
ω' is more than ω.

5.7 Velocity and Momentum Four-Vectors

Using the four-vector notation, let us denote the spacetime coordinates of a moving particle separated by an infinitesimal interval by x^μ and $x^\mu + dx^\mu$. As we have pointed out in Sect. 5.3, the quantity $d\tau$ defined through (5.30) is a Lorentz scalar and can be interpreted as the proper time in the frame of the moving particle. Since we know that dx^μ transforms as a four-vector, it is obvious that

$$V^\mu = \frac{dx^\mu}{d\tau} \qquad (5.63)$$

has to be a four-vector. We now use the convention that while writing four-vectors in a column, we shall write the three spatial components together as a three-dimensional vector. Following this convention and making use of (5.47), we can write

$$dx^\mu = \begin{pmatrix} c\,dt \\ d\mathbf{x} \end{pmatrix}. \qquad (5.64)$$

Using (5.30), it then follows from (5.63) that

$$V^\mu = \begin{pmatrix} \gamma c \\ \gamma \mathbf{v} \end{pmatrix}, \qquad (5.65)$$

where

$$\mathbf{v} = \frac{d\mathbf{x}}{dt}$$

is the usual velocity in three-dimensional space. The four-vector V^μ defined through (5.65) is called the *four-velocity*.

We now introduce the *four-momentum* P^μ of the particle by multiplying the four-velocity V_μ by the mass m of the particle, that is,

$$P^\mu = mV^\mu = \begin{pmatrix} m\gamma c \\ m\gamma \mathbf{v} \end{pmatrix}. \qquad (5.66)$$

Let us write the four-momentum in the form

$$P^\mu = \begin{pmatrix} E/c \\ \mathbf{p} \end{pmatrix}, \qquad (5.67)$$

where obviously

$$E = m\gamma c^2, \qquad (5.68)$$

$$\mathbf{p} = m\gamma \mathbf{v}. \qquad (5.69)$$

We can easily interpret **p** as given by (5.69) as the *relativistic momentum*. We shall now argue that E given by (5.68) can be interpreted as the *relativistic energy*. Making use of (5.10), we write

$$E = \frac{mc^2}{\sqrt{1 - v^2/c^2}}. \tag{5.70}$$

In the non-relativistic limit $v^2/c^2 \ll 1$, we can make a binomial expansion and write

$$E \approx mc^2 \left(1 + \frac{1}{2}\frac{v^2}{c^2}\right) \approx mc^2 + \frac{1}{2}mv^2. \tag{5.71}$$

We interpret the term mc^2 as the *rest mass energy* of the particle, since the particle will continue to have this energy even when it is at rest. We see from (5.71) that in the non-relativistic limit, the energy is given by the sum of the rest mass energy and the usual kinetic energy. Certainly (5.71) provides some justification for interpreting E given by (5.68) as the relativistic energy.

It is one of the most famous results of special relativity that the mass m of a particle contributes a term mc^2 to the non-relativistic expression of energy, provided we want the energy to be defined in such a way that it becomes a part of a four-vector, as shown in (5.67). As we shall discuss in Sect. 5.8, in order to have a formulation of the energy conservation law consistent with the Lorentz transformation, it is necessary to make the energy a part of a four-vector. It is not possible to do this if we consider only the kinetic energy. The rest mass energy has to be a part of the energy if we want the energy conservation law to be consistent with special relativity.

In earlier textbooks, it was a common practice to introduce the concept of the *relativistic mass $m\gamma$* varying with the speed of the particle. In this approach, the relativistic momentum given by (5.69) could be obtained by multiplying this relativistic mass by the usual three-dimensional velocity **v**. Since we do not find the concept of relativistic mass particularly illuminating, we shall not make use of this concept in this book.

Since the four-momentum given by (5.67) should transform as a four-vector under Lorentz transformations, we can introduce the following Lorentz scalar

$$\eta_{\mu\nu} P^\mu P^\nu = -\frac{E^2}{c^2} + p^2 \tag{5.72}$$

on using (5.54) and (5.67). This Lorentz scalar should have the same value in all inertial frames, including the frame in which the particle is instantaneously at rest. Now, from (5.66), it follows that in the frame in which the particle is at rest, we have

$$P^\mu = \begin{pmatrix} mc \\ 0 \end{pmatrix}. \tag{5.73}$$

From this expression of P^μ, it follows that

$$\eta_{\mu\nu} P^\mu P^\nu = - m^2 c^2. \tag{5.74}$$

Equating the values of this Lorentz scalar in two frames as given by (5.72) and (5.74), we get

$$-\frac{E^2}{c^2} + p^2 = -m^2 c^2,$$

from which we are led to the extremely important energy–momentum relationship

$$E^2 = m^2 c^4 + p^2 c^2. \tag{5.75}$$

5.8 Conservation of Momentum and Energy

We know the two famous laws of Newtonian mechanics that the total momentum and the total energy of particles involved in an elastic collision are conserved. We expect these fundamental laws to hold even for motions involving relativistic velocities. Before we discuss how to formulate these laws in the relativistic situation, we make a few comments on the following question.

Suppose we know the non-relativistic version of a law, which is expected to be valid in relativistic situations as well. How do we determine the relativistic version of this law?

We put down the following general procedure how we may proceed in this situation.

Write down an equation in which each term is a Lorentz tensor of the same order and which reduces to the known non-relativistic law in the non-relativistic limit. This equation would give the relativistic version of the law.

If all the terms are Lorentz tensors of the same order, then certainly all the terms would transform in exactly the same manner when we go from one inertial frame to another through a Lorentz transformation. This means that if the equation holds in one inertial frame, it will hold in all inertial frames. If we know that the non-relativistic version of the equation holds in a non-relativistic inertial frame, we can at once conclude that the relativistic version should hold in frames connected by Lorentz transformations. The relativistic form of an equation in which each term is a Lorentz tensor of the same rank is often called the *covariant* form of the equation.

Suppose we have an elastic collision involving a few particles. We know from Newtonian mechanics that the total linear momentum and the total energy of the particles have to be the same before and after the collision. We now want to write down the generalized relativistic versions of these laws by following the procedure outlined above. Let $P_{1i}^\mu, P_{2i}^\mu, P_{3i}^\mu, \ldots$ be the four-momenta of the particles before the collision and $P_{1f}^\mu, P_{2f}^\mu, P_{3f}^\mu, \ldots$ be the four-momenta of the particles afterwards. We write down

$$P_{1i}^\mu + P_{2i}^\mu + P_{3i}^\mu + \ldots = P_{1f}^\mu + P_{2f}^\mu + P_{3f}^\mu + \ldots. \tag{5.76}$$

From (5.67) and (5.71), it follows that in the non-relativistic limit, the 0-th component (i.e. the time-like component) of (5.76) should give the non-relativistic law of conservation of energy, and the other components (i.e. the space-like components) together should give the non-relativistic law of conservation of momentum. Since all the terms in (5.76) are Lorentz four-vectors which should transform the same way under a Lorentz transformation and the non-relativistic limit of (5.76) corresponds to the usual conservation laws of energy and momentum, we conclude that (5.76) must be the relativistic generalization of the energy–momentum conservation law. We thus see that we can write down the conservation laws of both energy and momentum in a compact manner through one four-vector equation.

To demonstrate the power and elegance with which the generalized energy–momentum conservation law (5.76) can handle collisions of particles moving relativistically, we consider two examples. Note that the particles before the collision and after the collision do not always have to be the same in order to apply (5.76). In other words, this equation can tackle creation and annihilation of particles.

Example 1. Bombarding a target
In accelerator experiments, we sometimes have an incident particle of mass m_i moving with four-momentum P_i^μ hitting a target particle of mass m_t at rest. If the moving particle has a specific energy E_i, the collision may produce a new particle of mass $M > m_i + m_t$. Let us determine the energy E_i of the incident moving particle for which this will happen.

Writing the four-momenta of the target particle and the product particle as P_t^μ and P^μ, respectively, we have the energy–momentum conservation law

$$P_i^\mu + P_t^\mu = P^\mu$$

applicable to this situation, from which

$$\eta_{\mu\nu}(P_i^\mu + P_t^\mu)(P_i^\nu + P_t^\nu) = \eta_{\mu\nu}P^\mu P^\nu.$$

Expanding the LHS of this equation, we get

$$\eta_{\mu\nu}P_i^\mu P_i^\nu + \eta_{\mu\nu}P_t^\mu P_t^\nu + 2\eta_{\mu\nu}P_i^\mu P_t^\nu = \eta_{\mu\nu}P^\mu P^\nu. \tag{5.77}$$

We can evaluate several terms of this equation by using (5.74), which leads to

$$-m_i^2 c^2 - m_t^2 c^2 + 2\eta_{\mu\nu}P_i^\mu P_t^\nu = -M^2 c^2.$$

To evaluate $2\eta_{\mu\nu}P_i^\mu P_t^\nu$, we note that P_i^μ is given by an expression of the kind (5.67), whereas P_t^μ is given by an expression of the kind (5.73). It is easy to see that (5.77) gives

$$-m_i^2 c^2 - m_t^2 c^2 - 2E_i m_t = -M^2 c^2,$$

from which

$$E_i = \frac{M^2 - m_i^2 - m_t^2}{2m_t} c^2. \tag{5.78}$$

This gives the energy with which the incident particle m_i should hit the target particle m_t in order to produce the new particle of mass M.

Example 2. Decay of particles
Now we consider a particle of mass M decaying into two particles of masses m_1 and m_2. We must have $m_1 + m_2 < M$ to make this decay energetically possible. The excess mass will be converted into energy so that the particles resulting from the decay will have some kinetic energies available to them. Let us try to find the energy of one of the decay particles in the frame in which the initial particle was at rest.

If P^μ is the four-momentum of the initial particle and P_1^μ, P_2^μ the four-momenta of the two decay particles, the energy–momentum conservation law suggests

$$P^\mu = P_1^\mu + P_2^\mu,$$

which can be written as

$$P_1^\mu = P^\mu - P_2^\mu.$$

As in the case of (5.77), it follows from the above equation that

$$\eta_{\mu\nu} P_1^\mu P_1^\nu = \eta_{\mu\nu} P^\mu P^\nu + \eta_{\mu\nu} P_2^\mu P_2^\nu - 2\eta_{\mu\nu} P^\mu P_2^\nu.$$

By making use of (5.74) and keeping in mind that P^μ in the frame in which the initial particle was at rest is given by an expression like (5.73), the above equation gives

$$-m_1^2 c^2 = -M^2 c^2 - m_2^2 c^2 + 2E_2 M,$$

from which

$$E_2 = \frac{M^2 + m_2^2 - m_1^2}{2M} c^2. \tag{5.79}$$

This is the energy of the decay particle with mass m_2. The energy of the decay particle with mass m_1 is obviously given by an analogous expression.

5.9 Covariant Formulation of Electrodynamics

We are now at last ready to discuss the relationship between electrodynamics and relativity. As we mentioned in Sect. 5.1 and showed explicitly in Sect. 5.2, if the speed of light is c in one inertial frame, it will be c in other inertial frames which are connected by Lorentz transformations. This suggests the possibility that Maxwell's equations—which lead to the conclusion that the speed of light is c in the frame in which these equations are valid—may be valid in all inertial frames. We shall now show this explicitly.

We shall proceed as in Sect. 5.8 where we presented the relativistic formulation of energy–momentum conservation laws. We shall attempt to write down equations in which each term is a Lorentz tensor of the same rank and which give the various electrodynamics laws in the non-relativistic limit. If we know that the electrodynamics laws are valid in some given inertial frame such as our laboratory frame, they have to be valid in all other inertial frames moving with uniform velocity with respect to this frame, provided we are able to write down the electrodynamics laws in the form of equations involving tensors of the same rank, ensuring that the equations would retain the same form under Lorentz transformations.

Let us begin with the charge conservation law (4.7), which is certainly expected to hold in all inertial frames. In the case in which the charge and the current make up a four-vector given by

$$j^\mu = \begin{pmatrix} \rho \\ \mathbf{j}/c \end{pmatrix}, \tag{5.80}$$

it is easy to check that the charge conservation law (4.7) can be put in the form

$$\frac{\partial j^\mu}{\partial x^\mu} = 0, \tag{5.81}$$

where x^μ is the four-vector defined in (5.47). If j^μ given in (5.80) is really a four-vector, then (5.81) will be a proper tensor equation valid in all inertial frames and giving charge conservation. Our requirement that charge conservation has to hold in all inertial frames leads to the conclusion that j^μ introduced in (5.80) must be a proper four-vector. This implies that the charge and current densities should transform between inertial frames according to the four-vector transformation law (5.52). The four-vector introduced through (5.80) is called the *charge-current four-vector* or the *four-current*. Thus, with very little effort, we have been able to put the charge conservation Eq. (4.7) in the form (5.81), which makes it explicit that this equation is valid all inertial frames connected by Lorentz transformations. In the process, we have realized that the charge and current densities make up the Lorentz four-vector given by (5.80). We remind the reader that we have used two conventions which we are going to use throughout our discussion of relativistic electrodynamics: (i) while writing a four-vector as a column, we may write the three space-like components together as an ordinary vector, as we have done in (5.80); (ii) an index repeated twice in contravariant and covariant places will imply summation as in (5.81), noting that μ appearing in x^μ in the denominator of a derivative has to be taken like a covariant index.

We next turn our attention to the Lorentz gauge condition (4.135). In the case in which the scalar and vector potentials make up a four-vector like

$$A^\mu = \begin{pmatrix} \Phi \\ c\mathbf{A} \end{pmatrix}, \tag{5.82}$$

we easily check that (4.135) can be written in the form

$$\frac{\partial A^{\mu}}{\partial x^{\mu}} = 0. \tag{5.83}$$

Since the Lorentz gauge condition (4.135) should hold in all inertial frames, we are driven to the conclusion that the scalar and vector potentials together make up the four-vector given by (5.82), known as the *four-potential*. We should point out that the scalar and vector potentials make up such a four-vector only if they satisfy the Lorentz gauge condition. We introduced the Lorentz gauge condition in Sect. 4.10 to simplify a few equations, even though we can have scalar and vector potentials not satisfying this condition which give the electric and magnetic fields correctly. We now realize that we need scalar and vector potentials satisfying the Lorentz gauge condition for a proper covariant formulation of electrodynamics.

We now take up the inhomogeneous wave Eqs. (4.138) and (4.139) for the scalar and vector potentials. Making use of (4.42), we can write these two equations together in the following form

$$\left(\nabla^2 - \frac{1}{c^2}\frac{\partial^2}{\partial t^2}\right)\begin{pmatrix} \Phi \\ c\mathbf{A} \end{pmatrix} = -\frac{1}{\epsilon_0}\begin{pmatrix} \rho \\ \mathbf{j}/c \end{pmatrix}. \tag{5.84}$$

Now, note that we can write

$$\nabla^2 - \frac{1}{c^2}\frac{\partial^2}{\partial t^2} = \frac{\partial}{\partial x^{\nu}}\frac{\partial}{\partial x_{\nu}} \tag{5.85}$$

by virtue of (5.47) and using (5.56) to get x_{μ} from it. It is clear that the differential operator in the LHS of (5.84) is a scalar operator. The combined inhomogeneous wave Eq. (5.84) implies that a scalar operator acting on a four-vector gives another four-vector. With the help of (5.80), (5.82) and (5.85), we can now put (5.84) in the following compact and elegant form:

$$\frac{\partial}{\partial x^{\nu}}\frac{\partial}{\partial x_{\nu}}A^{\mu} = -\frac{j^{\mu}}{\epsilon_0}. \tag{5.86}$$

This form (5.86) of the inhomogeneous wave equation makes it clear that it will transform under Lorentz transformations in such a manner that it will be valid in all inertial frames.

We now discuss how we handle electromagnetic fields in our relativistic formalism. Let us introduce the Lorentz tensor

$$F_{\mu\nu} = \frac{\partial A_{\nu}}{\partial x^{\mu}} - \frac{\partial A_{\mu}}{\partial x^{\nu}}. \tag{5.87}$$

It is obvious that $F_{\mu\nu}$ is an anti-symmetric rank-2 tensor, which can be represented by a 4×4 matrix. Such a matrix with real components can have six independent components. We leave it for the readers to determine from (4.128) and (4.130) that the different components of $F_{\mu\nu}$ correspond to different components of the electric and magnetic fields in the following way:

$$
F_{\mu\nu} = \begin{pmatrix} 0 & -E_x & -E_y & -E_z \\ E_x & 0 & cB_z & -cB_y \\ E_y & -cB_z & 0 & cB_x \\ E_z & cB_y & -cB_x & 0 \end{pmatrix}. \tag{5.88}
$$

While checking this, keep in mind that A_μ has to be obtained from A^μ given by (5.82) in accordance with (5.56). We see from (5.88) that the components of the electromagnetic fields make up the components of a rank-2 Lorentz tensor, called the *electromagnetic field tensor*. Sometimes, we shall have to use the contravariant version of this tensor given by

$$
F^{\mu\nu} = \eta^{\mu\sigma}\eta^{\nu\tau}F_{\sigma\tau}.
$$

Using (5.54), it is easy to show that

$$
F^{\mu\nu} = \begin{pmatrix} 0 & E_x & E_y & E_z \\ -E_x & 0 & cB_z & -cB_y \\ -E_y & -cB_z & 0 & cB_x \\ -E_z & cB_y & -cB_x & 0 \end{pmatrix}. \tag{5.89}
$$

One important corollary of the fact that the electromagnetic fields make up a rank-2 Lorentz tensor is the following: one can determine how electromagnetic fields transform from one inertial frame to another from the transformation laws of Lorentz tensors. We shall discuss this problem in the next section. Right now, let us focus our attention on Maxwell's equations.

From the definition (5.87) of $F_{\mu\nu}$, it readily follows that

$$
\frac{\partial F_{\mu\nu}}{\partial x^\sigma} + \frac{\partial F_{\sigma\mu}}{\partial x^\nu} + \frac{\partial F_{\nu\sigma}}{\partial x^\mu} = 0. \tag{5.90}
$$

It is easy to check from (5.88) that for the combination $\{1, 2, 3\}$ of μ, ν, σ, we get (1.2). Similarly, the combinations $\{0, 2, 3\}$, $\{0, 3, 1\}$ and $\{0, 1, 2\}$ of μ, ν, σ give the three components of (1.4). We had seen in Sect. 4.10 that when electromagnetic fields are defined through (4.128) and (4.130) using the scalar and vector potentials, Maxwell's equations (1.2) and (1.4) are automatically satisfied. Here, we are drawing essentially the same conclusion. We have introduced electromagnetic fields in terms of the four-potential by using (5.87), which is equivalent to (4.128) and (4.130). We find that (5.87) automatically leads to (5.90), which is equivalent to Maxwell's equations (1.2) and (1.4).

After having dealt with Maxwell's equations (1.2) and (1.4) which do not have source terms, we now come to the other Eqs. (1.1) and (1.3) involving the source terms ρ and $\mathbf{j}$. These equations can be written together compactly in the form

$$\frac{\partial F^{\mu\nu}}{\partial x^{\nu}} = \frac{j^{\mu}}{\epsilon_0}. \tag{5.91}$$

We leave for the readers the easy exercise of verifying, with the help of (5.80) and (5.89), that the 0-th component of (5.91) gives (1.1), whereas the other three components together make up (1.3).

All the equations in this section are nothing but the basic equations of electrodynamics cast in terms of Lorentz vectors and tensors. This formulation of electrodynamics is often referred to as the *covariant formulation*. The fact that all the basic equations of electrodynamics could be put in the form of equations involving Lorentz vectors and tensors so easily and elegantly makes one conclusion clear: if the laws of electrodynamics hold in one inertial frame, they should hold in all other inertial frames connected to it by Lorentz transformations. Especially, we find that all of Maxwell's equations can be put in the compact forms (5.90) and (5.91). As we go from one inertial frame to another, the values of all the quantities $\mathbf{E}$, $\mathbf{B}$, ρ and $\mathbf{j}$, as well as the spacetime coordinates, will change according to the transformation laws of Lorentz vectors and tensors. The fact that Maxwell's equations can be written in the forms (5.90) and (5.91) ensures that all the various quantities should transform between the frames in such a way that Maxwell's equations hold in all the frames.

Lastly, we should make one comment. Since all the laws of electrodynamics could be put in the covariant form so smoothly, we may not even realize that this is something truly extraordinary. In general, there is no compelling reason to expect that some physics laws arising out of non-relativistic considerations (Maxwell's equations were formulated several decades before special relativity) can always be put in the covariant form. Even for energy and momentum conservation laws, we saw in Sect. 5.8 that the standard non-relativistic laws (such as the law of kinetic energy conservation in elastic collisions) do not hold for relativistically moving particles, but we have to generalize them to suitable relativistic forms involving, for example, the replacement of the non-relativistic kinetic energy by the relativistic energy given by (5.68). It is amazing that Maxwell's equations can be put in covariant forms without even requiring such generalizations like the replacement of the non-relativistic kinetic energy by the relativistic energy. This shows the deep connection between electrodynamics and relativity, which should become even clearer in Sect. 5.12.

5.10 Transformation of Electromagnetic Fields Between Inertial Frames

In the previous section, various basic laws of electrodynamics were put in the forms of equations involving Lorentz vectors and tensors, which hold as we go from one inertial frame to another connected by a Lorentz transformation. However, while the equations may be the same in different frames, we expect various electromagnetic quantities themselves to change as we transform from one frame to another. Let us consider a primed frame K' moving with velocity $\mathbf{v}$ with respect to an unprimed frame K. We can choose the x axis in the direction of $\mathbf{v}$ without any loss of generality. Then the measurements of spacetime coordinates in the frames K' and K would yield values which are connected by (5.11–5.14). We want to determine how the electromagnetic fields $\mathbf{E}'$ and $\mathbf{B}'$ measured in the primed frame K' would be related to the fields $\mathbf{E}$ and $\mathbf{B}$ measured in the unprimed frame K.

In the covariant formulation of electrodynamics, the components of the electric and magnetic fields make up the Lorentz tensor $F^{\mu\nu}$ given by (5.89). Such a tensor transforms from one frame to another according to the transformation law (5.40). This being a Lorentz tensor, we have (5.51) with the transformation matrix Λ_ν^μ given by (5.48). From (5.40) and (5.51), it follows that $F'^{\mu\nu}$ in the primed frame K' has to be related to $F^{\mu\nu}$ in the unprimed frame K in the following manner:

$$F'^{\mu\nu} = F^{\alpha\beta}\Lambda_\alpha^\mu\Lambda_\beta^\nu. \tag{5.92}$$

If we follow the convention that the indices μ and ν in Λ_ν^μ correspond respectively to the rows and columns in the matrix representation (5.48) of Λ_ν^μ, then it is easy to check that the RHS of (5.92) corresponds to a multiplication of three matrices in the following order: (i) Λ_α^μ which is given by the matrix in the RHS of (5.48); (ii) $F^{\alpha\beta}$ which is given by the matrix in the RHS of (5.89); and (iii) the transpose of Λ_β^ν, which is a symmetric tensor, so that its transpose is also given by the matrix in the RHS of (5.48). On carrying out the multiplication of these three matrices, (5.92) leads to

$$F'^{\mu\nu} = \begin{pmatrix} 0 & E_x & \gamma(E_y - \beta c B_z) & \gamma(E_z + \beta c B_y) \\ -E_x & 0 & \gamma(c B_z - \beta E_y) & -\gamma(c B_y + \beta E_z) \\ -\gamma(E_y - \beta c B_z) & -\gamma(c B_z - \beta E_y) & 0 & c B_x \\ -\gamma(E_z + \beta c B_y) & \gamma(c B_y + \beta E_z) & -c B_x & 0 \end{pmatrix}. \tag{5.93}$$

We also expect on the basis of (5.89) that $F'^{\mu\nu}$ in the primed frame K' should be represented by the matrix

$$F'^{\mu\nu} = \begin{pmatrix} 0 & E_x' & E_y' & E_z' \\ -E_x' & 0 & c B_z' & -c B_y' \\ -E_y' & -c B_z' & 0 & c B_x' \\ -E_z' & c B_y' & -c B_x' & 0 \end{pmatrix}. \tag{5.94}$$

We can now equate the various terms in the matrices given in (5.93) and (5.94). We easily come to the conclusion that the electric field transforms in the following way:

$$E_x' = E_x, \quad E_y' = \gamma(E_y - v B_z), \quad E_z' = \gamma(E_z + v B_y), \tag{5.95}$$

whereas the magnetic field transforms in this way

$$B_x' = B_x, \quad B_y' = \gamma\left(B_y + \frac{v}{c^2}E_z\right), \quad B_z' = \gamma\left(B_z - \frac{v}{c^2}E_y\right). \tag{5.96}$$

We now try to write down these transformation laws in a more general form. We note in (5.95) and (5.96) that E_x and B_x do not change under the transformation. Since the x direction is the direction of motion of the primed frame K', we write

$$E'_\parallel = E_\parallel, \tag{5.97}$$

$$B'_\parallel = B_\parallel, \tag{5.98}$$

where we have used the symbol $\parallel$ to indicate the direction of motion of the primed frame K'. As for the perpendicular components in the y and z directions, they are found to transform in (5.95) and (5.96) in such a way that their transformation laws can be combined and written down in the following manner:

$$\mathbf{E}'_\perp = \gamma \left(\mathbf{E}_\perp + \mathbf{v} \times \mathbf{B}_\perp \right), \tag{5.99}$$

$$\mathbf{B}'_\perp = \gamma \left(\mathbf{B}_\perp - \frac{\mathbf{v}}{c^2} \times \mathbf{E}_\perp \right), \tag{5.100}$$

where we use the symbol $\perp$ to indicate the component of any vector perpendicular to the direction of motion of the primed frame K'. How an electric field transforms from one frame to another frame is specified by (5.97) and (5.99) together, while how a magnetic field transforms from one frame to another frame is specified by (5.98) and (5.100) together. We can draw one important conclusion. Even if either only an electric field or only a magnetic field is present in one frame, in general, both types of fields will be present in another frame moving with respect to it.

We now point out that we can construct some Lorentz scalars out of the electromagnetic field tensor. One such scalar is

$$F^{\mu\nu} F_{\mu\nu} = 2 \left(c^2 B^2 - E^2 \right) \tag{5.101}$$

from (5.88) and (5.89). Another scalar is

$$\det F^{\mu\nu} = c^2 \left(\mathbf{E} \cdot \mathbf{B} \right)^2. \tag{5.102}$$

We can conclude that this is a Lorentz scalar by taking determinants of both the sides of (5.92). It is instructive to show from the transformation laws (5.97–5.100) that the values of these quantities remain unchanged when we go from one inertial frame to another (Exercise 5.5).

So far, our discussion of the covariant formulation of electrodynamics has been rather formal and abstract. Let us now consider some practical examples in the next few subsections which may make the reader more comfortable with the subject.

5.10.1   From the Electric Field of a Line Charge to the Magnetic Field of a Line Current

One standard problem in electrostatics is to find the electric field of a line charge (Exercise 2.1), whereas one standard problem in magnetostatics is to find the magnetic field of a line current, which we worked out in Sect. 3.3, leading to (3.16). Now we want to argue that there is a deep connection between these problems which follows from the transformation law of electromagnetic fields.

Let us first consider a line charge along the x direction with Q charge per unit length, kept at rest in our frame of reference. We look at the electric field at a point in the y direction at a distance r away from the line charge, as indicated in Fig. 5.3. As we asked the reader to work out this electric field in Exercise 2.1, it follows easily from Gauss's law in electrostatics discussed in Sect. 2.6 that the electric field at this distance r is in the y direction and is given by

$$\mathbf{E} = \frac{1}{2\pi\epsilon_0} \frac{Q}{r} \hat{\mathbf{e}}_y. \tag{5.103}$$

We now consider an observer moving with a velocity $-v\,\hat{\mathbf{e}}_x$ in the negative x direction in Fig. 5.3 with respect to our frame. It would appear to this observer that the line charge is moving with a velocity $v\,\hat{\mathbf{e}}_x$. You may expect this to lead to a line current

$$I = Q\,v.$$

However, this turns out to be a non-relativistic approximation. The charge density ρ and the current density $\mathbf{j}/c$ together make up a four-vector, as indicated in (5.80). The reader should be able to argue from this that the charge per unit length Q and the current I/c should also behave like a four-vector when we consider transformations between frames moving parallel to the direction of the line charge. To find the current in the frame moving with velocity $-v\,\hat{\mathbf{e}}_x$, we should carry on a Lorentz transformation of this four-vector from the rest frame of the line charge with respect to which the we have charge Q per unit length and no current. An easy Lorentz transformation

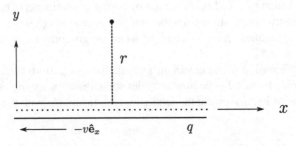

Fig. 5.3 A line charge extending along the x direction with a field point at a distance r in the y direction. We also consider an observer moving with a speed $-v\,\hat{\mathbf{e}}_x$ in the negative x direction

shows that the relativistically correct expression of the current in the frame moving with velocity $-v\,\hat{\mathbf{e}}_x$ should be

$$I = \gamma\,Q\,v. \tag{5.104}$$

We want to determine the magnetic field at the point where the electric field in the rest frame of the line charge is given by (5.103). The magnetic field at this point that will be inferred by the moving observer can be found from (5.100). Noting that there is no magnetic field in the rest frame of the line charge but only the electric field given by (5.103), we find on applying (5.100) that the magnetic field in the frame of the observer moving with velocity $-v\,\hat{\mathbf{e}}_x$ is given by

$$\mathbf{B} = \frac{\gamma}{c^2}\,v\,\frac{1}{2\pi\epsilon_0}\frac{Q}{r}\,\hat{\mathbf{e}}_z \tag{5.105}$$

on substituting $-v\,\hat{\mathbf{e}}_x$ for $\mathbf{v}$ and from (5.103) for $\mathbf{E}_\perp$. Making use of (5.104) and (4.42), we conclude

$$\mathbf{B} = \frac{\mu_0}{2\pi}\frac{I}{r}\,\hat{\mathbf{e}}_z.$$

Keeping in mind that this magnetic field is in the z direction perpendicular to the plane of Fig. 5.3, we note that this is exactly the same as the magnetic field due to a line current given by (3.16).

To the observer moving with uniform speed $-v\,\hat{\mathbf{e}}_x$ with respect to the frame in which the line charge is at rest, it would seem that there is a static line current I given by (5.104), and in that observer's frame, magnetostatic considerations would lead to the result (3.16). We have now shown that the same result follows from transforming the electric field in the frame in which the line charge is at rest. This example suggests that, in a certain sense, we can regard the magnetic field as the relativistic effect of the electric field. Normally, we expect a relativistic effect to be negligible when the speeds involved are non-relativistic. Why then is the magnetic field due to a current measurable with very ordinary instruments? To answer that question, keep in mind that inside a current-carrying metal wire, we normally have a background of positive charges at rest against which the negative charges move. Since there would be equal amounts of positive and negative charges within a certain portion of the wire, the electric fields produced by them would cancel out. On the other hand, if opposite charges move in opposite directions—which would be the case if charges of both signs are free to move under the influence of an electric field—the magnetic fields created by the opposite charges would add up. In the more realistic situation of a metal wire within which positive charges remain at rest while negative electrons move, only the moving electrons produce a magnetic field which is not screened. Thus, a current-carrying wire produces only a magnetic field, the electric fields due to the opposite charges being cancelled out. It is instructive to consider how strong the effect of the electric field would have been if it were not cancelled out in this manner. For the sake of a comparison, let us consider both the electric field and the magnetic field due to an unscreened line charge in the frame of the observer moving with speed $-v\,\hat{\mathbf{e}}_x$. It follows from (5.99) that the electric field in this frame also would be given approximately by (5.103) if the speed v of the observer is non-relativistic. Suppose we have a charged particle moving with a velocity $\mathbf{v}_{particle}$ in the observer's frame. The ratio of the magnetic force to the electrical force on this particle as seen in this frame would be

$$\frac{|\text{Magnetic force}|}{|\text{Electrical force}|} \approx \frac{|v_{particle}|\,|B|}{|E|} \approx \frac{|v|\,|v_{particle}|}{c^2} \tag{5.106}$$

on using (5.103) and (5.105). For non-relativistic motions, the magnetic effect would be completely negligible compared to the electric effect. Only because the electric field is screened off in normal circumstances, the much smaller magnetic effect has the chance of being discernible. If you are asked whether you can measure a relativistic effect in the laboratory with simple apparatus, you can say this in answer: you can measure the magnetic field of a current, and it is a relativistic effect!

We are now in a position to address the concern we expressed in Sect. 2.1. While the SI unit ampere of current is of the order currents we encounter in everyday life, the corresponding unit coulomb of charge is much larger than the charges which we come across in the laboratory. Why is this so? It is clear from our discussion that charges of one sign corresponding to a current, if not screened off by charges of opposite sign, would give rise to a much stronger electrical force compared to the magnetic force produced by the current, as indicated in (5.106). The non-occurrence of such strong electrical forces implies that it is unlikely for such amounts of non-screened charges to be found under normal laboratory circumstances. The charges we typically encounter in our laboratories are very small compared to 1 C—which would be the typical amount of unscreened charge associated with a current of 1 A (i.e. the amount of charge that would flow in 1 s through a wire carrying a current of 1 A). Since the magnetic field is a relativistic effect, it is not possible to introduce electrical units in a such a manner that a unit of current and the corresponding unit of charge produce magnetic and electrical forces of comparable strength in typical laboratory circumstances. This problem occurs even with Gaussian units, as we point out in Sect. A.3 of Appendix A discussing Gaussian units.

5.10.2 The Field of a Relativistically Moving Charged Particle

We now consider a particle having charge q moving relativistically with velocity $\mathbf{v} = v\,\mathbf{e}_x$ with respect to our frame and determine the electromagnetic fields as they would appear in our frame. We write various various physical quantities without primes in the frame K in which the charged particle is at rest at the origin of the coordinate system, whereas they are written with primes our frame K'. In the rest frame K of the charged particle, there will be only an electric field given by (2.6), which becomes

$$\mathbf{E} = \frac{q}{4\pi\epsilon_0} \frac{x\,\mathbf{e}_x + y\,\mathbf{e}_y + z\,\mathbf{e}_z}{[x^2 + y^2 + z^2]^{3/2}}. \tag{5.107}$$

The components of the electric field in our frame K' can be obtained by applying the transformation law (5.95) taking $\mathbf{B} = 0$ and with $-v$ replacing v (since the frame K' is now moving in the negative x direction with respect to the frame K). We also should use (5.6–5.9) with $-v$ in the place of v to express the electric field as a function of the coordinates in our frame K'. This procedure gives

$$E'_x = \frac{q}{4\pi\epsilon_0} \frac{\gamma\,(x' - vt')}{[\gamma^2(x' - vt')^2 + y'^2 + z'^2]^{3/2}}, \tag{5.108}$$

$$E'_y = \frac{q}{4\pi\epsilon_0} \frac{\gamma\,y'}{[\gamma^2(x' - vt')^2 + y'^2 + z'^2]^{3/2}}, \tag{5.109}$$

$$E'_z = \frac{q}{4\pi\epsilon_0} \frac{\gamma\,z'}{[\gamma^2(x' - vt')^2 + y'^2 + z'^2]^{3/2}}. \tag{5.110}$$

Now, in our frame, $(x' - vt', y', z')$ would be the coordinates of the field point with respect to the position of the charged particle at time t'. It should be obvious from (5.108–5.110) that the electric field would be directed radially outward from the position of the charged particle at time t'. However, the electric field is not going to be isotropic. To determine the nature of the electric field, consider the electric field at the point $(x', 0, 0)$ on the x axis and at the point $(0, y', 0)$ perpendicular to the x axis. At the point $(x', 0, 0)$, it follows from (5.108–5.110) that the only non-zero component of $\mathbf{E}'$ is E'_x given by

$$E'_x(x', 0, 0) = \frac{q}{4\pi\epsilon_0} \frac{1}{\gamma^2(x' - vt')^2}. \tag{5.111}$$

On the other hand, at the point $(0, y', 0)$, the only non-zero component of $\mathbf{E}'$ is E'_y given by

$$E'_y(0, y', 0) = \frac{q}{4\pi\epsilon_0} \frac{\gamma}{y'^2}. \tag{5.112}$$

Note that $x' - vt'$ and y' are respectively the distances of the points $(x', 0, 0)$ and $(0, y', 0)$ from the position of the charged particle at time t'. If we consider points on the x axis and perpendicular to the x axis at equal distances from the charged particle, it follows from (5.111) and (5.112) that the electric field in the perpendicular direction is enhanced by a factor γ^3 (which can be much larger than 1 for highly relativistic motion) with respect to the electric field in the direction of motion. Figure 5.4 shows a sketch of what the electric field lines due to a relativistically moving charged particle would be like. The field lines are crowded in the direction perpendicular to the motion because the component of the electric field perpendicular to the motion is enhanced.

Fig. 5.4 The electric field
lines of a relativistically
moving charged particle as
they appear in our frame

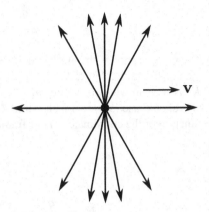

Fig. 5.4 The electric field lines of a relativistically moving charged particle as they appear in our frame

As the charged particle passes by us, we would experience a pulse of a strong electric field perpendicular to the direction of its motion (Fig. 5.4).

The moving charged particle also gives rise to a magnetic field, which can be found from (5.96), keeping in mind that we have to put $-v$ in the place of v. It follows from (5.95) and (5.96) that the magnetic field due to the moving charged particle is perpendicular to the direction of motion and is given by

$$\mathbf{B}' = \frac{\mathbf{v}}{c^2} \times \mathbf{E}'. \tag{5.113}$$

We wrote down an expression (3.72) for the magnetic field due to a moving charge by using the analogy between a moving charge and a current. If the charged particle is moving non-relativistically and we use the non-relativistic expression for $\mathbf{E}'$ due to a charge in (5.113), then we indeed end up with (3.72) on using (4.42). In the other limit of a highly relativistic charged particle moving with velocity $\mathbf{v}$ close to c, we find from (5.113) that the ratio of the value of the magnetic field to the value of the electric field would be about $1/c$, which is the same as the ratio in the case of an electromagnetic wave as given by (4.52). Thus, the pulse associated with the relativistic particle would appear to us like a pulse of an electromagnetic wave, with electric and magnetic fields perpendicular to each other in the plane perpendicular to the direction of motion and having their amplitudes related in the same way as in an electromagnetic wave.

There is another way of calculating the electromagnetic fields due to a uniformly moving charged particle. We have seen in (5.82) that the scalar potential Φ and the vector potential $\mathbf{A}$ (multiplied by c) make up a Lorentz four-vector. We can easily write down the scalar potential Φ in the frame K in which the charged particle is at rest ($\mathbf{A}$ is zero in that frame). By using the four-vector transformation law, we can determine the corresponding Φ' and $\mathbf{A}'$ in our frame K' with respect to which the particle is moving with speed $v\,\mathbf{e}_x$. Then we can use (4.128) and (4.130) to calculate the electromagnetic fields in our frame K' (consistently using primed variables for all quantities in our frame). This procedure also gives the same electromagnetic fields

(5.108–5.110) and (5.113) which we have found by the direct transformation of the fields. The reader is asked to verify this in Exercise 5.8.

Lastly, we should caution the reader that our discussion in this subsection holds only for uniformly moving charged particle. If the charged particle is moving non-uniformly (either relativistically or non-relativistically), there can be additional contributions to the electromagnetic fields due to the non-uniformity of motion, as we shall discuss in Sect. 6.4.

5.10.3   The Non-relativistic Limit

Suppose we have a frame S' moving with non-relativistic velocity $\mathbf{v}$ with respect to the frame S. We want to relate the electromagnetic fields $\mathbf{E}'$, $\mathbf{B}'$ found in the frame S' with the electromagnetic fields $\mathbf{E}$, $\mathbf{B}$ found in the frame S. This can be done by taking a non-relativistic limit of the transformation laws (5.97–5.100). Basically, the only thing we have to do is to take $\gamma = 1$ for the non-relativistic situation. Then we can combine (5.97) and (5.99) to write

$$\mathbf{E}' = \mathbf{E} + \mathbf{v} \times \mathbf{B}. \tag{5.114}$$

Similarly, (5.98) and (5.100) can be combined to give

$$\mathbf{B}' = \mathbf{B} - \frac{\mathbf{v}}{c^2} \times \mathbf{E}. \tag{5.115}$$

It follows that the electromagnetic fields measured in the two frames will be different even if the relative motion between them is non-relativistic and still the transformation law of the electromagnetic fields has to be obtained from relativistic considerations. When we study relativistic mechanics, we find that the relativistic effects can be neglected when only non-relativistic motions are involved. In our discussion of relativistic electrodynamics, in contrast, we arrive at the somewhat startling conclusion that, even when relative motions between frames are non-relativistic, the electromagnetic fields measured in them can be different—an essentially relativistic effect.

While it may appear surprising at first sight that electromagnetic fields transform even between frames moving non-relativistically by following transformation laws which arise from relativistic considerations, the transformation laws (5.114–5.115) help us understand many well-known phenomena from a deeper level. Take, for example, the Lorentz force equation (1.5). Suppose we go to the rest frame of the moving particle. Since the velocity is zero in that frame, there is no magnetic force, and we conclude that the force on the particle will be given by $q\,\mathbf{E}'$. If the electric field $\mathbf{E}'$ in that frame is related to the electromagnetic fields in our frame by (5.114), then the force would be the same as what is given by (1.5).

The transformation law (5.114) also helps us in solving a conceptual problem connected with electromagnetic induction which was mentioned towards the end of Sect. 4.1.2. Einstein began his first paper on special relativity published in 1905 by pointing out this problem. The first few lines of this revolutionary paper, titled 'On the Electrodynamics of Moving Bodies' (in English translation), are as follows (Lorentz et al. [1], p. 37):

> It is known that Maxwell's electrodynamics—as usually understood at the present time—when applied to moving bodies, leads to asymmetries which do not appear inherent in the phenomena. Take, for example, the reciprocal electrodynamic action of a magnet and a conductor. The observable phenomenon here depends only on the relative motion of the conductor and the magnet, whereas the customary view draws a sharp distinction between the two cases in which either the one or the other of these bodies is in motion.

When we have a moving magnet in the rest frame of a circuit, the magnetic field is changing with time, and the EMF induced in the circuit, given by the $\oint \mathbf{E} \cdot d\mathbf{l}$ term in the RHS of (4.9), can be calculated from (4.4). On the other hand, in the rest frame of the magnet, the magnetic field is not changing, and we now have to calculate the EMF induced in the moving circuit by using the second term in the RHS of (4.9), which follows from the assumption that there are charged particles inside the circuit acted on by the Lorentz force given by (1.5). This is the asymmetry in the explanations from the two frames which made Einstein dissatisfied. Once we have the transformation law (5.114), we do not have to rely on this assumption of charged particles inside the circuit. Transforming from the rest frame of the magnet in which the magnetic field is $\mathbf{B}$, the electric field in the frame of the circuit, according to (5.114), is given by $\mathbf{E}' = \mathbf{v} \times \mathbf{B}$ (noting that there is no electric field in the rest frame of the magnet), and we can obtain the induced EMF by integrating this electric field $\mathbf{E}'$ over the circuit. If you read the discussion in Sect. 4.1.2 and work out Exercise 4.3(a), then it should be clear to you that the EMF you calculate in this manner is the same as the EMF you would get from the rate of change of the magnetic flux passing through the circuit. If we want to avoid this asymmetry in the explanation from the two frames of reference, then the most reasonable approach would be to define the EMF in a circuit to be given by the line integral $\oint \mathbf{E}.d\mathbf{l}$ *in the frame of the circuit*. Then the Eq. (4.4) and the transformation law (5.114) together would enable us to calculate the EMF in any situation without the additional assumption about the existence of charged particles inside the circuit.

5.11   Covariant Formulation of the Lorentz Force Equation

We have been discussing various aspects of the covariant formulation of electrodynamics in the previous two sections. We pointed out at the very beginning in Sect 1.2 that, apart from Maxwell's equations, the other basic equation of electrodynamics is the Lorentz force equation (1.5). So far, we have not considered what happens to this equation when we make a covariant formulation of electrodynamics. Now we turn our attention to this equation. If **p** is the momentum of the charged particle on which the Lorentz force given by (1.5) operates, then we have

$$\frac{d\mathbf{p}}{dt} = q\,(\mathbf{E} + \mathbf{v} \times \mathbf{B}).\tag{5.116}$$

To make a covariant formulation of this equation, we proceed as discussed in Sect. 5.8. We try to write down an equation involving Lorentz vectors which would reduce to (5.116) in the non-relativistic limit. We expect that the LHS of (5.116) should now be generalized to

$$\frac{dP^{\mu}}{d\tau},$$

where P^{μ} is the momentum four-vector given by (5.66), and τ is the proper time given by (5.30). Since this is a Lorentz four-vector, the RHS of the equation we are trying to arrive at must also be a Lorentz four-vector. We now introduce the *Lorentz four-force* defined in the following way

$$f^{\mu} = \frac{q}{c}\,F^{\mu\nu}\,V_{\nu},\tag{5.117}$$

where V_{ν} is the four-velocity given by (5.65) along with (5.56). From (5.65) and (5.89), it is easy to show that the components of (5.117) are given as follows:

$$\frac{q}{c}\,F^{\mu\nu}\,V_{\nu} = q\left[\begin{array}{c}(\gamma/c)\mathbf{E}.\mathbf{v}\\ \gamma(\mathbf{E} + \mathbf{v} \times \mathbf{B})\end{array}\right],\tag{5.118}$$

It is obvious that the space-like part of this four-vector gives the Lorentz force (1.5) in the non-relativistic limit. This suggests that the RHS of the covariant version of (5.116) would have f^{μ} given by (5.117) so that we write

$$\frac{dP^{\mu}}{d\tau} = \frac{q}{c}\,F^{\mu\nu}\,V_{\nu}.\tag{5.119}$$

Both sides of this equation are Lorentz four-vectors which would transform similarly under Lorentz transformation so that (5.119) can be valid in all inertial frames. It is also easy to check from (5.30), (5.66) and (5.118) that the space-like part of this equation reduces to (5.116) in the non-relativistic limit. This makes it clear that (5.119) is the appropriate covariant generalization of the standard Lorentz force equation (5.116).

It is instructive to write down the time-like and space-like parts of (5.119) separately, which we proceed to do now. In Sect. 5.7, we used the symbol E for energy. To avoid confusion with the electric field, let us now use $\mathcal{E}$ for energy. Using (5.30), (5.67) and (5.118), we find that the time-like component of (5.119) is

$$\gamma\frac{d}{dt}\left(\frac{\mathcal{E}}{c}\right) = \gamma\frac{q}{c}\,\mathbf{E}.\mathbf{v},$$

from which

$$\frac{d\mathcal{E}}{dt} = q\,\mathbf{E}\cdot\mathbf{v}. \tag{5.120}$$

This signifies that the energy of the charged particle changes due to the work done by the electric field on it. This is expected because the magnetic field does not do any work on the charged particle and does not change its energy. We now look at the space-like part of (5.119). From (5.30), (5.66) and (5.118), it follows that

$$\gamma\frac{d}{dt}(\gamma\,m\mathbf{v}) = q\gamma\,(\mathbf{E} + \mathbf{v}\times\mathbf{B}).$$

Cancelling γ from both sides, we arrive at

$$\frac{d}{dt}(\gamma\,m\mathbf{v}) = q\,(\mathbf{E} + \mathbf{v}\times\mathbf{B}), \tag{5.121}$$

which is the relativistic generalization of (5.116). In the non-relativistic limit when $\gamma \approx 1$, (5.121) reduces to (5.116).

We work out one important example on the basis of (5.121). One of the well-known results of electromagnetic theory is that a charged particle moving in a plane perpendicular to a constant magnetic field traverses a circle. We expect all the readers to be familiar with this result discussed in many elementary textbooks. We now want to consider the generalization of this result when the charged particle moves relativistically. Let us consider a relativistic charged particle moving perpendicular to a constant magnetic field $\mathbf{B}$ in a region where there is no electric field. Since the magnetic field does not do any work on the charged particle, the energy of the charged particle does not change, as seen in (5.120), and γ for the particle remains constant. It then follows from (5.121) that

$$\gamma\,m\frac{d\mathbf{v}}{dt} = q\,\mathbf{v}\times\mathbf{B}. \tag{5.122}$$

Anticipating that the particle would traverse a circular path, the acceleration can be equated to the centripetal acceleration v^2/r directed towards the centre of the circular path so that (5.122) gives

$$\gamma\,m\frac{v^2}{r} = qvB,$$

from which the circular frequency of gyration of the relativistic particle is found to be

$$\omega_{c,r} = \frac{v}{r} = \frac{Bq}{\gamma\,m}. \tag{5.123}$$

If we put $\gamma = 1$ in (5.123), then we get the standard expression for the circular frequency of gyration of a non-relativistic charged particle moving in a constant magnetic field. This frequency is known as the *cyclotron frequency* and is given by

$$\omega_c = \frac{Bq}{m}.$$ (5.124)

From (5.123) and (5.124), we note that the relativistic gyration frequency is related to the non-relativistic cyclotron frequency in the following manner:

$$\omega_{c,r} = \frac{\omega_c}{\gamma}.$$ (5.125)

The interesting point to note is that the non-relativistic cyclotron frequency of a particle given by (5.124) is independent of its energy. However, if the particle is relativistic, then its frequency given by (5.125) decreases (i.e. its period increases) with increasing energy. If we keep in mind that a more energetic particle would traverse a circular path of larger radius, then this is indeed a result which we would expect.

5.12   Action Principle Formulation of Electrodynamics

One of the most remarkable facts about physics is that many of its dynamical laws can be obtained by extremizing some suitably defined action. We assume the readers to be familiar with the action principle in classical mechanics (also called *Hamilton's principle*). Let L be the Lagrangian of a system. Suppose the system evolves from an initial configuration at time t_i to a final configuration at time t_f. There are many possible paths the system could follow during its evolution from the initial configuration to the final configuration. Which particular path does the system follow? According to the action principle, the system follows the path which makes the action

$$S = \int_{t_i}^{t_f} L \, dt$$ (5.126)

an extremum. The dynamical equations which follow from this action principle turn out to be identical with the equations which we obtain from the usual laws of mechanics (i.e. Newton's second law for a system of particles). Since this subject is discussed in any standard textbook on advanced classical mechanics (see, for example, Goldstein [6], Chap. 2; Landau and Lifshitz [5], Chap. 1), we shall not provide any more explanation of this. However, we point out that a basic knowledge of Lagrangian mechanics is a prerequisite for reading this section.

   We now want to show that an action principle formulation for electrodynamics is possible. This is not merely a matter of intellectual curiosity. We shall see that this formulation provides a deep insight into the structure of electrodynamics and also gives us an indication of how electromagnetic fields may be handled in quantum mechanics.

   We have shown in Sects. 5.9–5.11 that the laws of electrodynamics hold in all inertial frames. The action principle for electrodynamics has to be formulated in such a manner that we shall get the same dynamics in any inertial frame. A requirement for this is that the action has to be a Lorentz-invariant scalar. If this is not the case, then the actions for various possible paths will appear different to observers in different inertial frames. If one particular observer finds that the action is extremized for a particular path, there would be no guarantee that observers in other frames also would conclude that the action is indeed extremized for this particular path. Only if the action is a Lorentz-invariant scalar, would we expect all the observers to find that the action is extremized for the same path even when observed from different frames.

How to make an action formulation of the dynamics of a charged particle in an electromagnetic field governed by (5.121) will be discussed in Sect. 5.12.1. Then we shall show in Sect. 5.12.2 that even Maxwell's equations follow from a suitable action principle.

5.12.1   Charged Particle in an Electromagnetic Field

Let us begin our discussion by first figuring out the action of a freely moving particle. As we pointed out, the action has to be Lorentz-invariant. We are going to argue that the appropriate action for a free particle of mass m is

$$S_m = - mc^2 \int d\tau,$$ (5.127)

where $d\tau$ is the proper time interval and is a Lorentz scalar. In the non-relativistic limit, it follows from (5.29) that

$$d\tau \approx \left(1 - \frac{1}{2}\frac{v^2}{c^2}\right) dt.$$

Substituting this in (5.127) and integrating from t_i to t_f, we have

$$S_m \approx -mc^2(t_f - t_i) + \int_{t_i}^{t_f} \frac{1}{2}mv^2 \, dt.$$ (5.128)

We do not need to consider the constant first term when finding the extremum of this action. The second term, which is the crucial term, is the time integral of the kinetic energy of the particle. Now, the Lagrangian of a freely moving particle in non-relativistic mechanics is given by its kinetic energy so that the definition (5.126) of action implies that the action of such a free particle is given by a time integral of the kinetic energy. In other words, the second term in the RHS of (5.128) is nothing but the usual non-relativistic action of a free particle. Since the action defined in (5.127) is Lorentz-invariant and gives the correct non-relativistic limit (apart from an unimportant constant), we are justified in taking (5.127) as the relativistic action of a free particle.

When the particle is moving in a field, we expect an additional term in the action resulting from the interaction between the particle and the field. Let us try to write down the simplest kind of Lorentz-invariant interaction term we can think of. We shall find that this simplest possible interaction term is going to be the *actual* interaction term! We expect the Lorentz-invariant interaction term to depend on some variable of the field and the variable of the particle motion. A Lorentz vector A^μ may be the simplest type of a field variable. If x^μ and $x^\mu + dx^\mu$ are the spacetime coordinates of the particle at the beginning and at the end of an interval, then clearly dx^μ is a variable associated with the motion of the particle. The simplest Lorentz scalar we can construct from A^μ and dx^μ is $A_\mu dx^\mu$. We are going to argue that the integral of this, multiplied by q/c, turns out to be our appropriate interaction term in the action

$$S_{int} = \frac{q}{c} \int A_\mu \, dx^\mu. \tag{5.129}$$

We shall now show that the combination of S_m and S_{int} given by (5.127) and (5.129) leading to the action

$$S = S_m + S_{int} = -mc^2 \int d\tau + \frac{q}{c} \int A_\mu \, dx^\mu \tag{5.130}$$

results in the dynamical equation (5.121) of a charged particle in an electromagnetic field, when the field variable A^μ is identified with the four-potential defined by (5.82).

It follows from (5.30), (5.63) and (5.65) that

$$dx^\mu = d\tau \, V^\mu = \frac{dt}{\gamma} \, V^\mu = dt \begin{pmatrix} c \\ \mathbf{v} \end{pmatrix}.$$

Obtaining A_μ from (5.82) with (5.56), we have

$$A_\mu dx^\mu = (-c \, \Phi + c \, \mathbf{A} \cdot \mathbf{v}) \, dt. \tag{5.131}$$

Substituting this in (5.130) and using (5.30), we get

$$S = \int \left[-\frac{mc^2}{\gamma} - q \, \Phi + q \, \mathbf{A} \cdot \mathbf{v} \right] dt. \tag{5.132}$$

If this is the action of a charged particle in an electromagnetic field, then a comparison with (5.126) suggests that we can take

$$L = -mc^2 \sqrt{1 - \frac{v^2}{c^2}} - q\Phi + q \, \mathbf{A} \cdot \mathbf{v} \tag{5.133}$$

to be the Lagrangian of the system.

According to the standard principles of the calculus of variations, the requirement
that the action (5.130)—or equivalently (5.132)—is extremized leads to the conclusion that the Lagrangian L defined in (5.133) should satisfy Lagrange's equation.
Now, Lagrange's equation connected with the x direction is

$$\frac{d}{dt}\left(\frac{\partial L}{\partial v_x}\right) - \frac{\partial L}{\partial x} = 0. \tag{5.134}$$

From (5.133), we have

$$\frac{\partial L}{\partial v_x} = -mc^2 \frac{\partial}{\partial v_x}\left(\sqrt{1 - \frac{v^2}{c^2}}\right) + qA_x = \gamma m v_x + qA_x. \tag{5.135}$$

On substituting this in (5.134), we get

$$\frac{d}{dt}(\gamma m v_x + qA_x) = \frac{\partial L}{\partial x}.$$

We can write down similar equations for the y and z directions. Combining them
together, we arrive at the vector equation

$$\frac{d}{dt}(\gamma m \mathbf{v} + q\mathbf{A}) = \nabla L. \tag{5.136}$$

To proceed further, we need to evaluate ∇L. Note that only Φ and $\mathbf{A}$ in (5.133) are
field variables of which we have to consider spatial derivatives. From (5.133), we
get

$$\nabla L = -q\nabla\Phi + q\nabla(\mathbf{A}.\mathbf{v}).$$

Using the vector identity (B.6) and noting that $\mathbf{v}$ is not a field variable (i.e. we do not
have to consider the spatial derivatives of $\mathbf{v}$), we get

$$\nabla L = -q\,\nabla\Phi + q\,[(\mathbf{v}.\nabla)\mathbf{A} + \mathbf{v}\times(\nabla\times\mathbf{A})].$$

Substituting this in (5.136), we get

$$\frac{d}{dt}(\gamma m\mathbf{v}) = -q\,\nabla\Phi + q\left[-\frac{d\mathbf{A}}{dt} + (\mathbf{v}.\nabla)\mathbf{A}\right] + q\,\mathbf{v}\times(\nabla\times\mathbf{A}). \tag{5.137}$$

The last step in our derivation involves figuring out the significance of the term
$d\mathbf{A}/dt$ in the above equation. This is basically the rate at which $\mathbf{A}$ changes at the position of the moving particle. This type of time derivative is often called the *Lagrangian
time derivative*. For any arbitrary field variable $Q(\mathbf{x}, t)$, the Lagrangian derivative can
be obtained in the following manner. If $\mathbf{x}$ and $\mathbf{x} + \mathbf{v}\,\delta t$ are the positions of the particle
at times t and $t + \delta t$, then the Lagrangian time derivative of $Q(\mathbf{x}, t)$ is given by

$$\frac{dQ}{dt} = \lim_{\delta t \to 0} \frac{Q(\mathbf{x} + \mathbf{v}\,\delta t, t + \delta t) - Q(\mathbf{x}, t)}{\delta t}. \tag{5.138}$$

Keeping the first-order terms in the Taylor expansion, we have

$$Q(\mathbf{x} + \mathbf{v}\,\delta t, t + \delta t) = Q(\mathbf{x}, t) + \delta t \frac{\partial Q}{\partial t} + \delta t\, \mathbf{v}.\nabla Q.$$

Putting this in (5.138), we get

$$\frac{dQ}{dt} = \frac{\partial Q}{\partial t} + \mathbf{v}.\nabla Q. \tag{5.139}$$

Note that $\partial Q/\partial t$ here is the time derivative at a fixed position and is often called the *Eulerian time derivative*. The important relationship between the Lagrangian and Eulerian time derivatives is given by (5.139). On the basis of (5.139), we expect

$$\frac{d\mathbf{A}}{dt} = \frac{\partial \mathbf{A}}{\partial t} + \mathbf{v}.\nabla \mathbf{A}.$$

Substituting from this in (5.137), we get

$$\frac{d}{dt}(\gamma m \mathbf{v}) = q\left(-\nabla \Phi - \frac{\partial \mathbf{A}}{\partial t}\right) + q\,\mathbf{v} \times (\nabla \times \mathbf{A}). \tag{5.140}$$

Introducing $\mathbf{B}$ and $\mathbf{E}$ as given by (4.128) and (4.130), we are finally led to the Eq. (5.121) giving the dynamics of a charged particle in an electromagnetic field. This completes our proof that the extremization of the action (5.130) gives us the dynamical equation of a charged particle in an electromagnetic field.

5.12.2 Dynamics of the Electromagnetic Field

So far, we have not considered any term corresponding only to the electromagnetic field in the action. While considering the dynamics of a charged particle in a specified electromagnetic field, such a term is not needed. But now we want to consider the dynamics of the electromagnetic field itself, for which a term corresponding to the electromagnetic field is needed in the action. Let us determine the kind of term we have to include in the action. We may expect the electromagnetic field to have some kind of Lagrangian density $\mathcal{L}$, so that the Lagrangian of the field is $\int \mathcal{L}\,dV$, and the action as given by (5.126) would be

$$S_f = \int \mathcal{L}\,dV\,dt = \int \mathcal{L}\,d^4x. \tag{5.141}$$

To proceed further, we should note that $dV dt$, which is often written as d^4x and is a volume element of the four-dimensional spacetime, behaves as a scalar under Lorentz transformation. To show this, let us consider a frame moving in the x direction. Certainly dy and dz would not change, as indicated in (5.16) and (5.17). We would have

$$dx\, dt = \begin{vmatrix} \left(\frac{\partial x}{\partial x'}\right)_{t'} & \left(\frac{\partial t}{\partial x'}\right)_{t'} \\ \left(\frac{\partial x}{\partial t'}\right)_{x'} & \left(\frac{\partial t}{\partial t'}\right)_{x'} \end{vmatrix} dx'\, dt'.$$

Since we find from (5.15) and (5.18) that

$$\begin{vmatrix} \left(\frac{\partial x}{\partial x'}\right)_{t'} & \left(\frac{\partial t}{\partial x'}\right)_{t'} \\ \left(\frac{\partial x}{\partial t'}\right)_{x'} & \left(\frac{\partial t}{\partial t'}\right)_{x'} \end{vmatrix} = \begin{vmatrix} \gamma & \gamma\frac{v}{c^2} \\ \gamma v & \gamma \end{vmatrix} = 1,$$

we conclude

$$dx\, dt = dx'\, dt', \tag{5.142}$$

suggesting that

$$d^4x = d^4x'. \tag{5.143}$$

We have argued in the beginning of Sect. 5.12 that any part of the action of a relativistically valid theory has to be a Lorentz-invariant scalar. Since we have shown that d^4x is a Lorentz-invariant scalar, it follows from (5.141) that the Lagrangian density $\mathcal{L}$ of the field also has to be a Lorentz-invariant scalar. Our job now is to determine the appropriate $\mathcal{L}$ of the electromagnetic field.

The dynamical Eq. (5.121) of the charged particle, which has been shown to follow from the action principle in Sect. 5.12.1, is the space-like part of the full relativistic equation (5.119). It is clear from (5.119) that the dynamics of the particle is determined by field variables like $F_{\mu\nu}$, as defined in (5.87) and (5.88), rather than the four-potential A_μ which is not unique due to the gauge freedom. We, therefore, expect the Lagrangian density $\mathcal{L}$ of the electromagnetic field to involve the tensor $F_{\mu\nu}$. We also have argued that $\mathcal{L}$ has to be a Lorentz-invariant scalar. The simplest kind of scalar we can construct from $F_{\mu\nu}$ is $F_{\mu\nu}F^{\mu\nu}$. We may be tempted to guess that the Lagrangian density $\mathcal{L}$ should be given by this, apart from some constant of proportionality. We shall now show that we can derive Maxwell's equations with source terms if we take

$$\mathcal{L} = -\frac{\epsilon_0}{4c} F_{\mu\nu} F^{\mu\nu}.$$

Then, according to (5.141), the part of the action due to the electromagnetic field becomes

$$S_f = -\frac{\epsilon_0}{4c} \int F_{\mu\nu} F^{\mu\nu} d^4x. \tag{5.144}$$

Adding this term to the action as given in (5.130), the full action becomes

$$S = S_m + S_{int} + S_f. \tag{5.145}$$

Substituting from (5.127), (5.129) and (5.144), we get

$$S = -mc^2 \int d\tau + \frac{q}{c} \int A_\mu dx^\mu - \frac{\epsilon_0}{4c} \int F_{\mu\nu} F^{\mu\nu} d^4x. \tag{5.146}$$

We know that two of Maxwell's equations (1.2) and (1.4) are automatically satisfied if the electromagnetic fields are introduced through the potentials as given in (4.128) and (4.130)—or their relativistic equivalent (5.87). So we do not have to do anything special for these two Maxwell equations. Our job now is to show that the other two Maxwell equations (1.1) and (1.3) with source terms can be obtained by making the action given by (5.146) an extremum.

Before proceeding to the mathematical analysis, we have to do one last thing. The interaction term as given by (5.129) pertains to a moving charge particle. We can think of such a moving charged particle to be equivalent to a charge density and current density in the local region. We now want to write down the interaction term (5.129) using the charge and current densities. Note that q is obtained by integrating ρ over the region where ρ is non-zero. So we can write (5.129) as

$$S_{int} = \frac{1}{c} \int \rho \, dV \int A_\mu \, dx^\mu.$$

Keep in mind that ρ and dV will both change while transforming from one inertial frame to another, but their product $\rho \, dV$ will be invariant to make sure that we get the same amount of charge in different frames. We now write the above equation in the form

$$S_{int} = \frac{1}{c} \int \rho A_\mu \frac{dx^\mu}{dt} dV \, dt = \frac{1}{c} \int \rho A_\mu \frac{dx^\mu}{dt} d^4x. \tag{5.147}$$

Since we know that S_{int} is a Lorentz-invariant scalar and so is d^4x, we conclude that

$$\rho \frac{dx^\mu}{dt} = j^\mu \tag{5.148}$$

has to be a Lorentz four-vector to make (5.147) consistent. We now give the following simple argument to show that j^μ introduced in (5.148) is the four-current arising out of the moving charge. Let ρ_0 be the charge density in the frame in which the charge is at rest. We know that ρ_0 for a charged particle will be given by a delta function as in (2.4). For the charge at rest, ρ_0 is the 0-th component of the four-current given by (5.80), whereas the space-like component is zero. It is easy to see that the charge density in another frame with Lorentz factor γ would be $\gamma\rho_0$. Substituting this for ρ in (5.148) and using (5.30), we get

$$j^\mu = \rho_0 \frac{dx^\mu}{d\tau}. \tag{5.149}$$

The space-like component of this four-vector in the non-relativistic limit is $\rho_0\mathbf{v}$, which is the standard non-relativistic expression of the current density arising out of the charge density ρ_0 moving with velocity $\mathbf{v}$. Clearly, j^μ given by (5.149) is the relativistic expression of the four-current arising out of the moving charge density. Using (5.148), which is equivalent to (5.149), we write (5.147) as

$$S_{\text{int}} = \frac{1}{c} \int j^\mu A_\mu \, d^4x. \tag{5.150}$$

Using this expression for S_{int} in (5.145), the full action given by (5.146) now becomes

$$S = -mc^2 \int d\tau + \frac{1}{c} \int j^\mu A_\mu \, d^4x - \frac{\epsilon_0}{4c} \int F_{\mu\nu} F^{\mu\nu} \, d^4x. \tag{5.151}$$

To find the dynamics of the electromagnetic field, we now have to consider the variation δS of the total action given by (5.151) and set it to zero to get the extremum. While considering the dynamics of the field, we have to take into account the variations of the field variables, while the particle variables can be held fixed. This means that we do not have to consider variations in the first term in the RHS of (5.151) or variations in j^μ in the second term. Proceeding in this manner, we conclude that

$$\delta S = \frac{1}{c} \int j^\mu \delta A_\mu \, d^4x - \frac{\epsilon_0}{2c} \int F^{\mu\nu} \delta F_{\mu\nu} \, d^4x. \tag{5.152}$$

We have used the fact that $F_{\mu\nu} \delta F^{\mu\nu}$ and $F^{\mu\nu} \delta F_{\mu\nu}$ can easily be shown to be equal. To simplify the second term in the RHS of (5.152), we use (5.87) to obtain

$$\int F^{\mu\nu} \delta F_{\mu\nu} \, d^4x = \int F^{\mu\nu} \delta \left(\frac{\partial A_\nu}{\partial x^\mu} - \frac{\partial A_\mu}{\partial x^\nu} \right) d^4x. \tag{5.153}$$

The first term in the RHS of (5.153) can be written as

$$\int F^{\mu\nu} \frac{\partial}{\partial x^\mu} (\delta A_\nu) \, d^4x = \int \frac{\partial}{\partial x^\mu} (F^{\mu\nu} \delta A_\nu) \, d^4x - \int \delta A_\nu \frac{\partial F^{\mu\nu}}{\partial x^\mu} \, d^4x. \tag{5.154}$$

The first term in the RHS is the integral of a four-dimensional divergence over the four-dimensional spacetime volume. Just as we can use Gauss's theorem (B.13) to convert a similar three-dimensional volume integral into a surface integral, this term also can be converted into an integral of $F^{\mu\nu} \delta A_\nu$ over the three-dimensional surface enclosing the four-dimensional spacetime volume. If j^μ and the fields arising out of it are confined in localized regions, then we can argue that this surface integral taken at a large distance goes to zero. So we are left with only the second term in the RHS of (5.154). We leave it for the reader to show that the other term in the RHS of (5.153) is also exactly equal to this (keep in mind that $F_{\mu\nu}$ is an anti-symmetric tensor). Then (5.153) leads to

$$\int F^{\mu\nu}\,\delta F_{\mu\nu}\,d^4x = -2\int \delta A_\nu\,\frac{\partial F^{\mu\nu}}{\partial x^\mu}\,d^4x = 2\int \delta A_\mu\,\frac{\partial F^{\mu\nu}}{\partial x^\nu}\,d^4x,$$

where we have interchanged the dummy variables μ and ν and have used the fact $F^{\mu\nu} = -F^{\nu\mu}$. Substituting this in (5.152), we get

$$\delta S = \frac{1}{c}\int \left(j^\mu - \epsilon_0\frac{\partial F^{\mu\nu}}{\partial x^\nu} \right)\delta A_\mu\,d^4x.$$

The condition for the extremum is $\delta S = 0$, which suggests

$$\frac{\partial F^{\mu\nu}}{\partial x^\mu} = \frac{j^\mu}{\epsilon_0}.$$

This is the same as (5.91), which is the relativistic combination of Maxwell's equations (1.1) and (1.3) with source terms.

To summarize, when we consider a charged particle in an electromagnetic field, the action has one term corresponding to the particle, one term corresponding to the field and one term corresponding to the interaction between the particle and the field, as we see in (5.145–5.146). By varying the particle variables while holding the field variables fixed, we can get the equation of motion of the particle given by (5.140). On the other hand, by varying the field variables while holding the particle variables fixed, we can get (5.91) which is the relativistic combination of Maxwell's equations with sources, whereas (5.90) combining the other Maxwell's equations is automatically satisfied due to the structure of the theory.

5.12.3   Some General Remarks

We can imagine a hypothetical brilliant theoretical physicist who has a mastery over all other areas of physics except electromagnetism. She somehow never had a chance to study electromagnetism. She would know that the action of a free particle would be given by (5.127). Suppose this brilliant theoretical physicist wants to determine whether particles can interact in a manner which is consistent with the principles of special relativity. Then she has to include a Lorentz-invariant interaction term in the action. For a moving particle, the simplest interaction term she could presumably think of is (5.129). If she includes such a term in the action, then she would be led to the equation of motion (5.140). Suppose then she wants to determine how the interaction field may arise from the particles—which would be necessary to understand how particles interact with each other (i.e. the particles themselves should produce the interaction field which acts on the particles). The brilliant theoretical physicist would note that A_μ appears in the equation of motion (5.140) in a manner which involves only $F_{\mu\nu}$ defined in (5.87). Identifying $F_{\mu\nu}$ as the appropriate field acting on the particles, this brilliant theoretical physicist would want to add a term in the action involving $F_{\mu\nu}$ that would be Lorentz-invariant. She is likely to figure out that the appropriate term would be of the form (5.144), finally leading to Maxwell's equations. Thus, this brilliant theoretical physicist without any prior knowledge of electromagnetism may arrive at all the basic principles of electromagnetism merely by trying to determine the simplest kind of interaction consistent with special relativity that particles may have among themselves. Landau and Lifshitz's *The Classical Theory of Fields* [2], perhaps the most beautiful textbook on electromagnetic theory ever written, develops the subject essentially by giving the arguments which we are ascribing to this hypothetical brilliant theoretical physicist.

When we first learn Maxwell's equations, we may think that Mother Nature just happens to obey these equations. We may wonder whether electromagnetism could possibly be governed by somewhat different equations. However, the action principle formulation of electromagnetism makes us feel that there is something inevitable and inexorable about the basic laws of electromagnetism. We realize that electromagnetism is the simplest possible theory of interaction among particles that is consistent with special relativity. We should also point out that these fundamental considerations do not lead to the requirement that there have to be two kinds of charges. We know that like charges repel each other. However, if we had taken a positive sign in (5.144) instead of a negative sign, then we would end up with the result that like charges attract each other. If we had only kind of charge—with particles either repelling or attracting each other—a little reflection would show that the Universe in which we find ourselves would not be possible. Perhaps our hypothetical brilliant theoretical physicist will also find that there must be two kinds of charges to make the Universe as we know it.

We are all familiar with one other long-range interaction: gravitational interaction. Since there is only one kind of mass in contrast to two kinds of charge, we often have the mistaken notion that the gravitational interaction must be simpler than the electromagnetic interaction. When we try to determine the simplest kind of possible interaction among particles, readers may wonder why we arrive at the electromagnetic interaction rather than the gravitational interaction. A short answer to the question is that, in reality, the gravitational interaction is much more complicated. It is not difficult to see that Newton's law of gravitation is not consistent with special relativity. Suppose we have two particles m_1 and m_2 at rest. The gravitational force between them can be found from Newton's law of gravitation. Now, suppose m_1 starts moving and we want to determine the force exerted by it on m_2. The instantaneous separation between two particles in relative motion is not something that can be readily defined within the framework of special relativity. Still, if we somehow define it in a suitable manner, Newton's law of gravitation would lead to the conclusion that the force on m_2 will change as soon as m_1 starts moving, implying an infinite propagation speed of the gravitational interaction, in contradiction with special relativity. It is clear that Newton's law of gravitation is a non-relativistic approximation. One has to generalize it suitably to handle relativistic situations. Einstein's efforts in doing this finally led to the general theory of relativity—the complete classical theory of gravitational interaction—which is much more complicated than the theory of electromagnetic interaction.

5.13 The Non-relativistic Hamiltonian for a Charged Particle in an Electromagnetic Field

One important topic in non-relativistic quantum mechanics is to study the quantum dynamics of a charged particle in an electromagnetic field (see, for example, Merzbacher [3], pp. 71–75; Landau and Lifshitz [4], Chap. XV). If we know the classical Hamiltonian $H(\mathbf{x}, \mathbf{p})$ for a system, the standard procedure in non-relativistic quantum mechanics is to find the wave function ψ from Schrodinger's equation written in the form

$$i\,\hbar\,\frac{\partial\psi}{\partial t} = H(\mathbf{x}, -i\,\hbar\,\nabla)\psi \qquad (5.155)$$

(see, for example, Landau and Lifshitz [5], Sect. 17). So the first step for us should be to find out the classical non-relativistic Hamiltonian for a charged particle in an electromagnetic field. We now point out how the appropriate Hamiltonian can be found from the Lagrangian given by (5.133). Note that we do not have to consider the term S_f in (5.145) because we are not interested in the dynamics of the electromagnetic field (this term becomes important in quantum field theory).

The standard procedure for obtaining the Hamiltonian from the Lagrangian (often called the *Legendre transformation*) is described in textbooks of classical mechanics (Goldstein [6], Sect. 8–1; Landau and Lifshitz [5], Sect. 40). We assume the reader to be familiar with this procedure and point out only the main steps without trying to explain the procedure in detail. In the non-relativistic limit, the Lagrangian given by (5.133) becomes

$$L = \frac{1}{2}\,m\,v^2 - q\,\Phi + q\,\mathbf{A}\,.\mathbf{v}, \qquad (5.156)$$

where we have not included the constant term $-mc^2$ because it plays no role when we differentiate the Lagrangian to study the dynamics. The standard procedure for obtaining the i-th component of the momentum $\mathbf{p}$ is

$$p_i = \frac{\partial L}{\partial v_i} = m\,v_i + q\,A_i$$

on using (5.156), the vectorial form of the equation being

$$\mathbf{p} = m\,\mathbf{v} + q\,\mathbf{A}. \qquad (5.157)$$

Once we have the expression of the momentum, the Hamiltonian is given by

$$H = \sum_i p_i v_i - L = \mathbf{p}\,.\mathbf{v} - L.$$

On substituting for L from (5.156) and for $\mathbf{p}$ from (5.157), we obtain

$$H = \frac{1}{2} m\, v^2 + q\, \Phi.$$

Finally, the standard convention is to write H as a function of $\mathbf{p}$ rather than $\mathbf{v}$. This can be done by eliminating $\mathbf{v}$ in favour of $\mathbf{p}$ by using (5.157). Then we are led to

$$H = \frac{1}{2m}\, (\mathbf{p} - q\mathbf{A})^2 + q\, \Phi. \tag{5.158}$$

This is the Hamiltonian which has to be used in Schrodinger's equation (5.155) with the operator $-i\,\hbar\,\nabla$ for momentum replacing $\mathbf{p}$. This gives

$$i\,\hbar\frac{\partial\psi}{\partial t} = \frac{1}{2m}\, (-i\hbar\,\nabla - q\mathbf{A})^2\psi + q\, \Phi\,\psi. \tag{5.159}$$

This is the equation which the wave function ψ of a non-relativist charged particle in an electromagnetic field has to satisfy.

We now point out a very curious fact. Unlike the classical equation (5.116) for the dynamics of a charged particle in which the electromagnetic fields $\mathbf{E}$ and $\mathbf{B}$ appear, the quantum equation (5.159) for the dynamics involves the potentials Φ and $\mathbf{A}$ explicitly. We remind the reader of the discussion in Sect. 3.4.2 whether the vector potential $\mathbf{A}$ has any observable consequences apart from the consequences due to the magnetic field $\mathbf{B}$. We concluded that, in classical physics, the vector potential is merely a mathematical trick to calculate the magnetic field and does not have any independent reality. Interestingly, we find that the quantum equation (5.159) involves $\mathbf{A}$ rather than $\mathbf{B}$. This suggests that the vector potential $\mathbf{A}$ may give rise to quantum phenomena which do not have any classical analogues. It is beyond the scope of the present book to discuss this provocative and deep question any further, although readers working out Exercise 5.11 will become aware of some of the important issues.

Exercises

5.1 A muon is produced 3 km about the surface of the Earth due to the collision of a cosmic ray particle with a particle in the atmosphere and moves downward. If the mean lifetime of the muon in its rest frame is 2.2×10^{-6} s, estimate the speed with which the muon has to move if it has to reach the surface of the Earth before its decay.

5.2 Show that the relativistic transformation law for accelerations is

$$a_x = \frac{a'_x}{\gamma^3\sigma^3},$$

$$a_y = \frac{a'_y}{\gamma^2\sigma^2} - \frac{u'_y v}{c^2}\frac{a'_x}{\gamma^2\sigma^3},$$

$$a_z = \frac{a_z'}{\gamma^2 \sigma^2} - \frac{u_z' v}{c^2} \frac{a_x'}{\gamma^2 \sigma^3},$$

where

$$\sigma = 1 + \frac{v u_x'}{c^2},$$

and the other symbols have usual meanings.

If the primed frame happens to be the instantaneous rest frame of an accelerated particle, then show that the acceleration of the particle satisfies

$$a_\parallel' = \gamma^3 a_\parallel, \quad a_\perp' = \gamma^2 a_\perp.$$

5.3 Consider rotations in a two-dimensional Cartesian system. Show that

$$A_{ij} = \begin{pmatrix} y^2 & -xy \\ -xy & x^2 \end{pmatrix}, \quad B_{ij} = \begin{pmatrix} -xy & -x^2 \\ -y^2 & xy \end{pmatrix}$$

transform as tensors, but

$$C_{ij} = \begin{pmatrix} y^2 & xy \\ xy & x^2 \end{pmatrix}, \quad D_{ij} = \begin{pmatrix} xy & y^2 \\ x^2 & -xy \end{pmatrix}$$

do not transform as tensors.

5.4 Consider a moving proton colliding with a proton at rest. If the moving proton is sufficiently energetic, then this collision can give rise to an additional proton–antiproton pair in the following way:

$$P + P \rightarrow P + P + P + \overline{P}.$$

Show that the threshold energy of the moving proton for this reaction to take place is $7Mc^2$, where M is the mass of a proton.

5.5 From the transformation law (5.97–5.100) of electromagnetic fields between inertial frames, show that $\mathbf{E} \cdot \mathbf{B}$ and $E^2 - c^2 B^2$ remain invariant between frames.

5.6 Although the charge density changes on transformation from one frame to another, argue that the total charge of an object remains invariant under Lorentz transformation.

5.7 Consider a straight infinite wire carrying current I as observed in the frame K in which it is at rest. Now, consider another frame K' moving with velocity v parallel to the direction of the current.

(a) Show that in the frame K' there would be a charge density Q' per unit length associated with the wire. Calculate the electric field $\mathbf{E}'$ produced by this charge density in the frame K'.

(b) There would be a magnetic field $\mathbf{B}$ in the frame K due to the current-carrying wire. By the Lorentz transformation of this magnetic field, find the electric field $\mathbf{E}'$ in the frame K' and show that it is the same as what you get in part (a) of this Exercise.

5.8 Suppose a charged particle is moving with velocity $\mathbf{v} = v\mathbf{e}_x$ with respect to our frame K'. In the frame K in which the particle is at rest, the electrostatic potential Φ is clearly given by

$$\Phi = \frac{1}{4\pi\epsilon_0} \frac{q}{[x^2 + y^2 + z^2]^{1/2}}.$$

Use the Lorentz transformation formula to write down the potentials Φ' and $\mathbf{A}'$ in our frame K'. From these potentials, calculate the electric field $\mathbf{E}'$ in our frame by using (4.130). Check whether you find the same expression of the electric field as (5.108–5.110). Calculate also the magnetic field and show that it is given by (5.113).

5.9 The thickness of our galaxy is about 100 parsecs (1 parsec $\approx 3 \times 10^{16}$ m), and it has a magnetic field of about 10^{-10} T. Determine the critical energy of a relativistic electron in eV such that electrons with higher energies would not remain confined in the galaxy.

5.10 Let ϕ be a scalar field having a local, Lorentz-invariant Lagrangian density

$$\mathcal{L} = \partial_\mu \phi \, \partial^\mu \phi + m^2 \phi^2.$$

Show that it satisfies the dynamical equation

$$\left(\frac{1}{c^2} \frac{\partial^2}{\partial t^2} - \nabla^2 \right) \phi + m^2 \phi = 0.$$

5.11 (a) Consider the quantum mechanical wave function of a particle with charge q in a magnetic field given by the vector potential $\mathbf{A}$ (i.e. assume the electric field and the scalar potential Φ to be zero). Suppose $\psi_0(\mathbf{x})$ is the time-dependent wave function when the magnetic field is zero. When the magnetic field is given by $\mathbf{A}$, argue on the basis of (5.159) that the wave function along a path is given by

$$\psi(\mathbf{x}) = \psi_0(\mathbf{x}) \, e^{i\frac{q}{\hbar} \int \mathbf{A}.d\mathbf{x}}.$$

(b) Suppose the charged particle q can pass through two symmetric paths above and below a solenoid S to reach the point P, as shown in Fig. 5.5. If we have magnetic flux Φ through the solenoid, argue that there would be a phase difference between the two paths of the charged particle given by

$$\frac{q}{\hbar} \Phi,$$

the phase difference becoming zero when the current producing the magnetic field in the solenoid is switched off (Fig. 5.5).

Fig. 5.5 The circular cross section of a solenoid S having an internal magnetic flux perpendicular to this page, with two possible symmetric paths of a charged particle on the two sides of the solenoid. Refer to Exercise 5.11

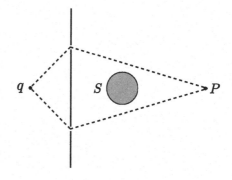

This is the celebrated *Aharonov–Bohm effect*, which has been experimentally studied in the laboratory. Even if the path of a particle passes only through regions where the magnetic field is zero, still the wave function may be affected by the presence of non-zero vector potential **A**. See the discussion in Sect. 3.4.2.

References

1. Lorentz, H.A., Einstein, A., Minkowski, H., Weyl, H.: The Principle of Relativity. Methuen and Co. (Reprinted by Dover) (1923)
2. Landau, L.D., Lifshitz, E.M.: The Classical Theory of Fields: Volume 2 (Course of Theoretical Physics), 4th edn. Butterworth-Heinemann (1980)
3. Merzbacher, E.: Quantum Mechanics, 3rd edn. Wiley (1998)
4. Landau, L.D., Lifshitz, E.M. Quantum Mechanics: Volume 3 (Course of Theoretical Physics), 3rd edn. Butterworth-Heinemann (1981)
5. Landau, L.D., & Lifshitz, E.M.: Mechanics: Volume 1 (Course of Theoretical Physics), 3rd edn. Butterworth-Heinemann (1976)
6. Goldstein, H.: Classical Mechanics, 2nd edn. Addison-Wesley (1980)
7. Kleppner, D., Kolenkow, R.J.: An Introduction to Mechanics. McGraw-Hill (1973)

Chapter 6
Electromagnetic Fields of Time-Varying Sources

6.1 A Few Remarks on Inhomogeneous Equations

So far we have not addressed the question of finding electromagnetic fields due to time-varying charges and currents. Now we need to turn our attention to this problem—especially if we want to understand the process of the emission of electromagnetic radiation. For this purpose, we have to solve Maxwell's equations allowing for all the terms to be non-zero. Since this is a somewhat complicated subject, it is a good strategy to give the first introduction to this subject without the additional complication of dealing with the polarization of the medium. The treatment of the subject in this chapter and the next will be based on the form (1.1–1.4) of Maxwell's equations rather than the form (4.1–1.4). This means that we either consider the charges and the currents in free space, or we assume that ρ and $\mathbf{j}$ include the charges and the currents induced in the medium. We have pointed out in Sect. 4.10 that the problem of solving Maxwell's equations (1.1–1.4) can be reduced to solving the inhomogeneous wave equations (4.138) and (4.139) for the scalar and vector potentials, from which the electromagnetic fields can be obtained by using (4.128) and (4.130). Let us point out what we exactly mean by solving these inhomogeneous wave equations. In many situations, we may have $\rho(\mathbf{x}, t)$ and $\mathbf{j}(\mathbf{x}, t)$ given to us. We then may want to solve (4.138) and (4.139) to obtain $\Phi(\mathbf{x}, t)$ and $\mathbf{A}(\mathbf{x}, t)$ for the given $\rho(\mathbf{x}, t)$ and $\mathbf{j}(\mathbf{x}, t)$. A standard technique of solving such inhomogeneous equations is to use *Green's function*. We now make a few comments about this technique.

Suppose we know a function $\chi(\mathbf{x}, t)$ and want to determine the function $\psi(\mathbf{x}, t)$ which satisfies the equation

$$\mathcal{L}\,\psi = \chi, \tag{6.1}$$

where $\mathcal{L}$ is a linear differential operator involving derivatives with respect to spatial and temporal coordinates. If χ is zero, then (6.1) would be a linear homogeneous equation for ψ. The presence of non-zero χ makes the equation inhomogeneous. The first step in solving for ψ from (6.1) is to obtain Green's function for the operator $\mathcal{L}$ that satisfies

© Springer Nature Singapore Pte Ltd. 2022
A. R. Choudhuri, *Advanced Electromagnetic Theory*, Lecture Notes in Physics 1009,
https://doi.org/10.1007/978-981-19-5944-8_6

$$\mathcal{L} G(\mathbf{x}, t; \mathbf{x}', t') = \delta(\mathbf{x} - \mathbf{x}') \, \delta(t - t'). \tag{6.2}$$

Usually, it turns out that it is easier to solve (6.2) to obtain $G(\mathbf{x}, t; \mathbf{x}', t')$ than to solve (6.1) to obtain $\psi(\mathbf{x}, t)$. Once we have $G(\mathbf{x}, t; \mathbf{x}', t')$ and we know $\chi(\mathbf{x}, t)$, we can at once write down the solution of (6.1). We now show that the appropriate solution happens to be

$$\psi(\mathbf{x}, t) = \int \chi(\mathbf{x}', t') \, G(\mathbf{x}, t; \mathbf{x}', t') \, dV' \, dt'. \tag{6.3}$$

To show that $\psi(\mathbf{x}, t)$ given by (6.3) is the solution of (6.1), we substitute (6.3) into (6.1) and keep in mind that the linear operator $\mathcal{L}$ operates only on unprimed variables $\mathbf{x}, t$ and not on primed variables $\mathbf{x}', t'$. We obtain from (6.3)

$$\mathcal{L} \psi(\mathbf{x}, t) = \int \chi(\mathbf{x}', t') \, \mathcal{L} G(\mathbf{x}, t; \mathbf{x}', t') \, dV' \, dt'.$$

Substituting from (6.2), we then get

$$\mathcal{L} \psi(\mathbf{x}, t) = \int \chi(\mathbf{x}', t') \, \delta(\mathbf{x} - \mathbf{x}') \, \delta(t - t') \, dV' \, dt'. \tag{6.4}$$

It is straightforward to carry on these integrations over the δ functions, the volume integral being in accordance with (1.15). We readily see that (6.4) leads to (6.1), which completes our proof that (6.3) is the solution of (6.1). In order to solve an inhomogeneous equation like (6.1), we then have to proceed in the following way. We first determine Green's function for the operator $\mathcal{L}$ appearing in (6.1)—by solving (6.2) that this Green's function would satisfy. Once we have Green's function, we can write down the solution (6.3) of (6.1).

We may point out that, while discussing electrostatics, without using the name Green's function, we have made use of the technique outlined above. By comparing (1.16) with (6.2), we can at once conclude that

$$G(\mathbf{x}; \mathbf{x}') = -\frac{1}{4\pi} \frac{1}{|\mathbf{x} - \mathbf{x}'|} \tag{6.5}$$

is the three-dimensional Green's function for the operator ∇^2 satisfying

$$\nabla^2 G(\mathbf{x}, \mathbf{x}') = \delta(\mathbf{x} - \mathbf{x}'). \tag{6.6}$$

Following (6.3), we can write down the solution of Poisson's equation (2.10) as

$$\Phi(\mathbf{x}) = -\frac{1}{\epsilon_0} \int \rho(\mathbf{x}') \, G(\mathbf{x}; \mathbf{x}') \, dV. \tag{6.7}$$

Substituting from (6.5) in this, we would get (2.2) giving the electrostatic potential for a specified charge distribution. Now, as we have seen in Chap. 2, in electrostatics, we often solve boundary value problems with conductors. A specified charge distribution in space causes induced charges on the surfaces of the conductors—as we saw when we applied the method of images in Sect. 2.9 to solve the problem of a point charge in front of a plane or spherical conductor. If we include the induced charges on the surfaces of the conductors along with the prescribed charges, then the solution of the boundary value problem would be given by (2.2). However, it happens very often that we a priori do not know the induced charges, which also have to come out as a part of the solution. It is possible to construct Green's function in such a manner that (6.7) gives the solution when only the specified prescribed charges (and not the induced charges) are included in ρ. Exercise 6.1 should make this point clear. Readers interested in a more general discussion of Green's function approach in boundary value problems may turn to Jackson ([1], Sect. 1.10).

6.2 Solving the Inhomogeneous Wave Equation with the Green's Function Technique

The inhomogeneous wave Eqs. (4.138) and (4.139) are of the form

$$\left(\nabla^2 - \frac{1}{c^2}\frac{\partial^2}{\partial t^2}\right)\psi(\mathbf{x}, t) = -f(\mathbf{x}, t), \tag{6.8}$$

which is exactly like (6.1). Note that the vector equation (4.139) can be written as three scalar equations, each of which is like (6.8). As we discussed in the previous section, the first step for solving this equation will be to determine Green's function satisfying

$$\left(\nabla^2 - \frac{1}{c^2}\frac{\partial^2}{\partial t^2}\right)G(\mathbf{x}, t; \mathbf{x}', t') = \delta(\mathbf{x} - \mathbf{x}')\,\delta(t - t'). \tag{6.9}$$

Comparing (6.8) with (6.1) and (6.9) with (6.2), we are led from (6.3) to the conclusion that the solution of (6.8) should be given by

$$\psi(\mathbf{x}, t) = -\int f(\mathbf{x}', t')\,G(\mathbf{x}, t; \mathbf{x}', t')\,dV'\,dt', \tag{6.10}$$

where $G(\mathbf{x}, t; \mathbf{x}', t')$ is the solution of (6.9).

It should be clear from (6.3) that the primed variables $\mathbf{x}', t'$ are the coordinates of the source point, whereas the unprimed variables $\mathbf{x}, t$ are the coordinates of the field point. Since we tacitly assume symmetry with respect to spatial and temporal translations, we expect $G(\mathbf{x}, t; \mathbf{x}', t')$ to be a function of only the differences $\mathbf{x} - \mathbf{x}'$ and $t - t'$. Since

$$\delta(t - t') = \frac{1}{2\pi}\int_{-\infty}^{\infty} e^{-i\omega(t-t')}\,d\omega \tag{6.11}$$

is a well-known representation of the δ function, we are tempted to look for a Green's function of the form

$$G(\mathbf{x}, t; \mathbf{x}', t') = \frac{1}{2\pi} \int_{-\infty}^{\infty} G_k(\mathbf{x} - \mathbf{x}') \, e^{-i\omega(t-t')} \, d\omega. \tag{6.12}$$

As we shall see below, it is indeed possible to satisfy (6.9) with a solution of form (6.12). On substituting (6.11) and (6.12) in (6.9), we are led to the equation

$$(\nabla^2 + k^2)G_k(\mathbf{x} - \mathbf{x}') = \delta(\mathbf{x} - \mathbf{x}'), \tag{6.13}$$

where

$$k = \frac{\omega}{c}. \tag{6.14}$$

For the purpose of solving (6.13), let us take the origin of the coordinate system at the position of the source point so that $\mathbf{x} - \mathbf{x}'$ should correspond to (r, θ, ϕ) in spherical coordinates. Now, if we have spatial isotropy, then $G_k(\mathbf{x} - \mathbf{x}')$ should be independent of $\theta \, \phi$, which leads us to write $G_k(\mathbf{x} - \mathbf{x}')$ as $G_k(r)$. When the operator ∇^2 operates on $G_k(r)$, we do not have to keep the terms which involve derivatives with respect to θ or ϕ. Substituting from (2.37) for ∇^2 and not keeping the terms which involve derivatives with respect to θ or ϕ, we are led from (6.13) to the following equation for $G_k(r)$ in spherical coordinates

$$\frac{1}{r^2}\frac{d}{dr}\left(r^2\frac{dG_k}{dr}\right) + k^2 G_k = \delta(r). \tag{6.15}$$

To proceed further, we make use of the mathematical identity

$$\frac{1}{r^2}\frac{d}{dr}\left(r^2\frac{dR}{dr}\right) = \frac{1}{r}\frac{d^2}{dr^2}(rR) \tag{6.16}$$

for any arbitrary function $R(r)$. The easiest way of proving this identity is to expand both of the sides and to show that each side is equal to

$$\frac{2}{r}\frac{dR}{dr} + \frac{d^2R}{dr^2}.$$

Substituting from (6.16) into (6.15), we conclude that the equation

$$\frac{d^2}{dr^2}(rG_k) + k^2 rG_k = 0$$

must be satisfied at all point except at the origin $r = 0$. It is trivial to write down the solution of the above equation, which is

$$r\,G_k = C\,e^{\pm ikr},$$

where C is the constant of integration. As we have been writing r for $|\mathbf{x} - \mathbf{x}'|$, this solution can be put in the form

$$G_k(\mathbf{x} - \mathbf{x}') = C\,\frac{e^{\pm ik|\mathbf{x}-\mathbf{x}'|}}{|\mathbf{x} - \mathbf{x}'|}. \tag{6.17}$$

Note that this is the solution of (6.13), which we had put in the form (6.15), provided r is not zero, that is, provided $\mathbf{x} \neq \mathbf{x}'$. Since $G_k(\mathbf{x} - \mathbf{x}')$ given by (6.17) blows up at $\mathbf{x} = \mathbf{x}'$ where the δ function on the RHS of (6.15) also blows up, we have the possibility that (6.17) may be made to satisfy (6.15) even at $\mathbf{x} = \mathbf{x}'$ by choosing the constant of integration C suitably. We now explore this possibility.

We point out that (6.13) reduces (6.6) when $k \to 0$. Noting that (6.5) is the solution of (6.6), we conclude that the function $G_k(\mathbf{x} - \mathbf{x}')$ we are trying to obtain should reduce to (6.5) when $k \to 0$. In other words, we require a function $G_k(\mathbf{x} - \mathbf{x}')$ which has the form (6.17) at all points except $r = |\mathbf{x} - \mathbf{x}'| = 0$ and which should reduce to (6.5) when $k \to 0$. The appropriate function satisfying these requirements is

$$G_k(\mathbf{x} - \mathbf{x}') = -\frac{1}{4\pi}\frac{e^{\pm ik|\mathbf{x}-\mathbf{x}'|}}{|\mathbf{x} - \mathbf{x}'|}. \tag{6.18}$$

This is the solution of (6.13), which we now have to substitute in (6.12) to obtain the full Green's function. This gives

$$G(\mathbf{x}, t; \mathbf{x}', t') = -\frac{1}{2\pi}\int_{-\infty}^{\infty}\frac{1}{4\pi}\frac{e^{\pm i\frac{\omega}{c}|\mathbf{x}-\mathbf{x}'|}}{|\mathbf{x} - \mathbf{x}'|}\,e^{-i\omega(t-t')}\,d\omega.$$

Taking those quantities which do not depend on ω out of the integral sign, we have

$$G(\mathbf{x}, t; \mathbf{x}', t') = -\frac{1}{4\pi|\mathbf{x} - \mathbf{x}'|}\frac{1}{2\pi}\int_{-\infty}^{\infty}e^{-i\omega\left[(t-t')\mp\frac{1}{c}|\mathbf{x}-\mathbf{x}'|\right]}\,d\omega.$$

Noting the representation of the δ function as given in (6.11), we can put the above equation in the form

$$G(\mathbf{x}, t; \mathbf{x}', t') = -\frac{1}{4\pi|\mathbf{x} - \mathbf{x}'|}\,\delta\!\left(t' - \left[t \mp \frac{|\mathbf{x} - \mathbf{x}'|}{c}\right]\right). \tag{6.19}$$

This completes our derivation of Green's function that would satisfy (6.9).

The last step in our analysis is to write down the solution of our starting equation (6.8) by using (6.10), in which we now substitute Green's function as specified by (6.19). This gives

$$\psi(\mathbf{x}, t) = \frac{1}{4\pi} \int \frac{f(\mathbf{x}', t')}{|\mathbf{x} - \mathbf{x}'|} \delta\left(t' - \left[t \mp \frac{|\mathbf{x} - \mathbf{x}'|}{c}\right]\right) dV' \, dt'. \qquad (6.20)$$

Because of the δ function, it is trivial to carry on the integration over t', which gives

$$\psi(\mathbf{x}, t) = \frac{1}{4\pi} \int \frac{f(\mathbf{x}', t \mp \frac{|\mathbf{x}-\mathbf{x}'|}{c})}{|\mathbf{x} - \mathbf{x}'|} dV'. \qquad (6.21)$$

The form of this solution will remind the reader of the solution (2.2) of Laplace's equation (2.10), the only difference being that we are now considering time-varying sources, and we have to use the value of the source function $f(\mathbf{x}', t')$ not at the time t at which we want the value of field function $\psi(\mathbf{x}, t)$, but at a time earlier or later by the time interval $|\mathbf{x} - \mathbf{x}'|/c$. What is the physical significance of this? Clearly $|\mathbf{x} - \mathbf{x}'|/c$ is the time taken by a signal travelling with speed c to propagate from the source point $\mathbf{x}'$ to the field point $\mathbf{x}$. Keeping in mind that c is the speed of electromagnetic waves, we can give a particularly transparent physical interpretation of our results if we assume that information within electromagnetic fields propagates at speed c, even when electromagnetic waves are not explicitly present. If we consider the source function at position $\mathbf{x}'$ at a time $|\mathbf{x} - \mathbf{x}'|/c$ earlier than the time t when we want to obtain the field function $\psi(\mathbf{x}, t)$ at position $\mathbf{x}$, then the information about the source function at this earlier time carried by a signal propagating at speed c would reach the field point exactly at time t when the value of the field function is required. Now, if we take the minus sign in (6.21), it is obvious that the source function at such an earlier time only would contribute to the field. On the other hand, if we take the plus sign, that would mean that the field would depend on the source at a *later* time, which certainly violates causality. So although the solution with the plus sign in (6.21) is a mathematical solution of our equations, it is certainly nonphysical. We need to keep only the minus sign in (6.21) to get a meaningful solution. If we keep the minus sign in (6.21), it means that the field at a time t is produced by the sources at earlier times in such a manner that the information about these sources travelling at speed c reaches the field point at time t.

We now write the physically admissible part of the solution in (6.21) in the following manner:

$$\psi(\mathbf{x}, t) = \frac{1}{4\pi} \int \frac{[f]}{|\mathbf{x} - \mathbf{x}'|} dV', \qquad (6.22)$$

where we have used the convention that the square bracket put around any quantity Q means

$$[Q] = Q\left(\mathbf{x}', t - \frac{|\mathbf{x} - \mathbf{x}'|}{c}\right). \qquad (6.23)$$

Comparing (4.138) and (4.139) with (6.8), we at once realize that the solutions of (4.138) and (4.139), which would be analogous to (6.22), can be readily written as

$$\Phi(\mathbf{x}, t) = \frac{1}{4\pi\epsilon_0} \int \frac{[\rho]}{|\mathbf{x} - \mathbf{x}'|} \, dV', \qquad (6.24)$$

$$\mathbf{A}(\mathbf{x}, t) = \frac{\mu_0}{4\pi} \int \frac{[\mathbf{j}]}{|\mathbf{x} - \mathbf{x}'|} \, dV'. \qquad (6.25)$$

At last, we have some really elegant solutions of (4.138) and (4.139). These are often called *retarded solutions* to indicate that we have to use values of the source functions at earlier times to calculate these solutions. The time

$$t' = t - \frac{|\mathbf{x} - \mathbf{x}'|}{c} \qquad (6.26)$$

is sometimes referred to as the *retarded time*, although this nomenclature is a little confusing because we are referring to a time earlier than t. It should be clear from the discussion in Sect. 4.10 that (4.138) and (4.139) encapsulate the same physics as Maxwell's equations (1.1–1.4). The electromagnetic fields calculated from (6.24) and (6.25), which are solutions of (4.138) and (4.139), would thus give the completely general solution of Maxwell's equations with time-varying sources when no term in the equations is assumed zero. However, as we shall see later in this chapter, calculating $\mathbf{E}$ or $\mathbf{B}$ from (6.24) and (6.25) in a realistic situation by applying (4.128) and (4.130) can be fairly non-trivial. Finally, we point out that (6.24) and (6.25) reduce to (2.2) and (3.2), respectively, in the static situation. This means that we can readily obtain electrostatics and magnetostatics as special cases of our general theory.

6.3   The Lienard–Wiechert Potentials

After introducing the expression (2.2) for the electrostatic potential due to a distribution of charges, one immediate application we considered was to write down the electrostatic potential due to a point charge at rest, which is given by (2.5). Now that we have the expressions (6.24) and (6.25) giving the scalar and vector potentials, we may want to determine what these potentials may be for a moving charge. This superficially simple problem actually turns out to be more complicated than what you may expect at first. Even for a 'point' charge, we have to consider the charge distribution inside its interior. Since the light travel time from various locations inside the charge to the field point would be different, we need to use different retarded times for different locations inside the charge!

One possible approach is to express the charge and current densities associated with the moving charge in terms of δ functions, as given in (3.70). One can carry out a formal mathematical manipulation by using such expressions for the charge and current densities to derive the scalar and vector potentials due to a moving charge. This approach has been followed in Zangwill ([2], Sect. 23.2.1) and is given as an

Fig. 6.1 Sketch showing the
two positions of the
information-gathering sphere
around the field point **x** at
times t' (with radius r) and
$t' + dt'$ (with radius $r - dr$).
This collapsing
information-gathering sphere
passes through a charge
distribution. The shaded
region indicates the volume
element $dV' = c \, dt' \, dS'$

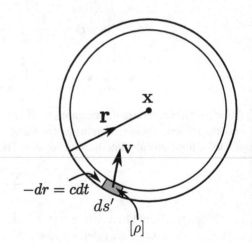

exercise problem (Exercise 6.2). Here, we present a different approach which makes
the physics of the derivation more transparent.

To determine how to proceed, let us carefully consider the physical significance of
the expressions (6.24) and (6.25). We can think of an information-gathering sphere
centred around the field point which is becoming smaller with inward radial velocity
c so as to collapse on the field point **x** exactly at time t. For a source point at **x'**,
we have to take the values of ρ and **j** as found by the information-gathering sphere
exactly when it passes through the source point **x'**. We have to substitute these values
of ρ and **j** in (6.24) and (6.25) to allow for the light travel time.

Figure 6.1 shows two successive positions of the information-gathering sphere
around the field point **x**, the outer sphere corresponding to its position at time t' and
the inner sphere corresponding to time $t' + dt'$. The radius of the outer sphere has to
be $r = c\,(t - t')$ to ensure that the information-gathering sphere collapses at the field
point **x** at time t. The radial separation between the two spheres is $dr = -c\,dt'$. We
now consider a volume $dV' = c\,dt'\,dS'$ between these two spheres having the cross
section dS' transverse to the radial direction. This volume is the shaded region in
Fig. 6.1. Suppose the charge density inside this volume is $[\rho]$ when the information-
gathering sphere passes through it. We have put square brackets around ρ to indicate
that we are considering its value at the retarded time t'. If the charge distribution is
at rest, then the information-gathering surface while passing from r to $r - dr$ would
'sample' the amount of charge

$$dq = [\rho]\, dV'$$

inside the volume dV' while traversing the volume. However, if the charge distribution is in motion, then it is possible for charges to leave or enter the volume dV'
while the information-gathering surface was traversing it. Let **v** be the velocity of
the charge distribution. Writing

$$\mathbf{r} = \mathbf{x} - \mathbf{x'}, \tag{6.27}$$

the component of velocity in the radially inward direction is $\mathbf{v}.\mathbf{r}/r$. The flux of charge across the surface element dS' of the inner sphere is

$$[\rho]\frac{\mathbf{v}.\mathbf{r}}{r}\,dS'.$$

Multiplying this by dt', we can get the amount of charge leaving the volume dV' while the information-gathering surface was traversing it. Hence, the amount of charge 'sampled' by the information-gathering surface while traversing the volume dV' would be

$$dq = [\rho]\,dV' - [\rho]\frac{\mathbf{v}.\mathbf{r}}{r}\,dS'\,dt'.$$

Noting that $dV' = c\,dt'\,dS'$, this can be written as

$$dq = [\rho]\,dV'\left(1 - \frac{\mathbf{v}.\mathbf{r}}{rc}\right),$$

from which

$$[\rho]\,dV' = \frac{dq}{1 - \frac{\mathbf{v}.\mathbf{r}}{rc}}. \tag{6.28}$$

Substituting this in (6.24) and noting that $|\mathbf{x} - \mathbf{x}'| = r$, we get

$$\Phi = \frac{1}{4\pi\epsilon_0}\int\frac{dq}{r - \frac{\mathbf{v}.\mathbf{r}}{c}}, \tag{6.29}$$

We now want to carry on the integration in (6.29) over one moving charge. While the information-gathering surface traverses the moving charge, the values of r and $\mathbf{v}$ will be almost constant. So we can just integrate over dq in (6.29) to obtain

$$\Phi = \frac{1}{4\pi\epsilon_0}\frac{q}{r - \frac{\mathbf{v}.\mathbf{r}}{c}}. \tag{6.30}$$

As we have pointed out in Sect. 3.9, a moving charge makes a contribution $q\mathbf{v}$ to the current. Applying considerations very similar to the considerations discussed above, we conclude that

$$\mathbf{A} = \frac{\mu_0}{4\pi}\frac{q\,\mathbf{v}}{r - \frac{\mathbf{v}.\mathbf{r}}{c}}, \tag{6.31}$$

The scalar and vector potentials due to a charged particle q moving with velocity $\mathbf{v}$ at a field point at a vectorial distance $\mathbf{r} = \mathbf{x} - \mathbf{x}'$ from the charged particle are given by (6.30) and (6.31). These potentials are called *Lienard–Wiechert potentials*.

In Exercise 6.3, readers are asked to show that the denominator in (6.30) or (6.31) is related to a Lorentz-invariant scalar. The significance of this is that we can combine (6.30) and (6.31) into a four-vector equation, as readers have to demonstrate in Exercise 6.3. Note that so far in this chapter, we have not explicitly introduced any relativistic considerations at all. We are carrying out our analysis in a frame of reference in which Maxwell's equations hold and we expect, as a consequence, that (4.138) and (4.139) also should hold. We are merely discussing solutions of these equations in this frame, without bothering about issues of transformation from one frame to another at all. It is remarkable that relativistic considerations still crop up in unexpected places, showing the deep connection between electrodynamics and relativity.

6.4   Electromagnetic Fields Due to a Moving Charged Particle

The electromagnetic fields of a moving charged particle can be obtained by substituting the Lienard–Wiechert potentials given by (6.30) and (6.31) into (4.128) and (4.130). Differentiating the Lienard–Wiechert potentials with respect to t or the spatial variables involves some subtleties and is quite messy. Since it is very important to obtain the electromagnetic fields due to a moving charged particle, we now go through some of the crucial steps in this messy and complicated derivation. It is possible to carry out the derivation in an explicitly relativistic manner to make it clear that the derivation should hold in any inertial frame. Interested readers are referred to Jackson ([1], Sect. 14.1) for this relativistic derivation. This derivation, though elegant, is rather formal and abstract. We believe that it is helpful for students to first learn the derivation with reference to a particular inertial frame in which the Lienard–Wiechert potentials hold, without bothering about relativistic considerations. Defining

$$s = r - \frac{\mathbf{v} \cdot \mathbf{r}}{c} \tag{6.32}$$

and using (4.42), we can write (6.30) and (6.31) as

$$\Phi = \frac{1}{4\pi\epsilon_0} \frac{q}{s}, \quad \mathbf{A} = \frac{1}{4\pi\epsilon_0 c^2} \frac{q\,\mathbf{v}}{s}. \tag{6.33}$$

Consider a charged particle moving in such a manner that its position at time t' is given by $\mathbf{x}'(t')$. The velocity of the charged particle is clearly given by

$$\mathbf{v} = \frac{d\mathbf{x}'}{dt'}. \tag{6.34}$$

We want to find the electromagnetic fields due to this charged particle at the field point $\mathbf{x}$ at time t. We first have to determine the retarded time we need to use for finding the Lienard–Wiechert potentials. Suppose information starting from the position $\mathbf{x}'(t')$ of the charged particle at time t' propagates at speed c and reaches the field point $\mathbf{x}$ at time t. In other words, we have

$$r = |\mathbf{x} - \mathbf{x}'(t')| = c\,(t - t'). \tag{6.35}$$

Certainly t' satisfying (6.35) should be the appropriate retarded time for us, and the value of r with respect to the position of the charged particle at this time t' should be put in (6.32). This means that the Lienard–Wiechert potentials as given by (6.33) depend on $\mathbf{x}'$ and t' which satisfy (6.35). Now, in order to obtain the electromagnetic fields by using (4.128) and (4.130), one has to operate on Φ and $\mathbf{A}$ with operators such as

$$\left(\frac{\partial}{\partial t}\right)_{\mathbf{x}}, \ (\nabla)_t. \tag{6.36}$$

While writing down (4.128) and (4.130), we had tacitly assumed that the differentiation with t has to be carried out holding $\mathbf{x}$ fixed, and the operation with ∇ has to be carried out holding t constant. Now that we have to deal with many variables, we are indicating these things explicitly. If we are given $\mathbf{x}$ where we want to find the fields at t and if $\mathbf{x}'(t')$ is specified to us, then (6.35) determines the retarded time t' and $\mathbf{x}'(t')$ at that retarded time. In other words, we can view t and $\mathbf{x}$ as our independent variables, from which the source variables t' and $\mathbf{x}'(t')$ can be obtained. Operations with the operators in (6.36) would be straightforward if we can express $\mathbf{r}$ appearing in (6.32) as a function of $\mathbf{x}$ and t alone by eliminating $\mathbf{x}'$ and t' with the help of (6.35). This is, however, possible to do in a closed form only for some specific kinds of simple motion like the motion with constant $\mathbf{v}$, which we shall discuss in Sect. 6.5. For more general kinds of motion, we may not be able to write $\mathbf{r}$ explicitly as a function of $\mathbf{x}$ and t. So it is not straightforward to operate with the operators in (6.36). How do we proceed?

We can think of t' and $\mathbf{x}$ as our independent variables. Once t' is known, we know $\mathbf{x}'(t')$. From knowledge of $\mathbf{x}'(t')$ and $\mathbf{x}$, the value of t follows simply from (6.35). When we regard t' and $\mathbf{x}$ as our independent variables, we have to consider operations with the operators

$$\left(\frac{\partial}{\partial t'}\right)_{\mathbf{x}}, \ (\nabla)_{t'} \tag{6.37}$$

rather than the operators in (6.36). As we shall see below, operations with the operators given in (6.37) are much simpler. The trick now is to determine the electromagnetic fields in the following three steps: (1) We first determine how to carry on operations with the operators shown in (6.37); (2) Then we shall relate the operators given in (6.36) with the operators given in (6.37); and (3) We finally calculate the electromagnetic fields by transforming (4.128) and (4.130) to expressions involving the operators in (6.37).

Step 1. Operations with operators in (6.37)

From (6.27) and (6.34), it is obvious that

$$\left(\frac{\partial \mathbf{r}}{\partial t'}\right)_{\mathbf{x}} = -\mathbf{v}, \ \left(\frac{\partial r}{\partial t'}\right)_{\mathbf{x}} = -\frac{\mathbf{r}.\mathbf{v}}{r}, \tag{6.38}$$

since the derivative of r with respect to t' is nothing but the component of the particle's velocity along $\mathbf{r}$. We also note that the acceleration of the charged particle is given by

$$\left(\frac{\partial \mathbf{v}}{\partial t'}\right)_{\mathbf{x}} = \dot{\mathbf{v}}. \tag{6.39}$$

From (6.32), (6.38) and (6.39), we easily arrive at

$$\left(\frac{\partial s}{\partial t'}\right)_{\mathbf{x}} = -\frac{\mathbf{r}.\mathbf{v}}{r} - \frac{\mathbf{r}.\dot{\mathbf{v}}}{c} + \frac{v^2}{c}. \tag{6.40}$$

Once we have determined how to operate on s with the operator $(\partial/\partial t')_{\mathbf{x}}$, it would be straightforward to operate on Φ or $\mathbf{A}$ given by (6.33) with this operator. We now turn our attention to the other operator listed in (6.37). It is easy to show from (6.32) that

$$(\nabla s)_{t'} = \frac{\mathbf{r}}{r} - \frac{\mathbf{v}}{c}. \tag{6.41}$$

Using this relation, we can operate on Φ or $\mathbf{A}$ given by (6.33) with the operator $(\nabla)_{t'}$.

Step 2. Relating the operators in (6.36) *to the operators in* (6.37)

We have explained how we can do mathematical manipulations with the operators listed in (6.37). The next step in our derivation is to relate these operators to the operators shown in (6.36), which we have to use in order to get the electromagnetic fields. While discussing the operators listed in (6.37), we treated t' and $\mathbf{x}$ as our independent variables. This makes sense because the other variables can be found from them. We have assumed $\mathbf{x}'(t')$ to be given and t can be found from (6.35). Considering t' and $\mathbf{x}$ as our independent variables, an infinitesimal variation in r can be written as

$$dr = \left(\frac{\partial r}{\partial t'}\right)_{\mathbf{x}} dt' + (\nabla r)_{t'}.d\mathbf{x}, \tag{6.42}$$

from which we have

$$\left(\frac{\partial r}{\partial t}\right)_{\mathbf{x}} = \left(\frac{\partial r}{\partial t'}\right)_{\mathbf{x}} \left(\frac{\partial t'}{\partial t}\right)_{\mathbf{x}} \tag{6.43}$$

and

$$(\nabla r)_t = (\nabla r)_{t'} + \left(\frac{\partial r}{\partial t'}\right)_{\mathbf{x}} (\nabla t')_t. \tag{6.44}$$

Readers are urged to provide the easy intermediate steps to arrive at (6.44) from (6.42). (Hint: you may substitute for dr and dt' the expansions that you would get if you consider r and t' to be functions of the independent variables t and $\mathbf{x}$.) From (6.38) and (6.43), we have

$$\left(\frac{\partial r}{\partial t}\right)_{\mathbf{x}} = -\frac{\mathbf{r}.\mathbf{v}}{r}\left(\frac{\partial t'}{\partial t}\right)_{\mathbf{x}}.$$

We also have from (6.35)

$$\left(\frac{\partial r}{\partial t}\right)_{\mathbf{x}} = c\left[1 - \left(\frac{\partial t'}{\partial t}\right)_{\mathbf{x}}\right].$$

Equating the above two expressions for $(\partial r/\partial t)_{\mathbf{x}}$, we get

$$\left(\frac{\partial t'}{\partial t}\right)_{\mathbf{x}} = \frac{1}{1 - \frac{\mathbf{v}.\mathbf{r}}{rc}} = \frac{r}{s}$$

using (6.32). This suggests the following relation between two differential operators of interest to us:

$$\left(\frac{\partial}{\partial t}\right)_{\mathbf{x}} = \frac{r}{s}\left(\frac{\partial}{\partial t'}\right)_{\mathbf{x}}. \tag{6.45}$$

Now, from (6.38) and (6.44), we get

$$(\nabla r)_t = \frac{\mathbf{r}}{r} - \frac{\mathbf{r}.\mathbf{v}}{r}(\nabla t')_t,$$

where we have made use of $(\nabla r)_{t'} = \mathbf{r}/r$. From (6.35), we get

$$(\nabla r)_t = -c\,(\nabla t')_t.$$

Equating the above two expressions for $(\nabla r)_t$, we easily arrive at the result

$$(\nabla t')_t = -\frac{\mathbf{r}}{s\,c}.$$

Substituting this in (6.44), we arrive at the following operator relationship

$$(\nabla)_t = (\nabla)_{t'} - \frac{\mathbf{r}}{s\,c}\left(\frac{\partial}{\partial t'}\right)_{\mathbf{x}}. \tag{6.46}$$

The operators appearing in (6.36) have now been expressed in (6.45) and (6.46) in terms of the operators in (6.37), which we already know how to handle.

Step 3. Calculation of the electromagnetic fields

We are now ready with the machinery we require for calculating the electromagnetic fields due to a moving charge. Substituting from (6.33) in (4.130), we get

$$\mathbf{E} = \frac{q}{4\pi\epsilon_0}\left[-\left(\nabla\left\{\frac{1}{s}\right\}\right)_t - \frac{1}{c^2}\left(\frac{\partial}{\partial t}\left\{\frac{\mathbf{v}}{s}\right\}\right)_{\mathbf{x}}\right],$$

from which

$$E = \frac{q}{4\pi\epsilon_0} \left[\frac{1}{s^2}(\nabla s)_t + \frac{\mathbf{v}}{s^2 c^2} \left(\frac{\partial s}{\partial t} \right)_x - \frac{1}{s\,c^2} \left(\frac{\partial \mathbf{v}}{\partial t} \right)_x \right].$$

We now substitute from (6.45) and (6.46) for the operators in this equation, which gives

$$E = \frac{q}{4\pi\epsilon_0} \left[\frac{1}{s^2}(\nabla s)_{t'} - \frac{\mathbf{r}}{s^3 c} \left(\frac{\partial s}{\partial t'} \right)_x + \frac{r\,\mathbf{v}}{s^3 c^2} \left(\frac{\partial s}{\partial t'} \right)_x - \frac{r\,\dot{\mathbf{v}}}{s^2 c^2} \right], \qquad (6.47)$$

where we have made use of (6.39) in the last term. We have already evaluated $(\partial s/\partial t')_x$ and $(\nabla s)_{t'}$ in (6.40) and (6.41), respectively. We now have to substitute these in (6.47) and rearrange the terms suitably. We leave it for the reader to carry on this straightforward, but slightly tedious, algebra and to show that we finally get

$$E = \frac{q}{4\pi\epsilon_0} \left[\frac{1}{s^3} \left(\mathbf{r} - \frac{r\,\mathbf{v}}{c} \right) \left(1 - \frac{v^2}{c^2} \right) + \frac{1}{c^2 s^3} \left[\mathbf{r} \times \left\{ \left(\mathbf{r} - \frac{r\,\mathbf{v}}{c} \right) \times \dot{\mathbf{v}} \right\} \right] \right]. \qquad (6.48)$$

This is the completely general expression of the electric field due to a charged particle moving in an arbitrary manner and occupying the position $\mathbf{x}'(t')$ at time t'. As the information-gathering sphere which would have to collapse on the field point $\mathbf{x}$ at t moves inward, it would intercept the moving charged particle only at one time. Taking that as the retarded time, we have to use the velocity $\mathbf{v}$ of the particle at that time in (6.48), and $\mathbf{r}$ is the position vector of the field point taken with respect to the position of the particle at that retarded time.

We leave it for the readers to calculate the magnetic field due to the moving charged particle by using (4.128) and to show that

$$B = \frac{\mathbf{r} \times E}{r\,c}. \qquad (6.49)$$

(Exercise 6.4).

The general expressions of the electric field and the magnetic field, as given by (6.48) and (6.49), seem quite complicated at the first sight. To obtain some insight into the nature of these fields, we shall now consider the field due to a uniformly moving charged particle and then determine what happens if the particle is suddenly brought to rest. These are taken up in the next section.

6.5   The Fields Due to a Uniformly Moving Charge

It is instructive to consider the electric field due to a charged particle moving with uniform velocity $\mathbf{v}$. This means we can put $\dot{\mathbf{v}} = 0$ in (6.48) leading to the electric field

Fig. 6.2 Sketch indicating
the direction of the electric
field produced at a field point
P by a charged particle
moving along the x axis,
which is at the positions
$x'(t')$ and $x'(t)$ at times t'
and t

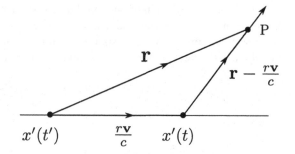

$$E = \frac{q}{4\pi\epsilon_0}\frac{1}{s^3}\left(\mathbf{r} - \frac{r\,\mathbf{v}}{c}\right)\left(1 - \frac{v^2}{c^2}\right). \tag{6.50}$$

This field is clearly in the direction of the vector $\mathbf{r} - (r\,\mathbf{v}/c)$. Let us determine the significance of this direction. Figure 6.2 shows the x axis by the horizontal line taken in the direction in which the charged particle q is moving, whereas P is the field point where we want the electric field. Remember that $\mathbf{r}$ gives the position of the field point with respect to the position of the charged particle at retarded time indicated by $x'(t')$. The vector $\mathbf{r}$ is shown in Fig. 6.2. A signal propagating at speed c starting from $x'(t')$ at time t' would reach the field point P at time t, the travel time of the signal being $t - t' = r/c$. During this travel time, the charged particle moves over the vectorial distance $\mathbf{v}(t - t') = r\,\mathbf{v}/c$ to reach the position $x'(t)$ as shown in Fig. 6.2. Obviously, $x'(t)$ is the position of the charged particle at time t when we want the field at the field point. It is clear from Fig. 6.2 that $\mathbf{r} - (r\,\mathbf{v}/c)$ is the vectorial distance of the field point with respect to the position $x'(t)$ of the particle at time t. This means that although information propagates within electromagnetic fields at speed c and the information that the charged particle has arrived at $x'(t)$ would not reach the field point yet at time t, the electric field at the field point is still directed outward from the position $x'(t)$ of the charged particle, as if the field arranges itself in anticipation that the charged particle would be at the point $x'(t)$ at time t.

We have here specialized for the case of the charged particle moving uniformly after obtaining the general result for an arbitrarily moving charged particle. As we saw in Sect. 6.4, the analysis to arrive at the general result is somewhat complicated. If the charged particle moves uniformly, then the electric field due to it can be calculated in a simpler way. It is instructive to show how to do this. We have pointed out in Sect. 6.4 that electromagnetic fields can be calculated by using (4.128) and (4.130) readily if it is possible to express s appearing in (6.33) as a function of the position $\mathbf{x}$ of the field point and the time t when the field is required. This cannot be done in a closed analytical form for an arbitrarily moving charged particle, necessitating the three-step process for calculating electromagnetic fields described in Sect. 6.4. However, for a uniformly moving charged particle, we now show that it is possible to write down s as a function of $\mathbf{x}$ and t.

Let us choose the origin of our coordinate system at the point where the charged particle is at time $t = 0$. Then the charged particle moving along x axis is at $(vt', 0, 0)$ at the retarded time t'. A signal starting from this point and propagating at speed c would reach the field point (x, y, z) at time t. This gives the condition

$$c^2(t - t')^2 = (x - vt')^2 + y^2 + z^2,$$

from which

$$(c^2 - v^2)t'^2 - 2(c^2 t - xv)t' + c^2 t^2 - x^2 - y^2 - z^2 = 0.$$

Solving this quadratic equation for t', we find

$$t' = \frac{c^2 t - vx \pm \sqrt{c^2(x - vt)^2 + (c^2 - v^2)(y^2 + z^2)}}{c^2 - v^2}. \tag{6.51}$$

Causality requires that we choose the minus sign in this solution, as we did for the solution (6.21) of the inhomogeneous wave equation. We now have to evaluate s as given by (6.32). Noting that r is given by (6.35) and $\mathbf{v} \cdot \mathbf{r} = v(x - vt')$, we have

$$s = c(t - t') - \frac{v(x - vt')}{c}.$$

Substituting for t' from (6.51) and carrying out some straightforward algebra, we arrive at

$$s = \sqrt{(x - vt)^2 + \left(1 - \frac{v^2}{c^2}\right)(y^2 + z^2)}. \tag{6.52}$$

Substituting this expression of s in (6.33), we get the potential Φ and $\mathbf{A}$ as functions of only the position $\mathbf{x}$ of the field point and the time t. It is then quite easy to obtain the electric field by using (4.130) and to show that it is the same as given by (6.50), keeping in mind that the charged particle is at the location $(vt, 0, 0)$ at time t and that the position vector of the field point with respect to this location is

$$\mathbf{r} - \frac{r\mathbf{v}}{c} = (x - vt)\,\hat{\mathbf{e}}_x + y\,\hat{\mathbf{e}}_y + z\,\hat{\mathbf{e}}_z \tag{6.53}$$

in this case. Readers are asked to carry on this calculation in Exercise 6.5

The electric field due to a uniformly moving charged particle can also be found out by first writing down the electric field in the frame in which the charged particle is at rest and then obtaining the electric field in the frame in which the charged particle is moving at speed $\mathbf{v}$ by applying the relativistic transformation laws. This derivation had been presented in Sect. 5.10.2, the electric field due to the uniformly moving charged particle being given by (5.108–5.110). Readers are asked in Exercise 6.5 to show that the method outlined in this section gives the same electric field. Keep in mind that we used primes in the discussion of Sect. 5.10.2 to indicate the frame of reference with respect to which the charged particle is moving with velocity $v\hat{e}_x$, whereas primes denote source variables throughout the present chapter. The variables (t', x', y', z') appearing in (5.108–5.110) should be replaced by (t, x, y, z) for comparison with results here. We emphasize again that all our derivations in this chapter were carried out with reference to one frame in which Maxwell's equations hold, without introducing relativistic considerations anywhere. The fact that the correct relativistic expression for the electric field of a relativistically moving charged particle can be obtained from the approach of this chapter makes the deep connection between electromagnetism and relativity clear once more. Unlike the laws of Newtonian mechanics which have to be corrected suitably when dealing with relativistically moving particles, Maxwell's equations require no such corrections.

For a uniformly moving charged particle, it is particularly easy to determine the magnetic field from the electric field. From (6.33), we have the general relation

$$\mathbf{A} = \frac{\mathbf{v}}{c^2}\, \Phi. \tag{6.54}$$

If $\mathbf{v}$ is constant, then we get

$$\mathbf{B} = \nabla \times \left(\frac{\mathbf{v}}{c^2}\, \Phi\right) = -\frac{\mathbf{v}}{c^2} \times \nabla\Phi$$

by using (B.5). It follows from (6.33) that $\mathbf{A}$ is in the direction of $\mathbf{v}$, and that would also be the case for $\partial\mathbf{A}/\partial t$ when $\mathbf{v}$ is constant. This means that we can put the equation above in the form

$$\mathbf{B} = \frac{\mathbf{v}}{c^2} \times \left(-\nabla\Phi - \frac{\partial\mathbf{A}}{\partial t}\right),$$

since the cross product of $\mathbf{v}$ and $\partial\mathbf{A}/\partial t$ is zero. It then follows from (4.130) that

$$\mathbf{B} = \frac{\mathbf{v}}{c^2} \times \mathbf{E}. \tag{6.55}$$

Readers can easily check that (6.49) and (6.55) will give the same magnetic field when we substitute for $\mathbf{E}$ from (6.50). It may be noted that (6.55) is the same as (5.113), which we obtained by very different considerations.

Lastly, before we get into a discussion in the next chapter of what happens when the moving charged particle has non-zero acceleration (i.e. when $\dot{\mathbf{v}}$ is not zero), we want to discuss the case of a uniformly moving charged particle which suddenly stops. This means that we have a sudden deceleration. Using the results of a uniformly moving charged particle that we already have, we can draw some conclusions about the effect of this sudden deceleration. We show in Fig. 6.3 that the charged particle has come to a halt at point S at time t_{stop}. Suppose we want to determine what the electric field would look like at a slightly later time t. Had the charged particle not

Fig. 6.3 The electric field
lines at time t due to a
uniformly moving charged
particle which has suddenly
stopped at S at an earlier
time t_{stop}. Had the charged
particle not stopped, it would
have reached the point P at
time t

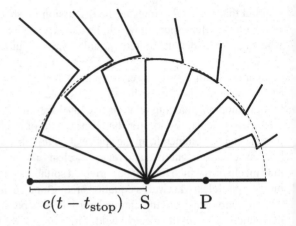

$$c(t - t_{\text{stop}}) \quad S \quad\quad P$$

suddenly stopped, it would have covered a distance $v(t - t_{\text{stop}})$ during the interval
from t_{stop} to t and would have come to the position P. Now, we know that information
within electromagnetic fields travel at speed c. So the information that the charged
particle had suddenly stopped at time t_{stop} would reach at time t up to the sphere
of radius $c(t - t_{\text{stop}})$ centred around S, as shown in Fig. 6.3. Beyond this spherical
surface, however, the information that the charged particle had stopped would not
reach. So in this outer region, the electric field would have been exactly like what
it would be if the charged particle kept moving and reached the point P at time
t. Since the electric field due to a uniformly moving charged particle is directed
outward from its present position, we expect the electric field outside the spherical
surface to be directed outward from P as the centre, as shown in Fig. 6.3. On the
other hand, the information that the charged particle is now at rest has reached the
spherical surface, and the electric field inside it would just be the electric field due
to the static charge at S, which must be directed outward from S. The electric field
lines between the regions interior to and exterior to the spherical surface are expected
to connect as indicated in Fig. 6.3. It is clear that we would get an electric field on
this spherical surface transverse to the radial direction. As the spherical information-
carrying surface expands radially outward with speed c, the transverse electric field
will also propagate away with speed c from the charged particle located at rest at S.
It follows from (6.49) that there would also be a magnetic field lying on the spherical
surface perpendicular to the electric field. It is clear that the sudden deceleration of
the charged particle would produce a disturbance which propagates outward with
speed c, having many characteristics of an electromagnetic wave which we derived
in Sect. 4.4. The disturbance would contain transverse electric and magnetic fields,
which are themselves transverse to each other. Thus, merely on the basis of the
results for a uniformly moving charged particle, we are able to argue that a sudden
deceleration would give rise to an electromagnetic wave pulse. The emission of
electromagnetic radiation by the acceleration of a charged particle will be discussed
in detail in Chap. 7.

Exercises

6.1 Suppose we have a static charge distribution $\rho(x' > 0, y', z')$ in front of an infinite conducting plane surface located at $x = 0$ and maintained at zero potential. Show that

$$-\frac{1}{4\pi}\left[\frac{1}{\sqrt{(x-x')^2+(y-y')^2+(z-z')^2}} - \frac{1}{\sqrt{(x+x')^2+(y-y')^2+(z-z')^2}}\right]$$

is the appropriate Green's function in this problem satisfying (6.6) and that the electrostatic potential in the region $x > 0$ is given by (6.7) with this expression appearing in the place of $G(\mathbf{x}; \mathbf{x}')$.

6.2 Write down expressions of $\Phi(\mathbf{x}, t)$ and $\mathbf{A}(\mathbf{x}, t)$ analogous to (6.20) in the form of four-dimensional integrals involving $\rho(\mathbf{x}, t)$ and $\mathbf{j}(\mathbf{x}, t)$. Substitute in these

$$\rho(\mathbf{x}', t') = q\delta(\mathbf{x}' - \mathbf{x}_q(t')), \quad \mathbf{j}(\mathbf{x}', t') = q\mathbf{v}\delta(\mathbf{x}' - \mathbf{x}_q(t'))$$

representing a moving charged particle which occupies position $\mathbf{x}'$ at time t'. Carry on the volume integration to show that

$$\Phi(\mathbf{x}, t) = \frac{q}{4\pi\epsilon_0}F(\mathbf{x}, t, \mathbf{x}_q(t'), t'), \quad \mathbf{A}(\mathbf{x}, t) = \frac{q\mathbf{v}}{4\pi\epsilon_0 c^2}F(\mathbf{x}, t, \mathbf{x}_q(t'), t'),$$

where

$$F(\mathbf{x}, t, \mathbf{x}_q(t'), t') = \int dt' \frac{\delta(t' - t + |\mathbf{x} - \mathbf{x}_q(t')|/c)}{|\mathbf{x} - \mathbf{x}_q(t')|}.$$

Evaluate $F(\mathbf{x}, t, \mathbf{x}_q(t'), t')$ to derive the Lienard–Wiechert potentials (6.30) and (6.31), where $\mathbf{r} = \mathbf{x} - \mathbf{x}_q(t')$ with t' corresponding to the retarded time.

Hint. You have to use the following well-known relation for the Dirac δ function:

$$\delta[f(x)] = \left|\left(\frac{df}{dx}\right)_{x=x_0}\right|^{-1}\delta(x - x_0),$$

where x_0 is the value of x for which $f(x_0) = 0$.

6.3 Suppose a signal propagating at speed c leaves the source point $\mathbf{x}'$ at time t' and reaches the field point $\mathbf{x}$ at time t. Argue that

$$r^\mu = \begin{bmatrix} c(t - t') \\ \mathbf{x} - \mathbf{x}' \end{bmatrix}$$

is a Lorentz four-vector. Assuming that the source corresponds to a moving charge occupying position $\mathbf{x}'(t')$ at time t', show that from s defined in (6.32) we can get the following Lorentz-invariant scalar

$$\gamma s = -\frac{r_\nu V^\nu}{c},$$

where V^μ is the four-velocity of the particle, and γ is the corresponding Lorentz factor. Argue that the four-potential has to be given by

$$A^\mu = -\frac{1}{4\pi\epsilon_0}\frac{qV^\mu}{r_\nu V^\nu}$$

and show that it is equivalent to (6.33).

6.4 Calculate the magnetic field due to a moving charged particle by taking the curl of $\mathbf{A}$ as given by (6.33) and show that it satisfies (6.49).

6.5 After substituting the expression of s given by (6.52) in the expressions of the potentials as given by (6.33), calculate the electric field due to a charged particle moving with uniform velocity $\mathbf{v}$. Show that this is the same as what you would get from (6.50). Verify also that the electric field you get by the method outlined here is the same as the electric field of a uniformly moving charged particle derived in Sect. 5.10.2 by carrying out a relativistic transformation from the frame in which the charged particle is at rest.

References

1. Jackson, J.D.: Classical Electrodynamics, 3rd edn. Wiley (1999)
2. Zangwill, A.: Modern Electrodynamics. Cambridge University Press (2013)

Chapter 7
Emission of Electromagnetic Radiation

7.1 Handling the Emission of Electromagnetic Radiation in Practical Situations

One important topic in classical electromagnetic theory is determining how electromagnetic waves are generated. We have studied various properties of electromagnetic waves in Chap. 4 on the basis of Maxwell's equations. However, the generation of electromagnetic waves is a topic which we avoided in that chapter. From our discussion of electrostatics and magnetostatics in Chaps. 2 and 3, we know that static charges and static currents do not give rise to electromagnetic waves. We clearly need charges and currents varying with time for producing electromagnetic waves. In Chap. 6, we obtained some general results about electromagnetic fields of time-varying charges and currents. These are going to be our basis for the discussion of the emission of electromagnetic waves by time-varying sources. Since some of the results obtained in Chap. 6 appear somewhat complicated, such as the expression (6.48) of the electric field due to a moving charge, one may think that the subject of emission of electromagnetic waves is going to be a mathematically involved subject. It is indeed true that the calculation of radiation emission in many situations can be fairly complicated. However, in many practical situations which are of great interest to us, it is possible to introduce some drastic simplifications, and the calculations involved in studying the emission of electromagnetic radiation are actually not too complicated in those situations. In this book, we shall restrict ourselves only to those situations in which the emission of electromagnetic waves can be studied without getting involved in very complicated mathematical calculations. In some important cases for which the detailed calculations are mathematically involved, we shall present approximate treatments which can bring out the essential physics quite clearly (especially in Sects. 7.9 and 7.10). In spite of our decision to stay away from mathematically involved calculations, we shall actually be able to cover some of the most important cases of practical relevance in the subject of the emission of electromagnetic radiation.

© Springer Nature Singapore Pte Ltd. 2022
A. R. Choudhuri, *Advanced Electromagnetic Theory*, Lecture Notes in Physics 1009,
https://doi.org/10.1007/978-981-19-5944-8_7

In the next two subsections, we show that we can arrive at reasonably simple formulae out of the general theory for calculating emission of electromagnetic radiation in two situations: (i) for non-relativistically moving charged particles and (ii) for periodically oscillating currents. Most of the practical problems that may interest a physicist can be worked out from the two not very complicated formulae (7.2) and (7.15) we are going to arrive at in the following two subsections. In the following two subsections, we merely discuss how to calculate the electromagnetic fields associated with the radiation arising in these two situations. Various applications of the formulae presented in these subsections will be taken up in the rest of this chapter.

It may be mentioned that the emission of electromagnetic radiation by atoms and molecules is an important topic in quantum physics. Much of the discussion in this chapter will be restricted to the emission of radiation by classical systems. However, in Sects. 7.7–7.10 towards the end of this chapter, we shall apply our results to study the emission and scattering of radiation by microscopic particles. While a full analysis may require a quantum treatment, we shall see that we can reach many important conclusions on the basis of a classical treatment.

7.1.1 Electromagnetic Fields Due to a Non-relativistically Moving Charged Particle

The general expression for the electric field due to a moving charged particle is given by (6.48). If the charged particle is moving non-relativistically, then we can neglect terms of order $|\mathbf{v}|/c$ and write r for s defined by (6.32). Then (6.48) reduces to

$$\mathbf{E} = \frac{q}{4\pi\epsilon_0}\frac{\mathbf{r}}{r^3} + \frac{q}{4\pi\epsilon_0 c^2}\frac{[\mathbf{r} \times (\mathbf{r} \times \dot{\mathbf{v}})]}{r^3}. \tag{7.1}$$

Before proceeding further, we point out some important differences between Newtonian mechanics and Maxwellian electrodynamics—the two pillars of classical physics. In Newtonian mechanics, many of the important results we derive from the basic equations—such as the laws of conservation of momentum and kinetic energy—normally hold in their Newtonian form only for particles moving at non-relativistic speeds. As we pointed out in Sect. 5.8, if some laws are expected to hold in relativistic situations, we have to write down Lorentz-covariant equations which reduce to the usual equations of Newtonian mechanics in the non-relativistic limit. However, the laws of Newtonian mechanics, in the forms in which they are usually written down in elementary textbooks, may not hold when the particles move relativistically. On the other hand, since Maxwellian electrodynamics is consistent with special relativity, results obtained *without neglecting any terms in Maxwell's equations* are valid even when relativistic motions are concerned. To get the non-relativistic versions of the results, we have to throw away terms of the order $|\mathbf{v}|/c$. That is what we had done to get (7.1) from (6.48). However, (6.48) itself has been derived without making any non-relativistic approximations and is expected to hold even when the particle motion is relativistic. Note that while deriving (6.48), we have never explicitly taken Lorentz transformations between frames of reference into account. We can begin with the ansatz that Maxwell's equations hold in one frame of reference and carry on the analyses of Sects. 4.10 and 6.1–6.4 in that frame to arrive at (6.48). Since Maxwellian electrodynamics is consistent with relativity, even without taking relativistic considerations explicitly into account, we have arrived at an equation like (6.48) that is valid for relativistic motions. As in the case of covariant formulation of electrodynamics presented in Sect. 5.9, it is possible to put (6.48) in a form which makes it manifestly clear that (6.48) would hold in all inertial frames connected by Lorentz transformation. We do not present that analysis in this book. We merely point out that within our frame of reference in which we are carrying out our analysis, (6.48) holds even when the motion of the charged particle is relativistic.

The first term in the RHS of (7.1) obviously gives the 'electrostatic' field produced by the charged particle, as in (2.6). If we substitute this term in (6.55), then we get the magnetic field produced by the moving charge, which is seen to be the same as (3.72). If we want to get the magnetic field due to a charged particle moving with a uniform relativistic speed, then we have to substitute (6.50) into (6.55), making it clear that (3.72) is an approximate expression giving the magnetic field of a non-relativistic particle.

We now turn to the second term on the RHS of (7.1), which is non-zero only when the motion of the charged particle is non-uniform, leading to a non-zero value of $\dot{\mathbf{v}}$. We have already pointed out at the end of Sect. 6.5 that the deceleration caused by the sudden stopping of a uniformly moving charged particle produces electromagnetic radiation, suggesting that electromagnetic radiation can be produced by the acceleration of a charged particle. We are now going to argue that the second term on the RHS of (7.1) indeed corresponds to the electric field associated with electromagnetic radiation. We write this electric field as

$$\mathbf{E}_{rad} = \frac{q}{4\pi\epsilon_0 c^2} \frac{[\mathbf{r} \times (\mathbf{r} \times \dot{\mathbf{v}})]}{r^3}. \tag{7.2}$$

From the nature of the triple vector product, one can easily determine that this electric field is transverse to the radial direction $\mathbf{r}$ and lies in the plane containing the vectors $\mathbf{r}$ and $\dot{\mathbf{v}}$, as shown in Fig. 7.1 where we indicate the electric field at the field point P due to the accelerated charge at S. If θ is the angle between $\mathbf{r}$ and $\dot{\mathbf{v}}$, then the magnitude of the electric field is easily found to be

$$|\mathbf{E}_{rad}| = \frac{q}{4\pi\epsilon_0 c^2} \frac{\dot{v}\sin\theta}{r}. \tag{7.3}$$

The important point to note here is that this field falls off as r^{-1}. We are now going to argue that this dependence r^{-1} of the field is an indication that this field may be associated with radiation.

The magnetic field corresponding to $\mathbf{E}_{rad}$ can be found from (6.49). We easily see that this magnetic field is going go be transverse to both $\mathbf{r}$ and $\mathbf{E}_{rad}$, which means that

Fig. 7.1 The 'radiation' part
of the electric field at point
P due to a charged particle at
S having acceleration $\dot{\mathbf{v}}$

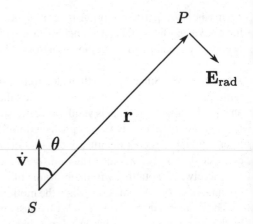

it is going to be perpendicular to the plane of Fig. 7.1 at P. We also find its amplitude
from (6.49), which gives

$$c\,|\mathbf{B}_{rad}| = |\mathbf{E}_{rad}|, \tag{7.4}$$

exactly the same as the relation (4.52) we have in the case of electromagnetic waves.
It is clear from (7.4) that the $\mathbf{B}_{rad}$ also has to fall off as r^{-1} like the electric field $\mathbf{E}_{rad}$.
Then the Poynting vector defined by (4.19), which gives the energy flux, falls off as
r^{-2}. This has a very important implication. Consider successive spherical surfaces
centred around the source. If the energy flux falls off as r^{-2}, then the same amount
of energy will pass through these successive spheres. This is what we expect in the
case of radiation.

We thus conclude that the electric and magnetic fields associated with radiation
have to fall off as r^{-1}. We have concluded in Sect. 2.13 that the electrostatic potential
due to a distribution of charges in a finite region cannot fall off slower than r^{-1}, which
means that the electric field cannot fall off slower than r^{-2}. Note that the electric
field can fall off faster as r^{-3} if the total electric charge is zero so that the first term
in (2.96) disappears. We stress again that these results hold only when the charges
are confined within a finite volume. The electric field due to an infinite line charge
falls off as r^{-1}. We then argued in Sect. 3.5 that the vector potential due to currents
confined in a finite region falls off as r^{-2} so that the magnetic field has to fall off as
r^{-3}. It is useful to keep the following rule of thumb in mind. Suppose we have some
localized charges and currents confined within a finite volume. The electrostatic and
magnetostatic fields would respectively fall off as r^{-2} and r^{-3}, whereas the emission
of radiation is indicated by the presence of electric and magnetic fields falling as r^{-1}
and satisfying (7.4).

We conclude from (7.1) that the ratio of the two terms on the RHS is approximately
given by

$$\frac{|\text{Radiation term}|}{|\text{Electrostatic term}|} \approx \frac{q\dot{v}/c^2 r}{q/r^2} \approx \frac{\dot{v}r}{c^2}. \tag{7.5}$$

At short distances

$$r \ll \frac{c^2}{\dot{v}},$$

the electrostatic term is the dominant term. On the other hand, at large distances

$$r \gg \frac{c^2}{\dot{v}},$$

the radiation term is the dominant term. So when we are considering the radiation going out from a source in a faraway region, we have to consider only the radiation field. While the general expression of the radiation field given by the second term of (6.48) is complicated, certainly (7.2) is not too difficult of an expression to handle. We shall present further discussions on the radiation from a non-relativistically moving accelerated charge in Sect. 7.2.

7.1.2 Electromagnetic Fields Due to Oscillating Currents

We now consider the oscillating current

$$\mathbf{j}(\mathbf{x}', t') = \mathbf{j}(\mathbf{x}') \, e^{-i\omega t'} \tag{7.6}$$

as a possible source of radiation. We have to substitute this in (6.25) to get the vector potential $\mathbf{A}(\mathbf{x}, t)$ at the field point $\mathbf{x}$ at time t. Following (6.23), we have

$$[\mathbf{j}] = \mathbf{j}\left(\mathbf{x}', t - \frac{|\mathbf{x} - \mathbf{x}'|}{c}\right) = \mathbf{j}(\mathbf{x}') \, e^{-i\omega t} \, e^{i\frac{\omega}{c}|\mathbf{x} - \mathbf{x}'|} \tag{7.7}$$

following from (7.6). Substituting this in (6.25) and writing $\omega/c = k$, we get

$$\mathbf{A}(\mathbf{x}, t) = \frac{\mu_0}{4\pi} e^{-i\omega t} \int \frac{\mathbf{j}(\mathbf{x}')}{|\mathbf{x} - \mathbf{x}'|} \, e^{ik|\mathbf{x} - \mathbf{x}'|} \, dV'. \tag{7.8}$$

Note that signals propagating with speed c starting from different regions of the source (with different $\mathbf{x}'$) will take different amounts of propagation time to reach the field point $\mathbf{x}$. As a result, when we want to obtain the field at $\mathbf{x}$ at a particular time t, the retarded time t' given by (6.26) will be different for different regions of the source. For an oscillatory source, this means that different regions of the source will contribute to the field when they are at different phases. This is the origin of the phase factor $e^{ik|\mathbf{x} - \mathbf{x}'|}$ in (7.8).

To get the magnetic field, we have to take the curl of (7.8), following (4.128). Keeping in mind that the operator $\nabla \times$ would operate only on the unprimed field variables and not on the primed source variables, we have

$$\mathbf{B}(\mathbf{x}, t) = \frac{\mu_0}{4\pi} e^{-i\omega t} \int \nabla \left(\frac{e^{ik|\mathbf{x}-\mathbf{x}'|}}{|\mathbf{x} - \mathbf{x}'|} \right) \times \mathbf{j}(\mathbf{x}') \, dV' \tag{7.9}$$

using the vector identity (B.5). Using (1.17), it is easy to show that

$$\nabla(e^{ik|\mathbf{x}-\mathbf{x}'|}) = ike^{ik|\mathbf{x}-\mathbf{x}'|} \frac{\mathbf{x} - \mathbf{x}'}{|\mathbf{x} - \mathbf{x}'|} = i\mathbf{k} \, e^{ik|\mathbf{x}-\mathbf{x}'|}, \tag{7.10}$$

where $\mathbf{k}$ is a vector in the direction of $\mathbf{x} - \mathbf{x}'$ having amplitude k. Using (1.18), (7.10) and (B.3), we have from (7.9) that

$$\mathbf{B}(\mathbf{x}, t) = \frac{\mu_0}{4\pi} e^{-i\omega t} \left[\int \frac{\mathbf{j}(\mathbf{x}') \times (\mathbf{x} - \mathbf{x}')}{|\mathbf{x} - \mathbf{x}'|^3} e^{ik|\mathbf{x}-\mathbf{x}'|} dV' + \int \frac{e^{ik|\mathbf{x}-\mathbf{x}'|}}{|\mathbf{x} - \mathbf{x}'|} i\mathbf{k} \times \mathbf{j}(\mathbf{x}') \, dV' \right]. \tag{7.11}$$

Comparing with (3.4), we realize that the first term in the RHS of (7.11) is reminiscent of the expression of the magnetic field due to a static current distribution. We can think of this term as the 'magnetostatic' term corresponding to the current (7.7) at the retarded time. For an oscillatory current, the magnetic field given by the first term in the RHS of (7.11) will be oscillatory and will have some properties of an electromagnetic wave. But it will not give rise to a finite energy flux at large distances from the source. This first term in the RHS of (7.11) does not correspond to the emission of radiation.

Anticipating that the second term in the RHS of (7.11) would correspond to the emission of radiation, we write

$$\mathbf{B}_{\mathrm{rad}}(\mathbf{x}, t) = i \frac{\mu_0}{4\pi} e^{-i\omega t} \int \frac{e^{ik|\mathbf{x}-\mathbf{x}'|}}{|\mathbf{x} - \mathbf{x}'|} \mathbf{k} \times \mathbf{j}(\mathbf{x}') \, dV'. \tag{7.12}$$

We now consider a field point at a distance much larger than the size L' of the source region. Then the variations in the value of $|\mathbf{x} - \mathbf{x}'|$ over different points of the source will be small compared to its overall value. This means that we can write $|\mathbf{x} - \mathbf{x}'| = r$ in the denominator of (7.11) and take it out of the integration. Also, $\mathbf{k}$, which is in the direction of $\mathbf{x} - \mathbf{x}'$ and will be different for different source points, can be taken to be constant when considering a distant point. We can then write (7.12) as

$$\mathbf{B}_{\mathrm{rad}}(\mathbf{x}, t) = i \frac{\mu_0}{4\pi} \frac{e^{-i\omega t}}{r} \mathbf{k} \times \int e^{ik|\mathbf{x}-\mathbf{x}'|} \mathbf{j}(\mathbf{x}') \, dV'. \tag{7.13}$$

Note that although we have taken $r = |\mathbf{x} - \mathbf{x}'|$ out of the integration in the denominator of (7.13), we have kept the phase term in which it appears inside the integration. The reason is this. If kL' is of order unity and larger, then phases of different points inside the source at their respective retarded times can be quite different, causing large variations in the phase factor $e^{ik|\mathbf{x}-\mathbf{x}'|}$ even when the field point is at a large distance. It should be clear from our discussion that $k = \omega/c$ should be interpreted as the wavenumber, which is related to wavelength by $k = 2\pi/\lambda$. The condition $kL' \geq 1$ translates into $L' \geq \lambda$. Thus, we assume different lengths in our problem to satisfy the relation

$$r \gg L' \geq \lambda. \tag{7.14}$$

Let us choose the origin inside the finite region within which currents are confined. Then $|\mathbf{x}'|$ can at most be of order L'. This suggests from (7.14) that $r \gg |\mathbf{x}'|$, and it is easy to show in this situation

$$k\,|\mathbf{x} - \mathbf{x}'| \approx kr - \mathbf{k}\,.\,\mathbf{x}',$$

if the condition (7.14) holds. Substituting this in (7.13), we get

$$\mathbf{B}_{\mathrm{rad}}(\mathbf{x}, t) = i\,\frac{\mu_0}{4\pi}\,\frac{e^{i(kr-\omega t)}}{r}\,\mathbf{k} \times \int e^{-i\mathbf{k}.\mathbf{x}'}\,\mathbf{j}(\mathbf{x}')\,dV'. \tag{7.15}$$

It is clear from this expression that $\mathbf{B}_{\mathrm{rad}}(\mathbf{x}, t)$ is produced by contributions from different regions of the source at different phases.

We see in (7.15) that $\mathbf{B}_{\mathrm{rad}}(\mathbf{x}, t)$ falls off as r^{-1}. We have already pointed out at the end of Sect. 7.1.1 that a field falling off as r^{-1} may be an indication that the field is associated with radiation. The $e^{i(kr-\omega t)}$ dependence in (7.15) makes it clear that this field varies as a wave propagating in the radially outward direction. Since $\mathbf{k}$ is the propagation vector at the field point, and the expression of $\mathbf{B}_{\mathrm{rad}}(\mathbf{x}, t)$ as given by (7.15) involves a cross product with $\mathbf{k}$, we conclude that this magnetic field at the field point is in a direction transverse to the propagation vector. We can find the electric field associated with this magnetic field by using (4.36), which involves taking a curl of $\mathbf{B}_{\mathrm{rad}}(\mathbf{x}, t)$. From (7.15) and the vector identity (B.5), we find that one term in the expression of the curl is going to be

$$(\nabla \times \mathbf{B}_{\mathrm{rad}})_1 = i\,\frac{\mu_0}{4\pi}\,\nabla[e^{i(kr-\omega t)}] \times \frac{\left[\mathbf{k} \times \int e^{-i\mathbf{k}.\mathbf{x}'}\mathbf{j}(\mathbf{x}')\,dV'\right]}{r}. \tag{7.16}$$

Using formula (C.5) and keeping in mind that $\mathbf{k}/k$ gives the unit vector in the radial direction, we have

$$\nabla[e^{i(kr-\omega t)}] = \frac{\partial}{\partial r}[e^{i(kr-\omega t)}]\,\frac{\mathbf{k}}{k} = e^{i(kr-\omega t)}\,i\,\mathbf{k}.$$

Substituting this in (7.16) and comparing with (7.15), we find that this term in the expression of $\nabla \times \mathbf{B}_{\mathrm{rad}}$ is given by

$$[\nabla \times \mathbf{B}_{\mathrm{rad}}]_1 = i\,\mathbf{k} \times \mathbf{B}_{\mathrm{rad}}. \tag{7.17}$$

We are now going to argue that only this term in the expression of $\nabla \times \mathbf{B}_{rad}$ is going to contribute to the emission of radiation. If we keep only this term in $\nabla \times \mathbf{B}_{rad}$, then (4.36) gives

$$-\frac{i\omega}{c^2}\mathbf{E}_{rad} = i\,\mathbf{k} \times \mathbf{B}_{rad}. \tag{7.18}$$

Since $\mathbf{B}_{rad}$ falls off as r^{-1}, the electric field $\mathbf{E}_{rad}$ given by (7.18) would also fall off as r^{-1}. With $\mathbf{B}_{rad}$ and $\mathbf{E}_{rad}$ falling off as r^{-1}, the Poynting vector introduced in (4.19) should fall off as r^{-2}, which we expect in the case of radiation emission. We leave it for the readers to argue that if we keep other terms in $\nabla \times \mathbf{B}_{rad}$ besides the term shown in (7.16), those terms would give rise to a part of the electric field falling faster than r^{-1}, which would not contribute to the energy flux emitted. However, although this other part of the electric field may not contribute to the energy flux, it may still be important in understanding the structure of the fields associated with the radiation emitted. An example discussed in Sect. 7.5 will illustrate these issues.

For the time being, we proceed with the understanding that the relevant part of the electric field we need to consider is given by (7.18). Then $\mathbf{E}_{rad}$, $\mathbf{B}_{rad}$ and $\mathbf{k}$ would make a triad of orthogonal vectors. It easily follows that the amplitudes of $\mathbf{E}_{rad}$ and $\mathbf{B}_{rad}$ would be related as in (7.4). We thus conclude that the electromagnetic fields produced by oscillating currents have parts which satisfy all the characteristics of electromagnetic waves we derived in Sect. 4.4.

To sum up, when we want to study radiation emission from a system of oscillating currents, the magnetic field associated with the radiation can be calculated by using (7.15). We shall work out an example in Sect. 7.3. Once we have $\mathbf{B}_{rad}$, we can determine $\mathbf{E}_{rad}$ from (7.4) and then the energy flux radiated as given by (4.19).

7.2   Larmor's Formula of Radiation Emission from an Accelerated Charge

Let us consider a particle with charge q moving non-relativistically with acceleration $\dot{\mathbf{v}}$. We have already discussed in Sect. 7.1.1 that such a charged particle emits electromagnetic radiation and that the electric field associated with this radiation at a field point at distance $\mathbf{r}$ from the charge is given by (7.2). This electric field $\mathbf{E}_{rad}$ is transverse to the radial direction $\mathbf{r}$ in which the electromagnetic radiation must be propagating. The magnetic field $\mathbf{B}_{rad}$ given by (6.49) is transverse to both $\mathbf{E}_{rad}$ and $\mathbf{r}$. It is easy to determine that the Poynting vector $\mathbf{N}$ at the field point $\mathbf{r}$ defined by (4.19) would be directed radially outward and would give the energy flux per unit area. The amplitude N of the Poynting vector is given by

$$N = |\mathbf{E}_{rad}||\mathbf{H}_{rad}| = \frac{|\mathbf{E}_{rad}||\mathbf{B}_{rad}|}{\mu_0} = \frac{|\mathbf{E}_{rad}|^2}{c\mu_0} \tag{7.19}$$

because of (7.4). Substituting from (7.3), we get

Fig. 7.2 A polar plot of the flux of radiation emitted in different directions by a non-relativistic accelerated charged particle

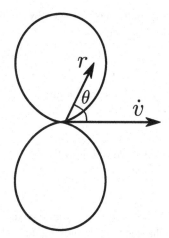

$$N = \frac{1}{c\mu_0}\left(\frac{q}{4\pi\epsilon_0 c^2}\right)^2 \frac{\dot{v}^2\sin^2\theta}{r^2} = \frac{1}{16\pi^2\epsilon_0 c^3}\frac{q^2\dot{v}^2}{r^2}\sin^2\theta. \qquad (7.20)$$

This expression gives the energy flux emitted from the accelerated charged particle in different directions. Keep in mind that θ is the angle between the acceleration $\dot{v}$ and the position vector $\mathbf{r}$ giving the direction of the field point. A polar plot of the flux in different directions is shown in Fig. 7.2. It is interesting to note that no radiation is emitted in the direction of acceleration $\dot{v}$. The maximum flux is in the direction $\theta = \pi/2$ perpendicular to $\dot{v}$.

Let us now determine the flux emitted into a solid angle $d\Omega$ from the accelerated charged particle. We consider a sphere of radius r (on which the field point lies) centred around the charged particle. An element of area $r^2 d\Omega$ of this surface will be within the solid angle $d\Omega$. Since N given by (7.20) is the flux per unit area, the flux through the element of area $r^2 d\Omega$ is obviously

$$dP = Nr^2 d\Omega = \frac{q^2\dot{v}^2}{16\pi^2\epsilon_0 c^3}\sin^2\theta \, d\Omega. \qquad (7.21)$$

Note that this expression is independent of r and should be considered as the energy flux emitted into the solid angle which will pass through an element of area inside this solid angle placed at any distance. We have already stressed the point that a requirement for the flux to be independent of distance is that both the electric and magnetic fields associated with the radiation should fall off as r^{-1} so that the Poynting vector falls off as r^{-2}.

To get the total energy radiated by the accelerated charged particle in unit time, we have to integrate (7.21) over all solid angles. Since the flux given in (7.21) is axisymmetric (i.e. independent of the azimuthal angle ϕ), we can consider the solid angle between two cones at θ and $\theta + d\theta$. Since this solid angle is $d\Omega = 2\pi\sin\theta \, d\theta$, the total energy emitted by the particle per unit time in all directions is found from (7.21) to be

$$P = \frac{q^2 \dot{v}^2}{16\pi^2 \epsilon_0 c^3} \, 2\pi \int_0^\pi \sin^3 \theta \, d\theta.$$

An easy integration gives

$$\int_0^\pi \sin^3 \theta \, d\theta = \frac{4}{3}$$

so that we end up with

$$P = \frac{q^2 \dot{v}^2}{6\pi \epsilon_0 c^3}. \tag{7.22}$$

This is *Larmor's formula* giving the energy radiated per unit time from an accelerated charged particle moving non-relativistically.

If the charged particle is moving relativistically, then it is much more complicated to calculate the emission of radiation from it. We have to start from the term involving $\dot{v}$ in the full expression of the electric field given in (6.48) rather than the term involving $\dot{v}$ in the non-relativistic expression (7.1). We shall not present a detailed discussion of this topic in this book. However, if one is interested only in figuring out the qualitative nature of the radiation emitted by a relativistically moving charged particle, one can apply some simple arguments. We have pointed out the relativistic beaming effect in Sect. 5.2. The relativistically moving particle would certainly emit radiation in its own frame in accordance with (7.20) shown as a polar plot in Fig. 7.2. In our frame, however, we would see the radiation beamed into the forward direction of relativistic motion in a cone with opening angle of order $1/\gamma$, as given in (5.27). Let us consider the two cases of a charged particle moving with relativistic velocity $\mathbf{v}$ and having acceleration $\dot{\mathbf{v}}$ (i) parallel to $\mathbf{v}$ and (ii) perpendicular to $\mathbf{v}$. We expect the polar plot in Fig. 7.2 to be distorted in such a fashion that much of the radiation would come out in the forward direction within an opening angle of order $1/\gamma$. A little reflection should convince the reader that the kinds of polar plots we may expect in these two cases are as shown in Fig. 7.3. Detailed calculations using the full term involving $\dot{v}$ in (6.48) show that radiation patterns in these two cases indeed turn out to be as shown

Fig. 7.3 Polar plots of the flux of radiation emitted in different directions by a charged particle moving relativistically with velocity $\mathbf{v}$ and having acceleration $\dot{\mathbf{v}}$ for the two cases: **a** $\dot{\mathbf{v}}$ is parallel to $\mathbf{v}$, and **b** $\dot{\mathbf{v}}$ is perpendicular $\mathbf{v}$

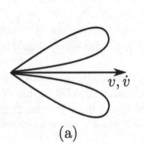

(a)

(b)

in Fig. 7.3. Readers interested in looking at the detailed calculations, which lead to fairly involved expressions, may turn to Panofsky and Phillips ([1], pp. 359–373) or Landau and Lifshitz ([2], Sect. 73). Here, we make only the following point. By combining the idea of relativistic beaming with the results of a non-relativistically moving particle, we can have a qualitative idea of the nature of the radiation emitted by a relativistically moving charged particle without doing the actual complicated calculations.

7.3 Radiation from a Centre-Fed Linear Antenna

On the basis of the discussion in Sect. 7.1.2, we expect alternating currents to cause emission of electromagnetic radiation. A configuration of conductors though which an alternating current is made to flow in order to produce electromagnetic radiation is called an *electrical antenna*. The formula (7.15) is the basic formula from which the radiation field due to an antenna can be calculated. We now give an illustration of the use of this formula by applying it to a simple antenna: made up of two conducting rods along the z axis—one extending from $z = -L/2$ to $z = 0$ and the other from $z = 0$ to $z = L/2$—as shown in Fig. 7.4. An alternating current is made to pass into the rods from the central region. The amplitude of the alternating current is maximum at some points near the centre and has to fall to zero at the extreme edges of the antenna. We can take the spatial part of the current to be of the form

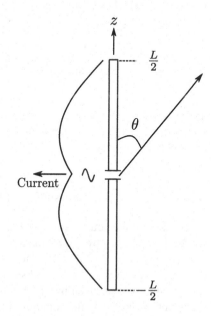

Fig. 7.4 The centre-fed linear antenna with an indication of the current flowing inside it. The current distribution shown is obtained from (7.23) for a typical case in which $kL = 3\pi/2$

$$I(z) = I_0 \frac{\sin\left(\frac{kL}{2} - k|z|\right)}{\sin\left(\frac{kL}{2}\right)}. \tag{7.23}$$

The full expression of the alternating current within the antenna can be obtained by multiplying this by $e^{-i\omega t}$.

We now have to evaluate the integral in (7.15) for this current in the antenna. Let us denote the integral by $\mathbf{X}$, that is,

$$\mathbf{X} = \int e^{-i\mathbf{k}.\mathbf{x}'}\, \mathbf{j}(\mathbf{x}')\, dV'. \tag{7.24}$$

We have been using primes to set apart source variables from the field variables. Now that we are dealing with an integral involving source variables alone, let us suppress the primes for simplicity. When we are dealing with currents through conductors, we know that $\mathbf{j}\, dV$ has to be replaced by $I\, d\mathbf{l}$ with $d\mathbf{l} = dz\, \hat{\mathbf{e}}_z$ in this case. Then we get from (7.24) that

$$\mathbf{X} = \hat{\mathbf{e}}_z \int_{-L/2}^{L/2} e^{-ikz\cos\theta} I(z)\, dz \tag{7.25}$$

on noting that $\mathbf{k}.\mathbf{x} = kz\cos\theta$ in this case, where θ is the angle which the direction of the field point makes with the z axis, as shown in Fig. 7.4. This integral is reminiscent of the integral we encounter when we calculate diffraction from a slit (see, for example, Born and Wolf [3], Sect. 8.5.1). We normally assume that all points in the slit contribute equally to the diffraction. Here, however, we have to use $I(z)$ given by (7.23). On substituting for $I(z)$ from (7.23) into (7.25) and carrying out the integration, we get

$$\mathbf{X} = \hat{\mathbf{e}}_z \frac{2I_0}{k\sin^2\theta\,\sin(kL/2)} \left[\cos\left(\frac{kL}{2}\cos\theta\right) - \cos\left(\frac{kL}{2}\right)\right]. \tag{7.26}$$

Using (7.24) and (7.26), we conclude from (7.15) that the magnetic field $\mathbf{B}_{\mathrm{rad}}(\mathbf{x}, t)$ associated with radiation at the field point is transverse to both $\mathbf{k}$ and $\hat{\mathbf{e}}_z$, which means that it is in the azimuthal direction perpendicular to the plane of Fig. 7.4. Its amplitude is given by

$$B_{\mathrm{rad}}(\mathbf{x}, t) = i\, \frac{\mu_0}{4\pi r} \frac{2I_0\, e^{i(kr-\omega t)}}{\sin\theta\,\sin(kL/2)} \left[\cos\left(\frac{kL}{2}\cos\theta\right) - \cos\left(\frac{kL}{2}\right)\right]. \tag{7.27}$$

We have pointed out in Sect. 7.1.2 that there would be an electric field $\mathbf{E}_{\mathrm{rad}}(\mathbf{x}, t)$ at the field point given by (7.18) transverse to both $\mathbf{k}$ and $\mathbf{B}_{\mathrm{rad}}(\mathbf{x}, t)$, which means that this electric field would lie in the plane of Fig. 7.4 perpendicular to the radial direction. The amplitude of this field would satisfy the relation (7.4). It is easy to check that the Poynting vector given by (4.19) will be in the radially outward direction. Note that when we are using the complex notation, the real parts in the expressions of the fields should be used for calculating the Poynting vector. Since both the electric and magnetic fields would vary sinusoidally, the Poynting vector would have a factor $\sin^2(kr - \omega t)$ of which the average value is $\frac{1}{2}$. Keeping this in mind, the amplitude of the average Poynting vector is

$$\overline{N} = \frac{|\mathbf{E}_{\text{rad}}||\mathbf{B}_{\text{rad}}|}{2\mu_0} = \frac{c\,|\mathbf{B}_{\text{rad}}|^2}{2\mu_0} \tag{7.28}$$

on using (7.4). We already pointed out in Sect. 7.2 that we have to multiply N by $r^2 d\Omega$ to get the energy flux dP radiated into the solid angle $d\Omega$. This means that

$$\frac{d\overline{P}}{d\Omega} = \overline{N}r^2 = r^2 \frac{c\,|\mathbf{B}_{\text{rad}}|^2}{2\mu_0}. \tag{7.29}$$

On substituting from (7.27), we finally arrive at

$$\frac{d\overline{P}}{d\Omega} = \frac{c\mu_0}{8\pi^2} I_0^2 \left[\frac{\cos\left(\frac{kL}{2}\cos\theta\right) - \cos\left(\frac{kL}{2}\right)}{\sin\theta \sin(\frac{kL}{2})} \right]^2. \tag{7.30}$$

This gives the radiation pattern from the antenna in different directions. We thus come to the end of this illustrative example to show how (7.15) can be used to determine the radiation emission from systems having alternating currents.

7.4 The Dipole Approximation

If the wavelength of radiation $\lambda = 2\pi/k = 2\pi c/\omega$ from a system having an alternating current turns out to be larger than the size L' of the system, then certain simplifications result. In Sect. 7.1.2, we assumed the condition (7.14). Let us now rather consider the condition

$$r \gg \lambda \gg L'. \tag{7.31}$$

In this case, $|\mathbf{k}.\mathbf{x}'|$ inside the radiating system will be much smaller than 1. Then, the exponential in (7.24) will give a series of increasingly smaller terms. If we keep only the first two terms, then we have

$$\int e^{-i\mathbf{k}.\mathbf{x}'} \mathbf{j}(\mathbf{x}')\,dV' = \int \mathbf{j}(\mathbf{x}')\,dV' - i\int (\mathbf{k}.\mathbf{x}')\,\mathbf{j}(\mathbf{x}')\,dV'. \tag{7.32}$$

If we were to keep only the first term in the RHS and substitute it in (7.15), then we get

$$\mathbf{B}_{\text{rad}}(\mathbf{x}, t) = i\,\frac{\mu_0}{4\pi}\frac{e^{i(kr-\omega t)}}{r}\,\mathbf{k} \times \int \mathbf{j}(\mathbf{x}')\,dV'. \tag{7.33}$$

Replacing (7.15) by (7.33) is called the *dipole approximation*. We shall argue below that (7.33) implies the source of radiation to be an electric dipole. On the other hand, if we substitute the second term in the RHS of (7.32) into (7.15), then what we get can be shown to be radiation from an electric quadrupole and a magnetic dipole. We shall make a few comments about this in Sect. 7.6. The physical significance of (7.33) is that the phase differences within different regions of the source can be neglected when considering their contributions to produce the magnetic field at the field point. We certainly expect the phase differences to be negligible when the source is much smaller than the wavelength of the radiation.

To show that (7.33) corresponds to radiation from an electric dipole, let us consider the volume integral of the quantity $\nabla'.[\mathbf{j}(\mathbf{x}', t')] x'_\alpha$, where $\mathbf{j}(\mathbf{x}', t')$ is as given in (7.6), and x'_α is a component of $\mathbf{x}'$. We have

$$\int \nabla'.[\mathbf{j}(\mathbf{x}', t')] x'_\alpha \, dV' = \int \nabla'.[x'_\alpha \mathbf{j}(\mathbf{x}', t')] \, dV' - \int \mathbf{j}(\mathbf{x}', t') . \nabla' x'_\alpha \, dV'$$

on using (B.4). The first term in the RHS can be transformed into a surface integral by Gauss's theorem and can easily be shown to be zero if the source is located completely within the region over which the integration is carried on. So we keep only the second term in the RHS, which gives

$$\int \nabla'.[\mathbf{j}(\mathbf{x}', t')] x'_\alpha \, dV' = - \int j_\alpha(\mathbf{x}', t') \, dV'.$$

As we shall have such an equation for each component α, we conclude

$$\int \mathbf{j}(\mathbf{x}', t') \, dV' = - \int \nabla'.[\mathbf{j}(\mathbf{x}', t')] \mathbf{x}' \, dV'.$$

Using the charge conservation equation (4.7), we get

$$\int \mathbf{j}(\mathbf{x}', t') \, dV' = \int \frac{\partial \rho}{\partial t'} \mathbf{x}' \, dV' = -i\omega \int \rho \mathbf{x}' \, dV', \qquad (7.34)$$

on using the fact that the charge density ρ at any point inside the source would vary as $e^{-i\omega t'}$ just like $\mathbf{j}(\mathbf{x}', t')$ as given in (7.6). It may be mentioned that while discussing the multipole expansion in magnetostatics in Sect. 3.5, we took $\int \mathbf{j}(\mathbf{x}', t') \, dV' = 0$, which is clearly the case in a static situation if the integration is done over the entire source region.

We introduced the electric dipole moment for an axisymmetric system defined through (2.94). For a more general charge distribution, the electric dipole moment is given by

$$\mathbf{p} = \int \rho \mathbf{x}' \, dV'. \qquad (7.35)$$

Given that ρ varies as $e^{-i\omega t'}$, we expect $\mathbf{p}$ also to vary that way and can write

$$\mathbf{p} = \mathbf{p}_0 \, e^{-i\omega t'}. \tag{7.36}$$

From (7.34), (7.35) and (7.36), we write

$$\int \mathbf{j}(\mathbf{x}', t') \, dV' = -i\omega \, \mathbf{p} = -i\omega \, \mathbf{p}_0 \, e^{-i\omega t'}. \tag{7.37}$$

On the basis of (7.6), we finally write

$$\int \mathbf{j}(\mathbf{x}') \, dV' = -i\omega \, \mathbf{p}_0. \tag{7.38}$$

Substituting this in (7.33), we get

$$\mathbf{B}_{\text{rad}}(\mathbf{x}, t) = \frac{\mu_0 \omega}{4\pi r} (\mathbf{k} \times \mathbf{p}_0) \, e^{i(kr - \omega t)}. \tag{7.39}$$

This gives the magnetic field associated with radiation at the field point. The energy flux radiated into the solid angle would be given by (7.29). Substituting for $\mathbf{B}_{\text{rad}}$ from (7.39), we get

$$\frac{d\overline{P}}{d\Omega} = \frac{\omega^4 p_0^2}{32\pi^2 \epsilon_0 c^3} \sin^2 \theta, \tag{7.40}$$

where θ is the angle which $\mathbf{k}$ makes with $\mathbf{p}_0$, and we have used (4.42). The angular dependence of the energy flux in different directions is given by the same $\sin^2 \theta$ factor which we encountered in Sect. 7.2 while discussing radiation from an accelerated charge. To get the total energy emitted per unit time in all directions, we have to integrate (7.40) over all solid angles, exactly the way we integrated (7.21). The result is

$$\overline{P} = \frac{\omega^4 p_0^2}{12\pi \epsilon_0 c^3}. \tag{7.41}$$

Let us now pause for a moment and think over the significance of what we have obtained. If we have oscillating currents given by (7.6) confined in a region much smaller than the wavelength of the radiation that would be emitted, then we can use (7.38) to get the equivalent electric dipole moment $\mathbf{p}_0$ of the system. Once we have the equivalent electric moment of the system, we can find the radiation emitted by the system by using (7.40) and (7.41). Even in the case of the centre-fed linear antenna discussed in Sect. 7.3, we can do this if the wavelength is much larger than the size L of the system. Let us now show this. We are considering the limit $kL \ll 1$. In this limit, the current given by (7.23) becomes

$$I(z) = I_0 \left(1 - \frac{2|z|}{L} \right). \tag{7.42}$$

Now, from (7.38), the equivalent electric dipole moment of the antenna is given by

$$-i\omega p_0 = I_0 \int_{-L/2}^{L/2} \left(1 - \frac{2|z|}{L}\right) dz = \frac{1}{2} I_0 L. \tag{7.43}$$

Substituting from (7.43) for p_0 into (7.40), we get

$$\frac{d\overline{P}}{d\Omega} = \frac{\omega^2 L^2 I_0^2}{128\pi^2 \epsilon_0 c^3} \sin^2 \theta. \tag{7.44}$$

It is easy to check that (7.30) reduces to this in the limit of small kL.

If we substitute from (7.43) for p_0 into (7.41), then we get

$$\overline{P} = \frac{\omega^2 L^2 I_0^2}{48\pi \epsilon_0 c^3}. \tag{7.45}$$

We can write this as

$$\overline{P} = \frac{1}{2} R_{\text{rad}} I_0^2, \tag{7.46}$$

where

$$R_{\text{rad}} = \frac{\omega^2 L^2}{24\pi \epsilon_0 c^3}.$$

Writing $\omega = kc = 2\pi c/\lambda$ (where λ is the wavelength of the emitted radiation) and using (4.42), this becomes

$$R_{\text{rad}} = \frac{\pi}{6} \sqrt{\frac{\mu_0}{\epsilon_0}} \left(\frac{L}{\lambda}\right)^2 = 197 \left(\frac{L}{\lambda}\right)^2 \text{ ohms.} \tag{7.47}$$

on substituting the values of ϵ_0 and μ_0 given in (2.8) and (3.11), respectively. Let us now interpret (7.47). When an alternating current is sent into a resistance R, some energy is dissipated inside this resistance, and the power supply source has to supply an average power

$$\frac{1}{2} R I_0^2$$

to provide the energy being dissipated. Looking at (7.46), we can interpret R_{rad} as the *radiative resistance* of the antenna. The power which is radiated from the antenna is supplied by the power supply as if the antenna has a resistance given by (7.47).

Lastly, we point out that oscillatory currents inside conductors mean that charges inside the conductors undergo oscillatory acceleration. Let us apply the formalism developed in Sect. 7.1.1 to a case of oscillatory charges. Let there be a few charged particles in a localized region, and let $\mathbf{x}_i$ and $\mathbf{v}_i = \dot{\mathbf{x}}_i$ be the position and velocity of the charge q_i. If we consider the field point far away and take $\mathbf{r}$ to be the distance of the field point from any charge, then from (7.2), we have

$$\mathbf{E}_{\text{rad}} = \frac{1}{4\pi\epsilon_0 c^2} \frac{[\mathbf{r} \times (\mathbf{r} \times \sum_i q_i \mathbf{v}_i)]}{r^3}.$$

The discrete version of (7.35) is

$$\mathbf{p} = \sum_i q_i \mathbf{x}_i \qquad (7.48)$$

so that

$$\mathbf{E}_{\text{rad}} = \frac{1}{4\pi\epsilon_0 c^2} \frac{[\mathbf{r} \times (\mathbf{r} \times \ddot{\mathbf{p}})]}{r^3}. \qquad (7.49)$$

If the charges are undergoing oscillatory motions given by $e^{-i\omega t'}$, then $\mathbf{E}_{\text{rad}}$ will be oscillatory with amplitude

$$|\mathbf{E}_{\text{rad}}| = \frac{\omega^2 p_0}{4\pi\epsilon_0 c^2 r} \sin\theta, \qquad (7.50)$$

where θ is the angle which $\mathbf{r}$ makes with $\mathbf{p}_0$. It may be noted that in this discussion we have not allowed for phase differences in the contributions from various charges, which means that we have tacitly assumed dipole approximation. If we obtain the magnetic field from this electric field by using (7.4), it is straightforward to check the magnetic field will be the same as given by (7.39).

In Sects. 7.1.1 and 7.1.2, we have discussed two procedures for calculating emission of radiation—one for accelerated charges and one for oscillatory currents. We expect them to give the same results in situations where both the procedures are applicable. Since (7.2) does not take account of the phase, it is properly applicable only in the situation of dipole approximation when we deal with several charges. If oscillatory charges or currents are confined within a region of small size compared to the wavelength of radiation, then we can introduce an equivalent electric dipole through (7.38) or (7.48). We have shown that both the procedures we have discussed give identical results in this situation.

7.5 Radiation Field from an Oscillating Electric Dipole

In a sense, an oscillating electric dipole is the prototype system emitting electromagnetic radiation. As we have discussed in Sect. 7.4, if oscillatory charges or currents are confined within a region small compared to the wavelength of the emitted radiation, then we can replace these charges or currents by an equivalent electric dipole, and the electromagnetic fields at large distances can be thought of as produced by this oscillating electric dipole. When we discussed emission of radiation by oscillatory currents in Sects. 7.3 and 7.4, we mainly focussed our attention on the energy fluxes and did not discuss the structure of the electromagnetic fields associated with the radiation. Now we shall point out how to find the full structure of the electromagnetic fields associated with the radiation emitted by an oscillating electric dipole. We repeat again that any system of oscillating charges or currents, if the dipole approximation holds, would produce exactly the same electromagnetic fields at large distances.

As we have already pointed out, the oscillating electric dipole (7.36) gives rise to the magnetic field (7.39). It is in the azimuthal direction and can be written as

$$\mathbf{B}_{\text{rad}}(\mathbf{x}, t) = \frac{\mu_0 \, \omega^2 p_0}{4\pi c r} \sin\theta \, e^{i(kr - \omega t)} \, \hat{\mathbf{e}}_\phi, \tag{7.51}$$

where θ is the angle between $\mathbf{p}_0$ and the direction of the field point. Since we shall now be concerned only with the radiation field, we are going to suppress the subscript 'rad' in the present discussion. The electric field associated with this magnetic field can be obtained from (4.36). In the discussion in Sect. 7.1.2, we considered only the transverse component of the electric field which was sufficient for calculating the energy flux given by the Poynting vector. Now we want to look at the full structure of the electric field. From the formula of the curl in spherical coordinates given by (C.7), we write down the components

$$-\frac{i\omega}{c^2} E_r = \frac{1}{r \sin\theta} \frac{\partial}{\partial\theta} (\sin\theta B_\phi), \tag{7.52}$$

$$-\frac{i\omega}{c^2} E_\theta = -\frac{1}{r} \frac{\partial}{\partial r} (r B_\phi). \tag{7.53}$$

With B_ϕ as given in (7.51) falling off as r^{-1}, it is easy check that E_θ given by (7.53) should have a term falling off as r^{-1}. It is only this term which will contribute to the Poynting vector. However, we are now interested in finding the nature of the full electric field. By substituting B_ϕ given by (7.51) in (7.52) and (7.53), we can get the full expressions of the electric field associated with radiation. We now show that the structure of the electric field can be found out in a simpler way without writing down the full expressions of its components.

The electric field lines would satisfy the equation

$$\frac{dr}{E_r} = \frac{r \, d\theta}{E_\theta}. \tag{7.54}$$

Substituting for E_r and E_θ from (7.52) and (7.53), we get

$$\frac{1}{r} \frac{\partial}{\partial r} (r B_\phi) \, dr + \frac{1}{r \sin\theta} \frac{\partial}{\partial\theta} (\sin\theta B_\phi) \, r \, d\theta = 0,$$

from which

$$\frac{\partial}{\partial r} (r \sin\theta B_\phi) \, dr + \frac{\partial}{\partial\theta} (r \sin\theta B_\phi) \, d\theta = 0.$$

Fig. 7.5 The electric field
lines at an instant associated
with the radiation emitted by
an oscillating electric dipole
at large distances

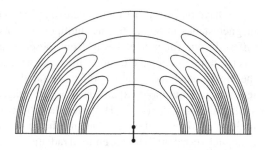

This means that

$$d(r \sin\theta B_\phi) = 0$$

along electric field lines. In other words, $r \sin\theta B_\phi$ has to be constant along these field lines. Using the real part of B_ϕ from (7.51), we conclude that

$$\sin^2\theta \cos(kr - \omega t) = \text{constant} \qquad (7.55)$$

along electric field lines. The electric field lines at a certain time, say $t = 0$, can be easily obtained by plotting the contours over which $\sin^2\theta \cos kr$ would have constant values. Figure 7.5 shows the electric field lines associated at an instant of time with the radiation emitted by the oscillating electric dipole.

7.6   A Short Note on Multipole Radiation

We have seen in (7.15) that the radiation field due to oscillating currents involves the integral $\int e^{-i\mathbf{k}.\mathbf{x}'}\, \mathbf{j}(\mathbf{x}')\, dV'$. For some simple current distributions, it is possible to do this integration in a closed form. This is the case for the centre-fed linear antenna discussed in Sect. 7.3.

If the size of the emitting system is much smaller then the wavelength of the emitted radiation, then $|\mathbf{k}.\mathbf{x}'| \ll 1$ throughout the source region, and we can use the series expansion for $e^{-i\mathbf{k}.\mathbf{x}'}$ in the integral, leading to the first two terms as shown in (7.32). In Sects. 7.4 and 7.5, our discussion was focussed on the first term, which is shown to correspond to the radiation from an oscillating electric dipole. Since the successive terms become smaller by the factor kL' (where L' is the size of the source of radiation), the higher-order terms are usually not of much practical importance, unless the lower terms turn out to be zero for some reason. Nevertheless, the higher-order terms, which lead to what is called *multipole radiation*, are sometimes of considerable theoretical interest. This subject has been discussed by Panofsky and Phillips ([1], pp. 260–270) and Jackson ([4], pp. 429–444) in some detail. Apart from making a few general remarks, we refrain from getting into a detailed discussion of this mathematically involved subject.

We have discussed the multipole expansion in electrostatics in Sect. 2.13 and in magnetostatics in Sect. 3.6. The multipole expansion of electromagnetic radiation differs in one fundamental way from these. In the multipole expansions both in electrostatics and magnetostatics, the successive terms go as higher powers of $1/r$, the higher-order terms falling off more rapidly with distance. In contrast, all the terms in the multipole expansion of electromagnetic radiation fall as $1/r$, which should be clear from (7.15). However, the different multipole terms involve different multipole moments of the current and charge distribution, leading to different kinds of angular distribution for the emitted radiation. The second term in the RHS of (7.32) can be shown to be a combination of the magnetic dipole radiation and the electric quadrupole radiation (Landau and Lifshitz [5], Sect. 71; Jackson [4], Sect. 9.3). While we shall not get into a detailed discussion of the mathematical analysis, let us give an indication how this result arises. One particular component (say the α component) of the second term in the RHS of (7.32) can be written as

$$-i\,\mathbf{k}.\int\, j_\alpha(\mathbf{x}')\,\mathbf{x}'\,dV'.$$

Note that the integral appearing here is the same as the integral in (3.25) in our discussion of the magnetic moment due to localized currents. Using (3.26), we can split this integral into symmetric and anti-symmetric parts. The anti-symmetric part corresponds to the magnetic moment as in Sect. 3.6. The symmetric part was shown to be zero in Sect. 3.6 by making use of the fact $\nabla'.\mathbf{j} = 0$, which is true in a static situation. Since we do not generally have $\nabla'.\mathbf{j} = 0$ in the dynamic situation we are now considering, the symmetric term is not going to be generally zero. It is this term which gives rise to the electric quadrupole radiation, while the anti-symmetric term gives the magnetic dipole radiation.

We have mentioned that the electric dipole radiation term may be zero in some particular situation. Then the next order term becomes the leading term. Readers may have an idea of how things go in such a situation by working out Exercise 7.3. One obvious example of such a situation is the radiation arising from an alternating current passing through a circular loop. It is easy to see that $\int \mathbf{j}(\mathbf{x}')\,dV'$ must be zero in this case, since elements of the circular loop located in opposite sides would make opposite contributions. As we have seen in Sect. 3.6, such a circular loop is equivalent to a magnetic dipole. It is, therefore, no surprise that an alternating current passing through a circular loop produces magnetic dipole radiation. This case is discussed in Griffiths ([6], Sect. 11.1.3).

7.7   Radiation Damping

We now turn our attention to a problem which does not have a proper resolution to this day. In fact, what we are going to discuss is one of the baffling puzzles which show that classical electrodynamics is not a fully self-consistent theory in spite of its many spectacular successes. When an accelerated charge radiates, it gives out some energy. Where does this energy come from? In some situations, there may be a source which is responsible for the acceleration of the charged particles, and we may expect the source to supply the energy which is radiated away. When we considered the radiation from an antenna in Sects. 7.3–7.4, we tacitly assumed that there must be some source of the alternating current which is supplying the energy. We pointed out in Sect. 7.4 that the energy radiated from the antenna introduces an effective resistance. However, there is not always such an identifiable source from which the energy radiated by a source can come. The only possibility in the case of an isolated accelerated charged particle is that the energy being radiated away must come from the kinetic energy of motion of the charged particle. If that happens, then we expect that the motion of the charged particle will slow down. In other words, we would expect the equation of motion of the charged particle to have a damping term to allow for the loss of energy due to radiation. However, the expression (1.5) for the electromagnetic force on a charged particle clearly does not have a damping term. One possibility is that there should be an additional damping force which is perhaps negligible in most practical situations, and (1.5) is an approximate expression of the electromagnetic force when we neglect this term. We can indeed argue that the damping term is going to be unimportant in practical situations. However, the problem is that it has not been possible to include a damping term to make electromagnetic theory self-consistent even at a conceptual level.

Let us first try to determine the conditions under which damping due to radiation emission is expected to be important. If P is the rate at which a particle moving at speed v is radiating away energy, then the time scale in which the particle will lose a significant amount of its kinetic energy is

$$T_{\text{rad}} \approx \frac{mv^2}{P}.$$

Using the expression of P as given by Larmor's formula (7.22), we get

$$T_{\text{rad}} \approx \frac{6\pi\epsilon_0 mc^3}{q^2} \left(\frac{v}{\dot{v}}\right)^2. \tag{7.56}$$

Keeping in mind that $v/\dot{v} \approx t_d$ is the dynamical time in which the speed of the particle changes significantly, we can write (7.56) in the form

$$\frac{T_{\text{rad}}}{t_d} \approx \frac{t_d}{\tau}, \tag{7.57}$$

where

$$\tau = \frac{q^2}{6\pi\epsilon_0 mc^3}. \tag{7.58}$$

We can estimate the value of τ for an electron by using the charge e and the mass m_e of the electron, which gives

$$\tau_e = \frac{e^2}{6\pi\epsilon_0 m_e c^3} \approx 10^{-23} \text{ s}. \tag{7.59}$$

Note that this is several orders of magnitude smaller than even the period of X-rays (X-rays of wavelength ≈ 1 Å having period of order 3×10^{-19} s). This means that, under reasonably expected circumstances, the dynamical time t_d for an electron is expected to be much larger than τ_e, implying that the RHS of (7.57) is much larger than 1. It is then clear from (7.57) that the time scale T_{rad} for radiation damping has to be much larger than the dynamical time t_d. In other words, an electron is expected to lose an insignificant part of its kinetic energy due to radiation damping during its dynamical time scale. When we study the dynamics of an electron over the dynamical time, the radiation damping is unimportant, and we can use (1.5) to study the dynamics of the electron.

Even if radiation damping is unimportant in most practical situations, self-consistency at a conceptual level demands that there should be a term arising out of radiation damping in the equation of motion. We now try to determine this term. We expect that the full equation of motion for a charged particle should be

$$m\dot{\mathbf{v}} = \mathbf{F} + \mathbf{F}_{rad} \tag{7.60}$$

with $\mathbf{F}$ given by (1.5). The radiation damping term $\mathbf{F}_{rad}$ implies that the charged particle would lose energy at the rate $-\mathbf{F}_{rad}.\mathbf{v}$, which must be equated with the energy loss rate due to radiation given by (7.22), that is, we should have

$$- \mathbf{F}_{rad}.\mathbf{v} = \frac{q^2 \dot{v}^2}{6\pi\epsilon_0 c^3} = m\tau\dot{v}^2 \tag{7.61}$$

on making use of (7.58). It is clearly not possible to write $\mathbf{F}_{rad}$ in such a way that (7.61) is satisfied at all times. If the charged particle is undergoing a periodic motion, the best that we can do is to demand that (7.61) averaged over time is satisfied. This requires

$$- \int_{t_1}^{t_2} \mathbf{F}_{rad}.\mathbf{v}\, dt = m\tau \int_{t_1}^{t_2} \dot{\mathbf{v}}.\dot{\mathbf{v}}\, dt.$$

An integration by parts of the RHS gives

$$- \int_{t_1}^{t_2} \mathbf{F}_{rad}.\mathbf{v}\, dt = m\tau \left[\dot{\mathbf{v}}.\mathbf{v}|_{t_1}^{t_2} - \int_{t_1}^{t_2} \ddot{\mathbf{v}}.\mathbf{v}\, dt \right]. \tag{7.62}$$

If t_1 and t_2 differ by a multiple of the period of the periodically moving particle, then $\dot{\mathbf{v}}.\mathbf{v}$ is expected to have the same value at times $t = t_1$ and $t = t_2$. In that case, (7.62) is satisfied if we take

$$\mathbf{F}_{rad} = m\tau\dddot{\mathbf{v}}. \tag{7.63}$$

This suggests a force which depends on the time derivative of acceleration. From (7.60), we now write down

$$m(\dot{\mathbf{v}} - \tau\ddot{\mathbf{v}}) = \mathbf{F}, \tag{7.64}$$

which is known as the *Abraham–Lorentz equation of motion*. This equation has many undesirable features. When $\mathbf{F} = 0$, certainly $\mathbf{v} = 0$ is a solution which we may expect. However, we also get a spurious solution growing exponentially

$$\mathbf{v} \propto e^{t/\tau}.$$

Since we had to assume the motion to be periodic in order to derive (7.64), we need to restrict ourselves only to periodic solutions of (7.64), and an exponentially growing solution cannot be considered.

There are other difficulties at the conceptual level. A term depending on the time derivative of acceleration cannot be incorporated within the theoretical structure of classical mechanics easily. An idea of central importance in classical mechanics is that the state of a dynamical system can be prescribed by giving the generalized coordinates and generalized momenta. If we have equations giving time derivatives of the coordinates and the momenta, then we can study how the state of the system evolves in time. In order to integrate the dynamical equations from a certain time, we need the values of the position and momentum coordinates at that time as initial conditions. If a dynamical equation has a time derivative of the acceleration, then that upsets this whole elegant scheme. In such a situation, we would need the initial value of acceleration also as an additional initial condition. Even Lagrangian or Hamiltonian formulations of such systems are not possible.

As we pointed out, the radiation damping may not be not very important in most of the practical applications of electromagnetic theory we may be interested in. However, our inability of coming up with a consistent formulation of this damping even at the fundamental level shows that the classical subject of electromagnetic theory still has problems at the conceptual level which we do not know how to fix. Not only are we unable to write down an expression of radiation damping which is valid at each instant, even the time-averaged expression we arrived at has many undesirable features such as introducing a time derivative of the acceleration in the dynamical theory. At the end of our discussion on the momentum of electromagnetic fields in Sect. 4.3, we mentioned the effort of building a model of the electron on the assumption that its mass is of electromagnetic origin—an effort pioneered by Lorentz. One aim of this model was to derive the radiation damping of an electron by considering forces on one part of the electron by other parts. Since this model was eventually abandoned as mentioned in Sect. 4.3, we shall not get into a detailed discussion of this model. We refer to Feynman, Leighton and Sands ([7], Chap. 28) for an illuminating discussion of this fascinating subject.

While developing the theory of optical dispersion in Sect. 4.9, we introduced a damping term $\gamma(d\mathbf{x}/dt)$ in (4.119). Now, we are concluding from (7.64) that the radiation damping term should be of the form $-\tau(d^3\mathbf{x}/dt^3)$. For oscillatory motions of the form $e^{-i\omega t}$, the two damping terms are going to be equivalent if

$$\gamma = \tau\omega^2. \tag{7.65}$$

In the theory of optical dispersion developed in Sect. 4.9, we assumed that γ is independent of t. However, a dependence on ω can be allowed for γ. The derivation presented in Sect. 4.9 will be fully valid even if γ is taken to be of the form (7.65).

7.8 Scattering of Electromagnetic Radiation by an Electron

We now discuss a phenomenon which turns out to be very important in many different situations. Consider a beam of electromagnetic radiation impinging on an electron. As we concluded in (4.53), the magnetic force due to this wave will be negligible compared to the electric force. Let us neglect the magnetic force and consider the effect of only the electric field of the wave on this electron. The oscillatory electric field will cause an oscillatory acceleration of the electron. The electron undergoing this oscillatory acceleration will definitely radiate out electromagnetic waves having the same frequency as the original electromagnetic wave. The energy of the electromagnetic waves radiating out from the accelerated electron must come from the energy of the original electromagnetic wave responsible for the acceleration of the electron. It would appear that some energy of the impinging beam of electromagnetic radiation is getting scattered by the electron in different directions. This phenomenon is called the *Thomson scattering*. If we can determine the average rate at which energy is radiated away from the accelerated electron and divide that by the intensity of the original beam, that should give us the cross section of the Thomson scattering. We now calculate this cross section first for a plane-polarized incident beam of electromagnetic radiation and then for an unpolarized beam—assuming the electron to be a free particle. Only in Sect. 7.8.3 shall we come to the question of how to treat this problem if the electron in bound inside an atom.

It may be pointed out that some of the results we are going to derive are quite general and should hold for any charged particle. However, we shall see that many of crucial formulae are going to have the mass of the particle in the denominator. As a result, other things being the same, the effects we are going to discuss are much more important for electrons due to the small mass of the electron.

7.8.1 The Case of Plane-Polarized Electromagnetic Radiation

Let us consider an electromagnetic wave propagating in the z direction with the electric field of the wave in the x direction. The electric field of the incident beam at the location of the electron would be of the form $E_0 \sin\omega t\, \hat{\mathbf{e}}_x$. The acceleration produced in the electron by this electric field is

Fig. 7.6 A sketch indicating an electromagnetic wave propagating in the z direction and the effect of the electric field of this wave in the x direction on an electron. We discuss in the text how we can calculate the energy flux in the direction making an angle θ with the propagation direction

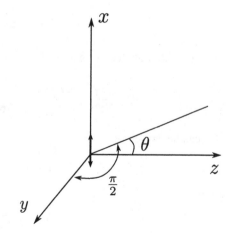

$$\ddot{\mathbf{x}} = -\frac{e}{m_e} E_0 \sin \omega t \, \hat{\mathbf{e}}_x. \tag{7.66}$$

We now want to determine the Poynting vector associated with the radiation emitted by the electron in a direction lying in the xz plane (i.e. the plane of polarization) and making an angle θ with the z direction (i.e. the propagation direction of the incident beam), as shown in Fig. 7.6. For this purpose, we can use (7.20), keeping in mind that θ in (7.20) was the angle made with the direction of acceleration which is now $\pi/2 - \theta$. Because of this, $\sin \theta$ appearing in (7.20) has now to be replaced by $\cos \theta$. Substituting the expression of acceleration given by (7.66) in (7.20), we get

$$N = \frac{1}{16\pi^2 \epsilon_0 c^3} \frac{e^4 E_0^2}{m_e^2 r^2} \cos^2 \theta \sin^2 \omega t.$$

On taking the time average, the mean amplitude of the Poynting vector is

$$\overline{N} = \frac{1}{16\pi^2 \epsilon_0 c^3} \frac{e^4 E_0^2}{m_e^2 r^2} \frac{1}{2} \cos^2 \theta.$$

The mean energy flux $d\overline{P}$ going into a solid angle $d\Omega$ in the direction we are considering is given by (7.21), from which

$$\frac{d\overline{P}}{d\Omega} = \overline{N} r^2 = \frac{e^4 E_0^2}{16\pi^2 \epsilon_0 m_e^2 c^3} \frac{1}{2} \cos^2 \theta. \tag{7.67}$$

Now, the intensity of the incident beam is given by the time-averaged amplitude of the Poynting vector associated with it. We already calculated this for a beam of electromagnetic radiation in Sect. 4.4.2. From (4.40) and (4.59), the intensity of the incident beam in free space is given by

$$\overline{N}_{\text{inc}} = \frac{1}{2}\epsilon_0 \, cE_0^2.$$

Dividing (7.67) by this, we get the differential scattering cross section

$$\frac{d\sigma_T}{d\Omega} = \frac{d\overline{P}/d\Omega}{\overline{N}_{\text{inc}}} = \frac{e^4}{16\pi^2\epsilon_0^2 m_e^2 c^4} \cos^2\theta.$$

This can be written as

$$\frac{d\sigma_T}{d\Omega} = r_0^2 \cos^2\theta, \tag{7.68}$$

where

$$r_0 = \frac{e^2}{4\pi\epsilon_0 m_e c^2} \tag{7.69}$$

has the dimension of length and is known as the *classical electron radius*. On substituting the numerical values of the various quantities appearing in its expression including the charge e and the mass m_e of the electron, we find

$$r_0 = 2.82 \times 10^{-15} \text{ m.} \tag{7.70}$$

We again stress that the expression (7.68) for the differential scattering cross section is valid only for scattering in the plane of polarization. When we consider scattering in a direction not in the plane of polarization, we have to consider the angle α which this direction would make with direction of the electric field in the wave, which is also the direction of acceleration of the electron. It is easy to check that $\cos^2\theta$ in (7.68) has to be replaced by $\sin^2\alpha$ to get the differential scattering cross section in this direction.

7.8.2 The Case of Unpolarized Electromagnetic Radiation

If the electromagnetic wave propagating in the z direction is unpolarized, then there would be electric fields in both x and y directions at the location of the electron on which this wave is impinging. As a result, the electron will have oscillatory accelerations in both x and y directions—both contributing to the scattering in a particular direction. We again try to find the differential scattering cross section in a direction in the xz plane making an angle θ with the z direction in which the wave is propagating. The contribution due to the acceleration of the electron in the x direction to the energy flux is still given by (7.67). However, to get the contribution due to the acceleration of the electron in the y direction, we need to keep in mind that θ appearing in (7.20) or (7.21) should be taken as $\pi/2$ in this situation, since a direction in the xz plane makes an angle $\pi/2$ with the y axis (see Fig. 7.6). This means that we have to take $\sin\theta = 1$ when we use (7.21) to find the flux due to the

y acceleration. Combining contributions due to accelerations of the electron both in x and y directions, the energy flux in the direction of consideration in the xz plane is given by

$$\frac{d\overline{P}}{d\Omega} = \frac{e^4}{16\pi^2\epsilon_0 m_e^2 c^3} \frac{1}{2}(E_{x,0}^2 \cos^2\theta + E_{y,0}^2).$$

Since $E_{x,0} = E_{y,0}$ for a fully unpolarized electromagnetic wave, we would have

$$\frac{d\overline{P}}{d\Omega} = \frac{e^4 E_{x,0}^2}{16\pi^2\epsilon_0 m_e^2 c^3} \frac{1}{2}(1 + \cos^2\theta). \tag{7.71}$$

The intensity of the beam of unpolarized radiation is clearly

$$\overline{N}_{\text{inc}} = \frac{1}{2}\epsilon_0 c(E_{x,0}^2 + E_{y,0}^2) = \epsilon_0 c E_{x,0}^2.$$

Dividing (7.71) by this, the differential scattering cross section for an unpolarized beam of electromagnetic wave due to Thomson scattering is

$$\frac{d\sigma_T}{d\Omega} = \frac{1}{2}r_0^2(1 + \cos^2\theta) \tag{7.72}$$

making use of (7.69).

Since all the directions around the propagation direction of a fully unpolarized beam would be symmetric, (7.72) is the expression for a differential scattering cross section in any direction making an angle θ with the propagation direction. To get the total scattering cross section, we merely have to integrate (7.72) over all directions, that is,

$$\sigma_T = \int \frac{d\sigma_T}{d\Omega} d\Omega = \pi r_0^2 \int_0^\pi (1 + \cos^2\theta) \sin\theta \, d\theta.$$

The easy integration gives the expression

$$\sigma_T = \frac{8\pi}{3}r_0^2. \tag{7.73}$$

for the Thomson scattering cross section. Readers are asked to show that differential scattering cross section (7.68) for plane-polarized electromagnetic radiation also leads to the same expression (7.73) for unpolarized radiation (Exercise 7.5).

It may be mentioned that a more complete theory of the interaction of a beam of radiation with an electron would require going beyond classical physics and treating the radiation as a beam of photons. If the energy $\hbar\omega$ of photons in the beam is comparable to or more than the rest mass energy $m_e c^2$ of the electron, then it is possible for a photon to transfer a part of its energy to the electron so that the scattered photon has less energy, leading to a lower frequency. This is called the *Compton scattering*. The cross section of the Compton scattering (known as the *Klein–Nishina formula*) can be calculated by the techniques of quantum electrodynamics (see, for example, Bjorken and Drell [10], Sect. 7.7). In the limit $\hbar\omega \ll m_e c^2$, the Compton cross section given by the Klein–Nishina formula reduces to the Thomson cross section, and the frequency shift of the photon also tends to zero. Since $\hbar\omega$ is much less than $m_e c^2$ for visible light, the quantum effects can be neglected, and the interaction of visible light with an electron can be treated classically.

Since r_0 given by (7.70) is several orders of magnitude smaller than the size of an atom, it may seem that the Thomson cross section given by (7.73) would be too small to have any relevance in a practical situation. A little reflection shows that this is not true. In a gaseous medium like air, the electrons are bound inside atoms and cannot respond to an electromagnetic wave in the same way as a free electron would respond. In Sect. 7.8.3, we shall present a phenomenological treatment of the effect of an electromagnetic wave on electrons bound inside atoms. However, if all the electrons inside all the air molecules could come out, then a beam of light propagating through the air could have been attenuated in a few metres due to the Thomson scattering (see Exercise 7.6).

7.8.3   The Case of Harmonically Bound Electrons

After discussing the scattering of electromagnetic radiation by free electrons, we now discuss how electrons bound inside atoms would respond to a beam of electromagnetic radiation. Although a proper treatment of this subject would require quantum mechanics, even a classical treatment by assuming that the electrons are harmonically bound inside atoms enables us to explain many observed phenomena. We assume that there is a restoring force $-m_e\omega_0^2\mathbf{x}$ acting on an electron when it is displaced by an amount $\mathbf{x}$ from its 'equilibrium position' inside the atom. Although we expect the radiation damping to be very small in most practical situations, we are going to include the radiation damping term (7.63) in our analysis. As we shall find, there will be a resonance at the normal frequency ω_0 of the electron, at which various physical quantities will become infinitely large if the damping is exactly zero. The equation of motion of the electron harmonically bound to the atom with restoring force $-m_e\omega_0^2\mathbf{x}$ and subject to the electric field $\mathbf{E}$ of the electromagnetic wave is

$$m_e \frac{d^2\mathbf{x}}{dt^2} - m_e\tau\frac{d^3\mathbf{x}}{dt^3} = -e\mathbf{E} - m_e\omega_0^2\mathbf{x}, \tag{7.74}$$

which follows easily from (7.64). If the electric field is taken to be of the form $\mathbf{E} = \mathbf{E}_0 e^{-i\omega t}$, then the solution of (7.74) is

$$\mathbf{x} = \frac{eE_0/m_e}{\omega^2 - \omega_0^2 + i\omega^3\tau}\, e^{-i\omega t}, \tag{7.75}$$

Fig. 7.7 A plot of the scattering cross section of electromagnetic radiation due to a harmonically bound electron, as given by (7.77) with $\omega_0\tau = 0.5$ (note that the actual value of $\omega_0\tau$ is expected to be much smaller, making the peak sharper)

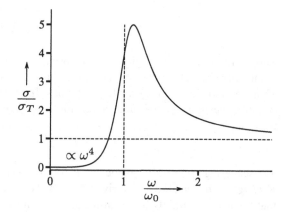

from which the acceleration is easily determined to be

$$\frac{d^2\mathbf{x}}{dt^2} = \frac{\omega^2 e E_0/m_e}{\omega^2 - \omega_0^2 + i\omega_0^3\tau}\, e^{-i\omega t}. \tag{7.76}$$

Note that the last term $i\omega^3\tau$ in the denominator is replaced by $i\omega_0^3\tau$, since the smallness of τ evident from (7.59) makes it clear that this term would make a significant contribution only when $\omega \approx \omega_0$ (making the remainder of the denominator close to zero). Using the expression (7.76) of the acceleration rather than (7.66), we have to repeat the analysis of Sects. 7.8.1 and 7.8.2 to obtain the scattering cross section, keeping in mind that we now have to use the acceleration given by (7.76) for $\dot{v}$ appearing in Larmor's formula (7.22). We leave this easy analysis to the reader, who should be able to show that the scattering cross section in this case is given by

$$\sigma = \sigma_T \frac{\omega^4}{(\omega^2 - \omega_0^2)^2 + (\omega_0^3\tau)^2}, \tag{7.77}$$

where σ_T is the Thomson scattering cross section given by (7.73).

Figure 7.7 shows a plot of σ/σ_T on taking $\tau = 1/2\omega_0$. The resonance peak becomes sharper on making τ smaller. Since scattering removes some radiation from the original beam, a plot of the scattering cross section can be thought of as a plot of the absorption coefficient. A sharp resonance at $\omega = \omega_0$ provides the classical theory of an absorption line.

It is instructive to consider the two limits away from the resonance. When $\omega \gg \omega_0$, (7.77) gives $\sigma \to \sigma_T$. It is not difficult to interpret this result. An electron responds like a free electron to an electromagnetic wave of sufficiently high frequency, and we get back the Thomson scattering cross section appropriate for free electrons. In the opposite limit, we get from (7.77) that

$$\omega \ll \omega_0 : \quad \sigma \to \sigma_T \left(\frac{\omega}{\omega_0}\right)^4 . \tag{7.78}$$

This corresponds to the region near the origin in Fig. 7.7. It is clear that the cross section of bound electrons for scattering light would be suppressed compared to the Thomson cross section σ_T for free electrons. In terms of the wavelength λ of the incoming beam of electromagnetic radiation instead of the frequency ω, we have in the limit

$$\sigma \propto \lambda^{-4}. \tag{7.79}$$

The scattering of electromagnetic waves following such a λ^{-4} dependence is called *Rayleigh scattering*. We have arrived at this λ^{-4} dependence by considering the limit that electrons are tightly bound inside atoms or molecules. However, we would like to point out that this λ^{-4} dependence can arise from more general considerations. When $\omega \ll \omega_0$, it follows from (7.75) that the displacement of charges would be approximately proportional to the instantaneous electric field, without the frequency ω appearing in the constant of proportionality. This implies that the oscillatory electric field of an electromagnetic wave would produce a polarization inside an object with bound electrons which is proportional to the electric field at an instant of time, but independent of frequency. The oscillatory polarization arising in this manner would cause the emission of electromagnetic waves, giving rise to a scattering cross section. It easily follows from (7.41) that the scattering cross section would have the λ^{-4} dependence if the polarization produced in the object is proportional to the electric field of the electromagnetic wave. You are asked to show this for a small dielectric sphere in Exercise 7.7. It is indeed found that the scattering of light in many situations approximately obeys the Rayleigh scattering law of λ^{-4} dependence.

We may expect molecules of air to scatter light (i.e. electromagnetic waves of low frequency compared to ω_0 corresponding to the binding of electrons inside atoms or molecules) by following the λ^{-4} dependence of the *Rayleigh scattering*. This dependence on wavelength implies that blue light with shorter wavelength should be scattered more than the red light. The blue colour of the sky can be explained on the basis of this. Molecules in the air scatter the blue colour of the sunlight more than the red colour into our eyes. The reddish colour of the rising or the setting Sun can also be explained on the basis of this wavelength dependence. When the Sun is near the horizon, a beam of sunlight has to traverse a longer path through the atmosphere. If more blue light is scattered away from this beam compared to red light, then the Sun would naturally appear redder.

Finally, we point out that the expression (7.77) of the scattering cross section is reminiscent of the expression (4.126) of the imaginary part of the refractive index derived in Sect. 4.9—especially if we assume that the damping used in the theory of dispersion is due to radiation damping and make use of (7.65). This is not an accident, but rather due to the fact that the theory of scattering has a deep connection with the theory of optical dispersion. The imaginary part of the refractive index implies that the intensity of a propagating beam of radiation would be attenuated as it passes through the medium. The scattering also has the effect of attenuating a beam of radiation by removing some energy from the incident beam. In fact, if (7.65) holds, one can show that the attenuation given by the theory of optical dispersion is exactly equivalent to what we would get on using the expression (7.77) of the scattering cross section (Exercise 7.8).

7.9   Cyclotron and Synchrotron Radiation from Charged Particle Moving in Magnetic Field

A standard result of electromagnetic theory is that a charged particle can move in a helical path in a uniform magnetic field. Since the Lorentz force $q\mathbf{v} \times \mathbf{B}$ does not have a component parallel to the magnetic field, the motion parallel to the magnetic field can only be uniform motion without acceleration and will not cause any emission of radiation. When we are interested in considering the emission of electromagnetic radiation by the charged particle, we need to consider only the circular motion in the plane perpendicular to the magnetic field, since the Lorentz force and the acceleration produced by it will lie in this plane. We have discussed the relativistic version of this problem in Sect. 5.11. Now we want to determine the kind of electromagnetic radiation which will be emitted by charged particles moving in circular paths perpendicular to the magnetic field, both when the motion is non-relativistic and when it is relativistic. The acceleration associated with a circular motion is always directed towards the centre of the circle. This is clearly a case of acceleration perpendicular to the velocity.

Let us first consider the charged particle to move non-relativistically. The gyration frequency is the cyclotron frequency ω_c given by (5.124), and the pattern of radiation with respect to the radially inward direction of acceleration should be as shown as in Fig. 7.2. The lobes of radiation will sweep across a stationary observer with the period of the gyrating charged particle. The observer will perceive this as radiation emitted at frequency ω_c. This is known as *cyclotron radiation*.

The situation changes considerably if the charged particle is moving relativistically. The full mathematical treatment of the problem leads to somewhat involved expressions derived in some books (Panofsky and Phillips [1], Sect. 20–24; Landau and Lifshitz [2], Sect. 74). We present here a heuristic discussion. Due to the relativistic beaming effect discussed in Sect. 5.2, a stationary observer will see an angular distribution of radiation as shown in Fig. 7.3b, with the radiation coming out mainly within a cone of opening angle $1/\gamma$ in the direction of motion of the particle. Only if the observer lies within this cone of angle $1/\gamma$ will the observer see the radiation from the electron. As shown in Sect. 5.11, the frequency of gyration of the charged particle, which is given by (5.123), can be written as ω_c/γ, where ω_c is the cyclotron frequency given by (5.124). Figure 7.8 shows a charged particle moving in a circular orbit, with the observer in the plane. When the charged particle is at position A, the observer comes within the cone of radiation and starts receiving the radiation. When the charged particle reaches B, the observer moves out of the cone and ceases to

Fig. 7.8 A sketch
illustrating how synchrotron
radiation arises. The observer
receives the radiation emitted
by the charged particle only
during its transit from A to B

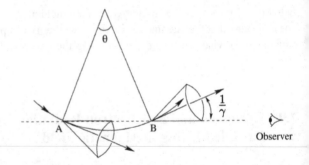

receive any more radiation. We now need to determine the duration of time during
which the observer receives the radiation.

Let L be the distance between A and B, which is also the arc length between them
if θ is small. The charged particle moving with speed v takes time L/v to travel from
A to B. This is the interval of time between the emissions of the earliest and latest
radiations which are seen by the observer. Keeping in mind that the radiation from
B takes time L/c less to travel to the observer compared to the radiation from A, it
should be clear that the time during which the observer receives the radiation is

$$\Delta t = \frac{L}{v} - \frac{L}{c} = \frac{L}{v}\left(1 - \frac{v}{c}\right). \tag{7.80}$$

Since $\theta \approx 2/\gamma$ in this case, we can write

$$\frac{L}{v} = \frac{\theta}{v/r} \approx \frac{2/\gamma}{\omega_c/\gamma} \approx \frac{2}{\omega_c}. \tag{7.81}$$

Also,

$$1 - \frac{v}{c} = \frac{1 - v^2/c^2}{1 + v/c} \approx \frac{1}{2\gamma^2} \tag{7.82}$$

if $v \approx c$. On making use of (7.81) and (7.82), we get from (7.80) that

$$\Delta t \approx \frac{1}{\gamma^2 \omega_c}. \tag{7.83}$$

Hence, as the charged particle gyrates around the magnetic field, the observer receives
a radiation pulse of this duration once every gyration period. If we take a Fourier
transform of this signal, the spectrum should peak at a frequency of about $\gamma^2 \omega_c$.
Clearly, the spectrum of radiation given out by highly relativistic charged particle's
gyrating in a magnetic field can be quite different from the spectrum due to non-
relativistic charged particle's. This type of radiation is called *synchrotron radiation*.

Although cyclotron or synchrotron radiation in principle can be emitted by any kind of charged particle gyrating around a magnetic field, the low mass of electrons makes them particularly efficient as sources of such radiation.

7.10 Bremsstrahlung

A hot gas cools by emitting radiation. The last application we want to consider on the subject of emission of electromagnetic radiation is to take a brief look at the physics of the emission of radiation from hot gases. If atoms in the hot gas are in excited states, then they make transitions to low-energy states by emitting photons. However, if the gas is very hot, then atoms would be broken into ions and electrons. The physics of this plasma state of a very hot gas will be discussed in the next chapter. In such a state in which a gas consists of many charged particles, these particles keep experiencing acceleration due to the forces exerted on them by the other particles. By Larmor's formula (7.22), this leads to emission of electromagnetic radiation from these particles, making the gas lose energy and cool. This process is called *Bremsstrahlung*. A lighter particle acquires more acceleration when subject to a force of particular strength. We thus expect the electrons in the hot gas to be the primary agents for energy loss. An electron can be accelerated by interacting with another electron of charge $-e$ or an ion of charge Ze. Due to the higher charge of the ion, an ion will be more effective in accelerating an electron. In an approximate treatment of Bremsstrahlung, we need to focus our attention on the acceleration of electrons caused by the electrostatic forces of ions.

Let us consider the 'collision' of an electron with an ion. The total emission of radiation by the electron can be obtained by integrating Larmor's formula (7.22) over time, that is,

$$W = \int P \, dt = \frac{e^2}{6\pi\epsilon_0 c^3} \int_{-\infty}^{\infty} |\dot{\mathbf{v}}(t)|^2 \, dt. \tag{7.84}$$

This emission will come out in different frequencies. To determine the spectrum of Bremsstrahlung, we need to find how the emitted energy would be distributed in different frequencies. When we considered the emission of electromagnetic radiation by a purely oscillatory system in Sect. 7.3–7.5, the radiation could be taken to be emitted at the frequency of the imposed oscillation. To find the frequency distribution in the present case, we need to express the acceleration in the form of a Fourier transformation

$$\dot{\mathbf{v}}(t) = \frac{1}{\sqrt{2\pi}} \int_{-\infty}^{\infty} \dot{\mathbf{v}}(\omega) \, e^{-i\omega t} \, d\omega, \tag{7.85}$$

of which the reverse transformation is

$$\dot{\mathbf{v}}(\omega) = \frac{1}{\sqrt{2\pi}} \int_{-\infty}^{\infty} \dot{\mathbf{v}}(t) \, e^{i\omega t} \, dt. \tag{7.86}$$

One standard result in the theory of Fourier transformations is the Parseval relation (see, for example, Arfken, Weber and Harris [8], p. 987):

$$\int_{-\infty}^{\infty} |\dot{\mathbf{v}}(t)|^2 \, dt = \int_{-\infty}^{\infty} |\dot{\mathbf{v}}(\omega)|^2 \, d\omega. \tag{7.87}$$

It is easy to check from (7.86) that $\dot{\mathbf{v}}(-\omega) = \dot{\mathbf{v}}(\omega)^*$, from which $|\dot{\mathbf{v}}(-\omega)|^2 = |\dot{\mathbf{v}}(\omega)|^2$. It then follows from (7.87) that

$$\int_{-\infty}^{\infty} |\dot{\mathbf{v}}(t)|^2 \, dt = 2 \int_0^{\infty} |\dot{\mathbf{v}}(\omega)|^2 \, d\omega. \tag{7.88}$$

Substituting this in (7.84), we get

$$W = \frac{e^2}{3\pi\epsilon_0 c^3} \int_0^{\infty} |\dot{\mathbf{v}}(\omega)|^2 \, d\omega.$$

This implies that the radiation coming out at the frequency ω is associated with energy

$$\frac{dW}{d\omega} = \frac{e^2}{3\pi\epsilon_0 c^3} |\dot{\mathbf{v}}(\omega)|^2. \tag{7.89}$$

It is clear that we need to calculate the Fourier transform $\dot{\mathbf{v}}(\omega)$ of the acceleration $\dot{\mathbf{v}}(t)$, as given by (7.86) in order to obtain the spectral distribution of the energy.

It is complicated to do a rigorous calculation of $\dot{\mathbf{v}}(\omega)$ in the case of a Coulomb collision of an electron with an ion inside a hot gas, since the Coulomb interaction is long-range, and, in principle, all the ions present in the gas can simultaneously exert forces on an electron. For a detailed mathematical treatment of this problem, readers are referred to other books (Landau and Lifshitz [2], Sect. 70). We follow Rybicki and Lightman ([9], Chap. 5) and present a rough order of magnitude estimate to give an idea of how things go. Let b be the impact parameter of an electron moving with velocity v which is scattered by an ion of charge Ze, as shown in Fig. 7.9. The electron would be close to the ion during an interval of time of order b/v, when it would have an acceleration predominantly perpendicular to its trajectory having magnitude of order

$$\dot{v}_{\perp}(t) \approx \frac{1}{4\pi\epsilon_0} \frac{Ze^2}{m_e b^2}. \tag{7.90}$$

Fig. 7.9 A sketch showing the 'collision' of an electron moving with speed v with an ion of charge Ze, the impact parameter being b

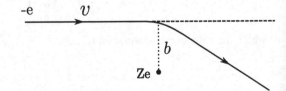

Substituting this in (7.86) and carrying on the integration over the time interval from $t = -b/v$ to $t = b/v$ when the electron is close to the ion (t being taken as 0 at the moment of the closest approach of the electron to the ion), we can get an approximate estimate of $\dot{v}_\perp(\omega)$. Let us now determine how we can do this integration approximately. If $\omega \gg v/b$, then the factor $e^{i\omega t}$ would oscillate very rapidly with time even during the interval from $t = -b/v$ to $t = b/v$ (the oscillations would be even more rapid outside this range of t). As a result of this, the integral in (7.86) would be close to zero when $\omega \gg v/b$. In the opposite limit of $\omega \ll v/b$, we can take $e^{i\omega t} \approx 1$ during the time when the electron is close to the ion and when $\dot{v}_\perp(t)$ is significantly non-zero. From such considerations, (7.86) and (7.90) lead to

$$\dot{v}_\perp(\omega) \approx \begin{cases} \frac{1}{\sqrt{2\pi}} \frac{1}{4\pi\epsilon_0} \frac{Ze^2}{m_e b^2} \frac{2b}{v} & \text{if } \omega \ll \frac{v}{b} \\ 0 & \text{if } \omega \gg \frac{v}{b} \end{cases} \tag{7.91}$$

By substituting this in (7.89), we can get the energy distribution in the emitted radiation resulting from a single collision of an electron with an ion having impact factor b.

Our aim should be to determine the radiation emitted from unit volume in unit time due to all the collisions taking place with different values of b. Let N_e and N_i be the number densities of electrons and ions per unit volume. An electron moving with velocity v would undergo $N_i v \times 2\pi b\, db$ collisions with ions per unit time having impact parameters in the range b, $b + db$. The spectral energy given out by one electron due to all the collisions it suffers in unit time is

$$\frac{dW_{\text{electron}}}{d\omega} = \int \frac{dW}{d\omega} N_i v\, 2\pi b\, db.$$

Substituting for $dW/d\omega$ from (7.89) with $\dot{v}(\omega)$ given by (7.91), we get

$$\frac{dW_{\text{electron}}}{d\omega} = \frac{1}{12\pi^3\epsilon_0^3} \frac{e^6}{m_e^2 c^3} \frac{Z^2 N_i}{v} \int_{b_{\min}}^{b_{\max}} \frac{db}{b}. \tag{7.92}$$

The integral in this equation gives $\ln b$, which would blow up at both the ends if we are to take $b_{\min} = 0$ or $b_{\max} = \infty$. The range of the impact parameter b has to be truncated at both the ends in order to yield a finite result. It follows from (7.91) that we need to consider $b \ll v/\omega$ when we are trying to find emission at frequency ω. The lower limit $b_{\min}$ arises from the breakdown of the classical treatment for very small b for which quantum corrections are necessary. We shall not get into a discussion of this here. In terms of the limits on b, we have

$$\frac{dW_{\text{electron}}}{d\omega} = \frac{1}{12\pi^3\epsilon_0^3} \frac{e^6}{m_e^2 c^3} \frac{Z^2 N_i}{v} \ln\left(\frac{b_{\max}}{b_{\min}}\right). \tag{7.93}$$

This is the radiation given out at frequency ω by one electron moving with speed v in unit time. In order to obtain the rate of radiant energy emission per unit volume per unit time, we need to integrate over all electrons in a unit volume moving with different speeds v. This is given as Exercise 7.10.

Although we have made several drastic approximations, even this over-simplistic derivation captures some essential aspects of Bremsstrahlung and gives the correct order-of-magnitude estimates. The spectrum of Bremsstrahlung is experimentally found to be fairly flat in a range of frequencies. This should be apparent from (7.93). The only place where ω appears in the RHS is through $b_{\max} \approx v/\omega$. This is a logarithmic dependence which would make the RHS of (7.93) to vary weakly with ω. On physical grounds, however, we do not expect the flat spectrum to continue indefinitely for very large ω. In a gas having temperature T, the typical electron has a kinetic energy $k_B T$, and we cannot expect this electron to produce photons with energies higher than $\hbar\omega \approx k_B T$. On including quantum corrections, the spectrum of Bremsstrahlung should be cut off for frequencies beyond $k_B T/\hbar$. To take account of this quantum consideration approximately in a classical calculation, we have to proceed as follows. To calculate total emission at frequency ω, we have to integrate over only emissions from electrons having kinetic energy larger than $\hbar\omega$, since only such electrons can emit at frequency ω. You are required to do this in Exercise 7.10.

Exercises

7.1 (a) Show that the energy radiated per unit time by a relativistic non-uniformly moving charged particle is

$$P = \frac{q^2}{6\pi\epsilon_0 c^3}(\gamma^6 a_\parallel^2 + \gamma^4 a_\perp^2),$$

where $a_\parallel$ is the component of acceleration in the direction of the instantaneous velocity of the particle, and $a_\perp$ is the acceleration in the perpendicular direction, whereas all the other symbols have their usual meanings.

Hint. You have to make use of the results concerning the transformation of acceleration between frames, which you were asked to work out in Exercise 5.2.

(b) Verify that the expression of energy radiated per unit time by a relativistically moving charged particle can also be written as

$$P = \frac{q^2\gamma^6}{6\pi\epsilon_0 c^3}\left(a^2 - \left|\frac{\mathbf{v} \times \mathbf{a}}{c}\right|^2\right),$$

where $\mathbf{a}$ is the acceleration of the charged particle having amplitude $a = \sqrt{a_\parallel^2 + a_\perp^2}$.

7.2 Two oscillating dipole moments (say radio antennas) $\mathbf{p}_1 e^{-i\omega t}$ and $\mathbf{p}_2 e^{-i\omega t}$ are oriented in the vertical direction and are placed one above the other at a vertical distance L apart. They oscillate in phase at the same frequency ω. Consider the radiation emitted at angle θ with respect to the vertical direction. Show that

$$\frac{dP}{d\Omega} = \frac{1}{32\pi^2} \left(\frac{\mu_0}{\epsilon_0}\right)^{1/2} \frac{\omega^4 \sin^2 \theta}{c^2} (p_1^2 + 2p_1 p_2 \cos \delta + p_2^2),$$

where

$$\delta = \frac{\omega L \cos \theta}{c}$$

and the other symbols have usual meanings.

7.3 When $L \ll \lambda$ in Exercise 7.2, show that the rate of radiation emission reduces to that from a single oscillating dipole of amplitude $p_1 + p_2$. Now consider the special case of $\mathbf{p}_2 = -\mathbf{p}_1$, which means that the two dipoles have the same strength and are exactly out of phase. Show that the dipole radiation term is zero and determine the lowest-order term in the expression for the emission of radiation from the system. This terms corresponds to the electric quadrupole radiation, which you will find to have $\sin^2 \theta \cos^2 \theta$ angular dependence.

7.4 If classical electrodynamics were valid at the atomic scale, make an order-of-magnitude estimate of the time in which a hydrogen atom would have collapsed due to the radiation damping of the orbital electron. Would it be justified to neglect the radiation damping term in the equation of motion?

7.5 Show that the total Thomson scattering cross section of a free electron for scattering a beam of a *plane-polarized* electromagnetic wave is the same as the cross section for scattering an unpolarized beam of an electromagnetic wave given by (7.73). What is the physical significance of this?

7.6 Consider an atmosphere of *completely ionized hydrogen* having the same density as the Earth's atmosphere. Estimate the length that a beam of light has to traverse before its intensity is reduced to half due to Thomson scattering.

7.7 Suppose an electromagnetic wave falls on a small dielectric sphere of radius a much smaller than the wavelength, that is, $ka \ll 1$, where k is the wavenumber of the electromagnetic wave. Since the oscillatory electric field of the electromagnetic wave will create an oscillatory magnetic moment in the dielectric sphere, some radiation will be scattered from the sphere. Show that the total scattering cross section of the sphere is

$$\sigma = \frac{8\pi}{3} k^4 a^6 \left(\frac{\kappa - 1}{\kappa + 2}\right)^2,$$

where κ is the dielectric constant. Note that this cross section obeys the law of Rayleigh scattering.

Hint. Make use of the relation (2.120).

7.8 We showed in Sect. 4.9 that the imaginary part of the refractive index given by (4.126) gives rise to the attenuation of a beam of electromagnetic radiation as indicated in (4.117). Show that you would get exactly the same amount of attenuation due to scattering given by the scattering cross section (7.77) on making use of (7.58) and (7.65).

7.9 Consider a collection of relativistic electrons spiralling around a magnetic field and having a power-law distribution of energy, that is, the number density of electrons in the range E to $E + dE$ is

$$dN \propto E^{-p}\, dE.$$

Assuming that synchrotron radiation is predominantly emitted at the frequency $\gamma^2 \omega_c$ (where ω_c is the cyclotron frequency), show that the spectrum of the synchrotron radiation also will have a power-law dependence

$$I(\omega)\, d\omega \propto \omega^{-\alpha}\, d\omega,$$

where

$$\alpha = \frac{p - 1}{2}.$$

Hint. You have to use the fact that the rate at which a relativistic spiralling electron gives out energy is proportional to γ^2 or E^2, which you can easily argue on the basis of the expression of P you have to derive in Exercise 7.1 (a).

7.10 We have derived the expression (7.93) for the energy loss rate due to Bremsstrahlung by a single electron moving with velocity v. Suppose we have an ionized gas of temperature T containing N_e free electrons obeying the Maxwellian distribution of velocities. Only electrons with kinetic energy $m_e v^2/2$ higher than $\hbar \omega$ can produce Bremsstrahlung with frequency ω. Show that the energy loss rate per unit volume due to Bremsstrahlung is

$$\frac{dW}{d\omega\, dV} = \frac{1}{6\sqrt{2}\pi^{7/2}\epsilon_0^3} \frac{e^6}{m_e^{3/2} c^3} \frac{Z^2 N_i N_e}{\sqrt{k_B T}} \exp\left(-\frac{\hbar \omega}{k_B T}\right) \ln\left(\frac{b_{max}}{b_{min}}\right),$$

where k_B is the Boltzmann constant.

Hint. To calculate the energy loss rate at frequency ω, you have to integrate over electrons moving with velocities in the range from $\sqrt{(2\hbar\omega/m_e)}$ to ∞.

References

1. Panofsky, W.K.H., Phillips, M.: Classical Electricity and Magnetism, 2nd edn. Addison–Wesley (reprinted by Dover) (1962)

2. Landau, L.D., Lifshitz, E.M.: The Classical Theory of Fields (Course of Theoretical Physics), 4th edn. Butterworth-Heinemann (1980)
3. Born, M., Wolf, E.: Principles of Optics, 6th edn. Pergamon Press (1980)
4. Jackson, J.D.: Classical Electrodynamics, 3rd edn. Wiley (1999)
5. Landau, L.D., Lifshitz, E.M.: The Classical Theory of Fields (Course of Theoretical Physics), vol. 2, 4th edn. Butterworth-Heinemann (1980)
6. Griffiths, D.J.: Introduction to Electrodynamics, 4th edn. Cambridge University Press (2017)
7. Feynman, R.P., Leighton, R.B., Sands, M.: The Feynman Lectures on Physics: Volume II, Mainly Electromagnetism and Matter. Addison-Wesley (1964)
8. Arfken, G.B., Weber, H.J., Harris, E.E.: Mathematical Methods for Physicists, 7th edn. Elsevier (2013)
9. Rybicki, G.B., Lightman, A.P.: Radiative Processes in Astrophysics. Wiley (1979)
10. Bjroken, J.D., Drell, S.D.: Relativistic Quantum Mechanics. McGraw. Hill (1964)

Chapter 8
Basics of Plasma Physics and Magnetohydrodynamics

8.1 Introductory Remarks

This is a textbook on a classical subject of which most of the fundamental principles were discovered decades ago. In fact, large parts of the book (apart from the relativistic topics) have been devoted to topics which had reached fairly finished forms by the end of the nineteenth century. In this final chapter, we turn to one application of electromagnetic theory which has become very important in recent years.

When a gas is heated to a very high temperature, many of the atoms break into positively charged ions and negatively charged electrons. The well-known *Saha ionization equation* discussed in standard textbooks of statistical mechanics gives the level of ionization at a given temperature and pressure (Fermi [1], pp. 151–155; Reif [2], pp. 363–365; Landau and Lifshitz [3], 313–314). An ionized gas, often called a plasma, consists of many randomly moving charged particles. We expect these particles to interact with each other through electromagnetic forces. The study of plasmas is the many-body problem of classical electrodynamics. The molecules in a dilute neutral gas interact with each other only when they 'collide', i.e. when they are separated by distances comparable to the size of the molecules. In contrast, since electromagnetic forces are long-range, a particle in a plasma, in principle, interacts with all the other plasma particles at all times. This makes the theory of plasmas much more complicated than the theory of neutral gases. Within the regime of validity of classical physics (which is the case when the de Broglie wavelengths of the particles are much smaller than typical inter-particle distances), the equation of motion of the i-th particle of mass m_i and charge q_i moving with velocity $\mathbf{v}_i$ should clearly be

$$m_i \frac{d\mathbf{v}_i}{dt} = q_i \left(\sum_{j \neq i} \mathbf{E}_{ji} + \mathbf{v}_i \times \sum_{j \neq i} \mathbf{B}_{ji} \right), \qquad (8.1)$$

where $\mathbf{E}_{ji}$ and $\mathbf{B}_{ji}$ are respectively the electric and magnetic fields produced by the j-th particle of the plasma at the location of the i-th particle. If we could evaluate $\mathbf{E}_{ji}$,

© Springer Nature Singapore Pte Ltd. 2022
A. R. Choudhuri, *Advanced Electromagnetic Theory*, Lecture Notes in Physics 1009,
https://doi.org/10.1007/978-981-19-5944-8_8

$\mathbf{B}_{ji}$ for all the pairs of plasma particles and then simultaneously solve equations like (8.1) for all the plasma particles, that would give a complete theory of the plasma. Since this is not feasible, we have to develop reduced simplified models to deal with the plasma. We shall introduce some of these reduced models in this chapter. All the reduced models have various limitations: each model can handle certain properties of the plasma and not the other properties.

Due to electrical attraction between the opposite charges, the positively charged and the negatively charged particles in a plasma remain well mixed. In other words, if we consider a volume element which is small by macroscopic standards, but large enough to contain many particles, that volume element is likely to have nearly equal amounts of positive and negative charges. One makes the statement that a plasma is a *quasi-neutral* medium. One may naively expect that a plasma may behave like a neutral gas in many ways. However, the difference between the two becomes evident as soon as we introduce an electric field. An electric field introduced in a neutral gas like air does not have much effect, unless it is strong enough to cause an electrical breakdown. On the other hand, an electric field introduced in a plasma would make positive charges move along it and negative charges move in the opposite direction, giving rise to a current. We easily see that a plasma would be a good conductor of electricity, even weak electric fields driving large currents. Clearly, magnetic fields can be associated with these currents, giving rise to magnetic forces. The study of the interaction between plasmas and magnetic fields is an important topic in plasma physics.

It is this property of quasi-neutrality that makes the study of plasmas manageable. Although an electric field is in principle long-range, the balancing effects of opposite charges can screen off an electric field in a plasma. We shall, however, introduce a parameter called the *plasma parameter* in Sect. 8.2. When this parameter is small, many charged particles can interact together, giving rise to various kinds of collective behaviour, in which nearby charged particles take part in a coherent manner.

The existence of currents shows that it is possible for the positive and negative charges in the plasma to move with respect to each other. We can study many aspects of plasma behaviours by treating a plasma as a combination of two inter-penetrating electric fluids of opposite charges occupying the same region of space. Each of these fluids will have a number density (from which the mass density or the charge density can be found) and will be associated at any point in space with a bulk velocity of the fluid. We shall show in the next two sections that charge separations arising out of relative displacements between the two fluids have to be confined over small length scales and short time scales. When we consider phenomena having larger length scales and longer time scales, we can neglect the charge separation and treat the plasma as a single fluid. The single-fluid model of the plasma is called *magnetohy-drodynamics*, abbreviated as MHD. In this model, we assume the charge density ρ to be always zero, but we allow non-zero current density $\mathbf{j}$. Certainly we can have positively charged and negatively charged particles in a local region of the plasma move in opposite directions giving rise to a current $\mathbf{j}$, but move in such a manner that the amount of positive and negative charges in a volume element balance each other such that $\rho = 0$. We shall show in this chapter that the two-fluid and the one-fluid (MHD) models of the plasma can be used to study many plasma phenomena. However, these two very useful reduced models are not sufficient to cover all plasma phenomena that may be of interest to us. We shall point out in Sect. 8.3.3 that there are phenomena which require us to consider microscopic energy distributions of plasma particles and which cannot be handled with either of these two reduced models which we discuss in this chapter.

Before starting our discussion of the two reduced models of the plasma, let us make a few comments about the scope of plasma physics. A major thrust area of plasma research is to study how we can generate energy in a controlled manner through thermonuclear fusion reactions taking place in a hot plasma. We shall give an introduction to this subject in Sect. 8.6. If we succeed in our goal of commercial production of energy in this manner, that will solve the energy problem of the human race for all times. Additionally, we have realized that most of the celestial objects which we observe through our telescopes are made up of materials in the plasma state. It is no wonder that we need to apply the principles of plasma physics to understand and explain many astrophysical phenomena. We shall consider some examples of such phenomena in Sect. 8.8. Lastly, an electron gas in a solid metal displays many properties of a plasma, although many phenomena connected with this electron gas require a quantum treatment (Fermi–Dirac statistics) for their explanation. Such a wide range of applications makes plasma physics a very important modern branch of physics research.

We end our introductory remarks with a comment on the use of Maxwell's equations in plasma physics. Much of this book has been devoted to applications of electromagnetism to situations in which we have conductors surrounded by non-conducting material media with some specified dielectric and magnetic properties. For handling such problems, it is often useful to distinguish between free charges/currents which are in the conductors and the induced charges/currents in the medium, leading to two kinds of electric vectors **E**, **D** and two kinds of magnetic vectors **B**, **H**, as shown in Sects. 2.14 and 3.6. When we consider charges/currents in plasmas, such distinctions are not meaningful. Therefore, whenever we have to apply Maxwell's equations in our discussion of plasmas, we shall use them in the original forms (1.1–1.4).

8.2 Debye Shielding. The Plasma Parameter

To have an idea of the length scale over which an electric field arising from a charge separation in a plasma can exist, let us consider the charge separation produced by introducing a charge q inside a plasma. We expect plasma particles of the opposite charge to be attracted and the plasma particles of the same charge to be repelled by this charge q, thereby producing an electric layer around q. If N_e is the number density of electrons and N_i that of ions (assumed singly ionized), then the charge density at a point in space is given by $(N_i - N_e)e$. Denoting the electrostatic potential by Φ, the Poisson equation (2.10) becomes

$$\nabla^2 \Phi = -\frac{(N_i - N_e)e}{\epsilon_0}. \tag{8.2}$$

If the plasma is in thermodynamic equilibrium, then we expect

$$N_i = N \exp\left(-\frac{e\Phi}{k_B T}\right), \qquad N_e = N \exp\left(\frac{e\Phi}{k_B T}\right), \tag{8.3}$$

where N is the undisturbed number density of either electrons or ions (which have to be equal for singly charged ions). Substituting from (8.3) in (8.2), we get

$$\nabla^2 \Phi = \frac{Ne}{\epsilon_0}\left[\exp\left(\frac{e\Phi}{k_B T}\right) - \exp\left(-\frac{e\Phi}{k_B T}\right)\right].$$

We now expand in Taylor series and neglect quadratic or higher-order terms in $(e\Phi/k_B T)$, which are expected to be small in usual circumstances. This gives

$$\nabla^2 \Phi = \frac{\Phi}{\lambda_D^2}, \tag{8.4}$$

where

$$\lambda_D = \left(\frac{\epsilon_0 k_B T}{2Ne^2}\right)^{1/2} \tag{8.5}$$

is known as the *Debye length*. The potential around the charge q satisfies (8.4). Using the representation of the Laplacian in spherical coordinates as given by (2.37), it can be easily shown that the solution of (8.4) in a spherically symmetric situation is

$$\Phi = \frac{q}{4\pi\epsilon_0} \frac{\exp(-r/\lambda_D)}{r} \tag{8.6}$$

on requiring that the solution should correspond to (2.5) in the limit when the plasma particle density N goes to zero and λ_D given by (8.5) becomes infinite. It thus appears that the effect of the charge q is screened beyond a distance λ_D. A plasma can therefore be considered charge-neutral when distances larger than the Debye length are considered.

Although the electric field of a charged particle in principle extends to infinity, the influence of a charged particle in a plasma is effectively felt only to a distance λ_D, i.e. within a volume of the order λ_D^3 called the *Debye volume*. Hence, the number of particles on which this charged particle would exert an influence is of order $N\lambda_D^3$. We have already mentioned in Sect. 8.1 that a plasma can exhibit collective phenomena arising out of mutual interactions of many charged particles. The number $N\lambda_D^3$ gives a measure of the number of particles which can interact simultaneously. The inverse of this number is known as the *plasma parameter* denoted as

$$g = \frac{1}{N\lambda_{\mathrm{D}}^3} = \frac{2\sqrt{2}e^3 N^{1/2}}{(\epsilon_0 k_{\mathrm{B}} T)^{3/2}} \tag{8.7}$$

on substituting from (8.5). When the plasma parameter g is smaller, there is more collective interaction in the plasma. It should be noted that g is smaller for smaller N. Therefore, the number of particles interacting collectively is greater for a low-density plasma. In such a plasma, the Debye shielding is less effective, so the Debye volume λ_{D}^3 is much larger. Hence, even if the number of particles per unit volume is less, the total number of particles in the Debye volume is larger.

The average distance between the particles of a plasma is of the order $N^{-1/3}$. Therefore, the average potential energy of electrostatic interaction between a pair of nearby particles is of the order $e^2 N^{1/3}/4\pi\epsilon_0$. Hence, the ratio of average potential to kinetic energy is

$$\frac{\langle \text{P.E.} \rangle}{\langle \text{K.E.} \rangle} \approx \frac{e^2 N^{1/3}}{4\pi\epsilon_0 k_{\mathrm{B}} T} \propto g^{2/3}. \tag{8.8}$$

Another interpretation of the plasma parameter, therefore, is that it is a measure of the potential energy of interactions compared to the kinetic energy. When g is small (as for a low-density plasma), the interaction amongst particles is weak, but a large number of particles interact simultaneously. On the other hand, a larger g implies few particles interacting collectively, but interacting strongly. The limit of small g is referred to as the *plasma limit*. Most textbooks on plasmas take the smallness of the plasma parameter g as a condition for the definition of a plasma.

8.3 Electromagnetic Oscillations in Cold Plasmas

After working out in the previous section the length scale over which a charge separation can exist, we now turn our attention to the time scale over which a charge separation in the plasma would evolve. For this purpose, we need to study the evolution of a charge separation in a plasma. We shall see that two important kinds of plasma phenomena will come out of our general analysis of charge separation. Firstly, when an electromagnetic wave propagates through a plasma, the electric field associated with the wave makes ions and electrons move in opposite directions. A mathematical analysis of this charge separation will give us a theory of electromagnetic wave propagation inside a plasma. The second phenomenon that we shall get out of our analysis is what is called a *plasma oscillation*. If electrons are displaced with respect to ions in a certain region of the plasma, we expect the electrostatic field arising out of this charge separation to pull back the electrons to their original positions. This type of electrostatic oscillation in the plasma will also be an outcome of our analysis.

Since the ions are much heavier, any electric field will produce a much smaller acceleration in an ion compared to an electron, and we shall neglect the motions of ions in our analysis. The ions are merely assumed to provide a background of positive charge to keep the plasma neutral. We further assume the plasma to be *cold*—which means that the electrons have no thermal motions and move only under the influence of the electric field of the wave.

Let $\mathbf{v}_e$, $\mathbf{E}$ and $\mathbf{B}$ respectively denote the velocity of the electron fluid, the electric field and the magnetic field. The equation of motion of an electron within the electron fluid is given by

$$m_e \frac{\partial \mathbf{v}_e}{\partial t} = -e\,\mathbf{E}. \tag{8.9}$$

As we concluded in (4.53) that the magnetic force is much smaller compared to the electric force in an electromagnetic wave, we have not included the magnetic force in (8.9). We need to combine (8.9) with the two Maxwell equations (1.3) and (1.4) giving the time evolutions of the fields, which we write again, keeping in mind that the current density arising out of the velocity $\mathbf{v}_e$ of the electron fluid would be $\mathbf{j} = -N_e e \mathbf{v}_e$ (where N_e is the number density of electrons):

$$\nabla \times \mathbf{B} = -\mu_0 N_e e\,\mathbf{v}_e + \epsilon_0 \mu_0 \frac{\partial \mathbf{E}}{\partial t}, \tag{8.10}$$

$$\nabla \times \mathbf{E} = -\frac{\partial \mathbf{B}}{\partial t}. \tag{8.11}$$

Since we expect an electromagnetic wave to be an outcome of our analysis, let us assume that the time dependence of all the quantities is of the form $\exp(-i\omega t)$ so that we can everywhere replace $\partial/\partial t$ by $-i\omega$. We then get from (8.9) that

$$\mathbf{v}_e = \frac{e}{i\omega m_e}\,\mathbf{E}. \tag{8.12}$$

On substituting this for $\mathbf{v}_e$ in (8.10), we have

$$\nabla \times \mathbf{B} = -\frac{i\omega}{c^2}\left(1 - \frac{\omega_p^2}{\omega^2}\right)\mathbf{E}, \tag{8.13}$$

where we have written $1/c^2$ for $\epsilon_0 \mu_0$ in accordance with (4.42), and

$$\omega_p = \sqrt{\frac{N_e e^2}{\epsilon_0 m_e}} \tag{8.14}$$

is known as the *plasma frequency*. On taking a time derivative of (8.13) and using (8.11), we end up with

$$\frac{\omega^2}{c^2}\left(1 - \frac{\omega_p^2}{\omega^2}\right)\mathbf{E} = \nabla \times (\nabla \times \mathbf{E}). \tag{8.15}$$

Assuming the background plasma to be spatially homogeneous, we may look for solutions of the perturbed quantities which are sinusoidal in space. In other words, we assume all perturbations to be of the form $\exp(i\mathbf{k}.\mathbf{x} - i\omega t)$. On substituting in (8.15), we get

$$\mathbf{k} \times (\mathbf{k} \times \mathbf{E}) = -\frac{\omega^2}{c^2}\left(1 - \frac{\omega_p^2}{\omega^2}\right)\mathbf{E}. \tag{8.16}$$

Without any loss of generality, we can choose our z axis in the direction of the propagation vector $\mathbf{k}$, i.e. we write $\mathbf{k} = k\,\mathbf{e}_z$. On substituting this in (8.16), we obtain the following matrix equation

$$\begin{pmatrix} \omega^2 - \omega_p^2 - k^2c^2 & 0 & 0 \\ 0 & \omega^2 - \omega_p^2 - k^2c^2 & 0 \\ 0 & 0 & \omega^2 - \omega_p^2 \end{pmatrix} \begin{pmatrix} E_x \\ E_y \\ E_z \end{pmatrix} = \begin{pmatrix} 0 \\ 0 \\ 0 \end{pmatrix}. \tag{8.17}$$

It is clear from (8.17) that the x and y directions are symmetrical, as we expect. The z direction, being the direction along $\mathbf{k}$, is distinguishable. This indicates that we may have two physically distinct types of oscillatory modes. They are discussed below.

8.3.1 Electromagnetic Waves

The matrix equation (8.17) allows a mode in which

$$E_z = 0, \qquad \omega^2 = \omega_p^2 + k^2c^2. \tag{8.18}$$

Since only the components E_x or E_y perpendicular to the propagation direction are allowed to be non-zero, this clearly corresponds to a transverse wave. It is actually nothing more than the ordinary electromagnetic wave modified by the presence of the plasma. If $\omega \gg \omega_p$, then we are led to limiting relation $\omega^2 = k^2c^2$, which is the usual dispersion relation for electromagnetic waves in a vacuum, as we can see from (4.45). In other words, if the frequency of the wave is too high, even the electrons, which are much more mobile than the ions, are unable to respond sufficiently fast so that the plasma effects are negligible. Only the propagation of low-frequency electromagnetic waves is affected significantly by the presence of the plasma.

It is also to be noted from (8.18) that if $\omega < \omega_p$, then k becomes imaginary so that the wave is evanescent. If an electromagnetic wave of frequency ω is sent towards a volume of plasma with a plasma frequency ω_p greater than ω (if $\omega < \omega_p$), then the electromagnetic wave is not able to pass through this plasma, and the only possibility is that it is reflected back.

The plasma frequency of the Earth's ionosphere is about 30 MHz. Radio waves from cosmic sources can penetrate through the ionosphere only if the frequency is higher than 30 MHz (or the wavelength is less than 10 m). Hence, radio telescopes have to be operated at higher frequencies if we are to receive radio signals from cosmic sources. On the other hand, if we want to communicate with faraway regions of the Earth's surface, then we may want to use radio waves of frequency less than 30 MHz which would be reflected back from the ionosphere. Although the electron gas inside a metal behaves as a quantum gas in many circumstances, a classical treatment is adequate for the propagation of electromagnetic waves. The typical plasma frequencies of electron gases in metals lie in the ultraviolet range. That is why metals are opaque to ordinary visible light with frequency lower than the plasma frequency. A polished metallic surface appears shiny because it reflects electromagnetic waves. However, it is found that many metals are transparent to ultraviolet radiation if the frequency of that radiation is higher than the plasma frequency of the metal.

8.3.2 Plasma Oscillations

The other solution of the matrix equation (8.17) is

$$E_x = E_y = 0, \qquad \omega^2 = \omega_p^2. \tag{8.19}$$

Here, the electric field is completely in the direction of the propagation vector $\mathbf{k}$, and it follows from (8.12) that all the displacements are also in the same direction. We also note that the group velocity $(\partial\omega/\partial k)$ is zero. We therefore have a non-propagating longitudinal oscillation with its frequency equal to the plasma frequency ω_p. Such oscillations are known as *plasma oscillations*.

It is not difficult to understand the physical nature of these oscillations. Against a background of nearly immobile and hence uniformly distributed ions, this solution will correspond to alternate layers of compression and rarefaction of the electron gas (unless $\mathbf{k} = 0$ so that the wavelength is infinite). The electrostatic forces arising out of such a charge imbalance drive these oscillations.

8.3.3 Landau Damping

In a physical system in which dissipative processes are present, oscillations usually damp out. For example, acoustic waves in air are damped by viscosity arising out of molecular collisions. If a layer of electrons is slightly displaced in a direction perpendicular to the layer, then the layer is attracted towards its undisturbed position due to the electrostatic forces arising out of the displacement. The analysis of Sect. 8.3.2 suggests that the layer would overshoot its undisturbed position and would undergo undamped oscillations. The plasma oscillations are driven by long-range interactions among plasma particles which do not 'collide', and it superficially seems that this system will not have any dissipation.

If the plasma is treated through the two-fluid model, it indeed appears that there is no sink to which the energy of the plasma oscillation can go, and there cannot be any dissipation in the system. However, the plasma particles would have an energy distribution (like the Maxwellian distribution in thermodynamic equilibrium), and it is possible for the energy of plasma oscillations to go into changing the energy distribution of the particles, thereby giving rise to dissipation. When one goes beyond the two-fluid model and formulates the problem in terms of the energy distribution of the plasma particles, it is possible to show that a plasma oscillation would be damped in a few periods if the wavelength is comparable to the Debye length. This was derived by Landau and is known as *Landau damping*. This example shows that even the two-fluid model of the plasma has important limitations and cannot handle many phenomena involving plasmas.

8.4 Basic Equations of MHD

As we saw in Sects. 8.2 and 8.3, the typical length scale associated with charge separation is the Debye length, and the typical time scale is the inverse of the plasma frequency, i.e. $1/\omega_p$. When we deal with plasma phenomena with longer length and time scales, charge separation can be neglected. In that situation, we can treat the electron and ion fluids to move together, giving rise to the one-fluid model. The one-fluid model is an even more reduced and restrictive model of the plasma compared to the two-fluid model. Important plasma phenomena such as propagation of electromagnetic waves through plasmas, which was treated in Sect. 8.3 with the two-fluid model, cannot be handled with the one-fluid model. Still, the one-fluid MHD model turns out to be an extremely useful model for studying the macroscopic dynamics of the plasma in many situations where charge separation can be neglected. The rest of this chapter will be devoted to developing the one-fluid MHD model and applying it to various situations.

Let us begin our discussion of the MHD model by first considering the question of how we provide a mathematical description of a state of the plasma at an instant of time in this model. We know that the thermodynamic state of a gas in a container is given by a pair of thermodynamic coordinates, say density ρ and pressure p. It may be noted that in this book we have been using ρ for charge density, which also happens to be the standard symbol for mass density in textbooks on fluid mechanics and MHD. Since we neglect charge separation in the one-fluid model of the plasma, the charge density will not occur anywhere in our discussion of this model. Thus, rather than inventing an unorthodox symbol for mass density, we shall use ρ for mass density in our discussion of MHD. When we consider a plasma in an extended region, we have to allow for the variations of the two thermodynamic variables within the plasma. Therefore, we take $\rho(\mathbf{x}, t)$ and $p(\mathbf{x}, t)$ as two thermodynamic variables for describing a state of the plasma. What are the other necessary variables? When we allow for motions inside the plasma, we have to allow for the velocity variable $\mathbf{v}(\mathbf{x}, t)$. Keep in mind that the one-fluid MHD model is a model in which the plasma is treated as a continuum, and we do not consider individual plasma particles separately. The velocity $\mathbf{v}(\mathbf{x}, t)$ introduced in our discussion is the macroscopic velocity with which an infinitesimally small volume of the plasma at position $\mathbf{x}$ moves at time t. Since we shall have to consider electromagnetic phenomena in the plasma, we have to introduce additional variables to treat them. At first, it may seem that we have to introduce the electric field $\mathbf{E}(\mathbf{x}, t)$ and the magnetic field $\mathbf{B}(\mathbf{x}, t)$ at various points inside the plasma. However, when charge separation is neglected, the electric field is negligible—at least, in the frame of reference in which the plasma is at rest in the local region. We shall later show that the very small electric field that may arise in the MHD model can be related to other variables, so we do not have to take the electric field as an additional variable in the MHD model. The magnetic field $\mathbf{B}(\mathbf{x}, t)$ is the only electromagnetic variable we have to deal with.

To summarize the discussion of the previous paragraph, the variables we need for specifying a state of the plasma at the time t are $\rho(\mathbf{x}, t)$, $p(\mathbf{x}, t)$, $\mathbf{v}(\mathbf{x}, t)$ and $\mathbf{B}(\mathbf{x}, t)$. Note that $\mathbf{v}(\mathbf{x}, t)$ and $\mathbf{B}(\mathbf{x}, t)$ are vector fields, each equivalent to three scalar fields corresponding to their three spatial components. We thus conclude that eight scalar fields are needed to describe the plasma at an instant of time in the MHD model. If we want to study the dynamics of the plasma, then we need time evolution equations for all these variables. If we have such equations, then we can calculate how the state of the plasma changes with time. Our aim now is to arrive at these time evolution equations.

We point out that a state of an ordinary neutral fluid is given by two thermodynamic variables like density $\rho(\mathbf{x}, t)$ and pressure $p(\mathbf{x}, t)$ along with the velocity field $\mathbf{v}(\mathbf{x}, t)$. The dynamics of a neutral fluid would be given by time derivative equations of all these variables. The MHD equations are an extension of the fluid dynamical equations due to the presence of the additional variable $\mathbf{B}(\mathbf{x}, t)$. We shall first write down the equations of fluid mechanics in Sect. 8.4.1 and then discuss in Sect. 8.4.2 how they are extended to give the MHD equations.

8.4.1 The Equations of Fluid Mechanics

Let us begin by drawing attention to the two different kinds of time derivatives: *Eulerian* and *Lagrangian*. The *Eulerian* derivative denoted by $\partial/\partial t$ implies differentiation with respect to time at a fixed point. On the other hand, one can think of moving with an infinitesimal element of the fluid with the velocity $\mathbf{v}$ and time-differentiating some quantity associated with this moving plasma element. This type of time derivative is called the *Lagrangian* derivative and is denoted by d/dt. If $\mathbf{x}$ and $\mathbf{x} + \mathbf{v}\,\delta t$ are the positions of the fluid element at times t and $t + \delta t$, then the Lagrangian time derivative of some quantity $Q(\mathbf{x}, t)$ is given by

$$\frac{dQ}{dt} = \lim_{\delta t \to 0} \frac{Q(\mathbf{x} + \mathbf{v}\,\delta t, t + \delta t) - Q(\mathbf{x}, t)}{\delta t}. \tag{8.20}$$

Keeping the first order terms in the Taylor expansion, we have

$$Q(\mathbf{x} + \mathbf{v}\,\delta t, t + \delta t) = Q(\mathbf{x}, t) + \delta t \frac{\partial Q}{\partial t} + \delta t \, \mathbf{v} . \nabla Q.$$

Putting this in (8.20), we have the very useful relation between the Lagrangian and the Eulerian derivatives:

$$\frac{dQ}{dt} = \frac{\partial Q}{\partial t} + \mathbf{v} . \nabla Q. \tag{8.21}$$

While developing the dynamical theory, if we can arrive at one type of time derivative of any quantity, we can easily get the other type of time derivative by the application of (8.21).

We now derive the first dynamical equation giving the time derivative of $\rho(\mathbf{x}, t)$. The mass $\int \rho \, dV$ inside a small volume can change only due to the motion of matter across the surface bounding this volume. Since the mass flux across an element of surface $d\mathbf{S}$ is $\rho \mathbf{v} . d\mathbf{S}$, we must have

$$\frac{\partial}{\partial t} \int \rho \, dV = - \oint \rho \mathbf{v} . \, d\mathbf{S},$$

where the minus sign implies that a mass flux out of the volume reduces the mass inside the volume. Transforming the RHS of the above equation by Gauss's theorem (B.13), we have

$$\int \left[\frac{\partial \rho}{\partial t} + \nabla . (\rho \mathbf{v}) \right] dV = 0.$$

Since this equation must be valid for any arbitrary volume dV, we must have

$$\frac{\partial \rho}{\partial t} + \nabla . (\rho \mathbf{v}) = 0. \tag{8.22}$$

This is known as the *equation of continuity*, which is essentially the equation of mass conservation. We point out that this is very similar to the charge conservation Eq. (4.7), which was derived in Sect. 4.1 by very similar arguments.

To obtain the time evolution equation of the velocity field $\mathbf{v}$, we consider an element of fluid with volume δV. The mass of this plasma element is $\rho \, \delta V$, and its acceleration is given by the Lagrangian derivative $(d\mathbf{v}/dt)$. Hence, it follows from Newton's second law of motion that

$$\rho \, \delta V \frac{d\mathbf{v}}{dt} = \delta \mathbf{F}_{\text{body}} + \delta \mathbf{F}_{\text{surface}}, \tag{8.23}$$

where we have split the force acting on the plasma element into two parts: the body force $\delta \mathbf{F}_{\text{body}}$ and the surface force $\delta \mathbf{F}_{\text{surface}}$. A body force is something which acts at all points within the body of the fluid. Gravity is an example of such a force. It is customary to denote the body force per unit mass as $\mathbf{F}$ so that

$$\delta \mathbf{F}_{\text{body}} = \rho \, \delta V \, \mathbf{F}. \tag{8.24}$$

The surface force on the element of the fluid is the force acting on it across the surface bounding the fluid element. Let $d\mathbf{S}$ be an element of area on the bounding surface. The pressure p would give rise to the force

$$d\mathbf{F}_{\text{surface}} = -p \, d\mathbf{S} \tag{8.25}$$

acting across this surface element. We add the minus sign because we want to consider the force acting on the fluid element inside the bounding surface (we use the usual convention that $d\mathbf{S}$ is in the outward direction). If there are viscous forces inside the moving fluid, then they also would contribute to the surface force. In the present discussion, we shall neglect the viscous forces and assume the surface force to be given entirely by (8.25). The total surface force acting across the whole bounding surface is then given by a surface integral

$$\mathbf{F}_{\text{surface}} = -\oint p \, d\mathbf{S}.$$

The right hand side can be transformed into the volume integral $-\int \nabla p \, dV$. For the small volume δV, we can write

$$\delta \mathbf{F}_{\text{surface}} = -\nabla p \, \delta V. \tag{8.26}$$

Substituting (8.24) and (8.26) into (8.23), we have

$$\rho \frac{d\mathbf{v}}{dt} = \rho \mathbf{F} - \nabla p. \tag{8.27}$$

If we use (8.21) to change from the Lagrangian derivative to the Eulerian derivative, then we get

$$\frac{\partial \mathbf{v}}{\partial t} + (\mathbf{v}.\nabla)\mathbf{v} = -\frac{1}{\rho}\nabla p + \mathbf{F}. \tag{8.28}$$

This equation is the central equation of fluid mechanics and is known as the *Euler equation*. If viscosity is included, then, in the place of the Euler equation, we have a more complicated equation known as the *Navier–Stokes equation*, which we shall not discuss in this book.

To complete our set of equations governing the dynamics of fluids, we need a time evolution equation of another thermodynamic variable beside density $\rho(\mathbf{x}, t)$. In general, one has to consider heat transfer inside the fluid in order to arrive at this equation. Without getting into a general discussion of this equation, we shall consider only the special case of a fluid which behaves like a perfect gas and the dynamics of which is assumed to be adiabatic. In other words, we assume that a small element of fluid does not exchange heat with its surroundings as it evolves with time. For a fluid obeying the perfect gas law, this means that p/ρ^γ will not change with time inside any element of the fluid, and we can write

$$\frac{d}{dt}\left(\frac{p}{\rho^\gamma}\right) = 0. \tag{8.29}$$

The Eqs. (8.22), (8.28) and (8.29) together constitute the basic equations of gas dynamics, i.e. the dynamical equations of a fluid satisfying the perfect gas law. If a state of such a fluid is given by specifying the values of $\rho(\mathbf{x}, t)$, $p(\mathbf{x}, t)$ and $\mathbf{v}(\mathbf{x}, t)$ at one instant of time, these equations enable us to calculate how these variables change with time so that we can determine the state of the fluid at some later time. It may be mentioned that (8.22) and (8.28) hold for different kinds of fluids, including nearly imcompressible liquids like water—for which we may need a different equation in the place of (8.29) for a complete dynamical theory. The dynamical theory of such fluids is more complicated than gas dynamics, and we shall not discuss it here.

8.4.2 Extension to MHD

We shall now discuss how the equations of gas dynamics can be extended to obtain the equations of MHD. Since we shall have to deal with another variable $\mathbf{B}(\mathbf{x}, t)$ inside the plasma, it is obvious that we shall need a time evolution equation for $\mathbf{B}(\mathbf{x}, t)$. Additionally, the magnetic field can give rise to a Lorentz force which has to be included in Euler's equation (8.28). We shall now show how these things can be done. Our treatment of MHD will be based on the assumption that the velocity $\mathbf{v}$ is non-relativistic everywhere, which introduces some elegant simplifications.

Suppose we are watching the dynamics of a plasma (treated as a one-fluid continuum) from the frame of reference of our laboratory. Let $\mathbf{E}$ and $\mathbf{B}$ be the electric and magnetic fields at the location of an element of plasma as seen from the laboratory frame, whereas $\mathbf{E}'$ and $\mathbf{B}'$ are the values of these fields in the frame of the plasma element moving at velocity $\mathbf{v}$. If $\mathbf{v}$ is non-relativistic, then the electromagnetic fields in the two frames are related in accordance with (5.114) and (5.115). We write these equations again:

$$\mathbf{E}' = \mathbf{E} + \mathbf{v} \times \mathbf{B}, \tag{8.30}$$

$$\mathbf{B}' = \mathbf{B} - \frac{\mathbf{v}}{c^2} \times \mathbf{E}. \tag{8.31}$$

We have already pointed out that, when we neglect charge separation, the electric field $\mathbf{E}'$ in the frame of the plasma would be very small. The electric field $\mathbf{E}$ in the laboratory frame arises mainly due to the motion of the plasma and (8.30) suggests that

$$|E| \approx |v||B|. \tag{8.32}$$

It then follows that the second term in the RHS of (8.31) is of order v^2/c^2 smaller compared to the first term. The spirit of MHD is that we keep terms of the order $|v|/c$, but neglect terms of the order v^2/c^2. Neglecting the second term in the RHS of (8.31), we write

$$\mathbf{B}' = \mathbf{B}. \tag{8.33}$$

The transformation laws of the electromagnetic fields between the laboratory frame and the frame of the plasma element are given by (8.30) and (8.33).

We now turn our attention to (1.3), one of Maxwell's equations. We shall show that the displacement current term is negligible compared to the $\nabla \times \mathbf{B}$ term within the framework of MHD. The ratio of these two terms is given by

$$\frac{\epsilon_0 \mu_0 |\frac{\partial \mathbf{E}}{\partial t}|}{|\nabla \times \mathbf{B}|} \approx \frac{|E|/\tau}{c^2 |B|/L},$$

where we have used (4.42) and have written the typical time scale and the typical length scale as τ and L, respectively. On taking $L/\tau \approx |v|$ and using (8.32), we find that the ratio of these two terms is of order v^2/c^2. Hence, we can neglect the displacement current term, when we neglect terms of order v^2/c^2. The approximation of neglecting the displacement current is often called the *MHD approximation*, which leads to

$$\nabla \times \mathbf{B} = \mu_0 \mathbf{j}. \tag{8.34}$$

We saw in Sect. 1.3 that this is one of the basic equations of magnetostatics. Now, this equation has a new significance. It is valid even in dynamic situations within the framework of MHD. We saw in Sect. 4.4 that the displacement current term plays a crucial role in the derivation of electromagnetic waves. When this term is neglected, we cannot get electromagnetic waves out of our equations. We conclude that MHD—the one-fluid model of the plasma—cannot handle the propagation of electromagnetic waves through plasmas, which was discussed in Sect. 8.3 with the help of the two-fluid model.

The current density and the electric field in a conducting medium are related by Ohm's law (4.61). When we allow for motions inside the medium like a plasma, this equation has to written down in a local frame of rest. We can write

$$\mathbf{j}' = \sigma \mathbf{E}', \tag{8.35}$$

where $\mathbf{j}'$ and $\mathbf{E}'$ are the current density and the electric field in the frame in which the element of plasma we are considering is at rest. For non-relativistic motions (i.e. $\gamma = 1$) with no charge separation in the rest frame of the plasma element (i.e. $\rho' = 0$), it easily follows from the discussion in Sect. 5.9 that the current density $\mathbf{j}$ in the laboratory frame should be equal to $\mathbf{j}'$, whereas the electric field $\mathbf{E}$ is given by (8.30). It then follows from (8.35) that

$$\mathbf{j} = \sigma(\mathbf{E} + \mathbf{v} \times \mathbf{B}). \tag{8.36}$$

Note that $\mathbf{v}$ in this equation is the velocity $\mathbf{v}(\mathbf{x}, t)$ at the point $\mathbf{x}$ inside the plasma treated as one fluid and *not* the velocity of individual charged particles. By combining (8.34) and (8.36), we can write the electric field as

$$\mathbf{E} = \frac{\nabla \times \mathbf{B}}{\mu_0 \sigma} - \mathbf{v} \times \mathbf{B}. \tag{8.37}$$

It is clearly seen in this equation that the electric field will be very small in the rest frame (where $\mathbf{v} = 0$) of a plasma with high electrical conductivity σ. Even when we allow for motions inside the plasma, (8.37) shows that the electric field $\mathbf{E}$ can be found from $\mathbf{v}$ and $\mathbf{B}$. This justifies our earlier assertion that the electric field does not have to be treated as an independent dynamical variable in the MHD model of the plasma.

One of the requirements of the MHD model is that we need a time evolution equation of the magnetic field $\mathbf{B}$. We can now readily get it from (1.4), which is one of Maxwell's equations incorporating electromagnetic induction. Substituting from (8.37) for $\mathbf{E}$ in (1.4), we get

$$\frac{\partial \mathbf{B}}{\partial t} = \nabla \times (\mathbf{v} \times \mathbf{B}) + \eta \nabla^2 \mathbf{B}, \tag{8.38}$$

where

$$\eta = \frac{1}{\mu_0 \sigma}, \tag{8.39}$$

and we have assumed that σ does not vary with position. The Eq. (8.38) is the central equation of MHD and is known as the *induction equation*. We shall discuss the significance of this equation in the next section.

To complete our derivation of the basic equations of MHD, the last question we need to address is how we can extend Euler's equation (8.28) to incorporate the Lorentz force which must arise in a plasma when magnetic fields are present. We basically have to add a term to this equation giving the Lorentz force per unit mass. As we pointed out in (1.6), the Lorentz force per unit volume is given by $\mathbf{j} \times \mathbf{B}$. To obtain the Lorentz force per unit mass, we merely have to divide this by ρ. Using (8.34) to substitute for $\mathbf{j}$, the Lorentz force per unit mass follows from the second term in the RHS of (1.6) and is

$$\mathbf{F}_{\text{Lorentz}} = \frac{1}{\mu_0 \rho} (\nabla \times \mathbf{B}) \times \mathbf{B}.$$

Adding this term to Euler's equation (8.28), we get

$$\frac{\partial \mathbf{v}}{\partial t} + (\mathbf{v}.\nabla)\mathbf{v} = \mathbf{F} - \frac{1}{\rho}\nabla p + \frac{1}{\mu_0 \rho}(\nabla \times \mathbf{B}) \times \mathbf{B}. \tag{8.40}$$

On making use of the vector identity

$$(\nabla \times \mathbf{B}) \times \mathbf{B} = (\mathbf{B}.\nabla)\mathbf{B} - \nabla\left(\frac{B^2}{2}\right),$$

which follows from (B.6), we can put (8.40) in the form

$$\frac{\partial \mathbf{v}}{\partial t} + (\mathbf{v}.\nabla)\mathbf{v} = \mathbf{F} - \frac{1}{\rho}\nabla\left(p + \frac{B^2}{2\mu_0}\right) + \frac{(\mathbf{B}.\nabla)\mathbf{B}}{\mu_0\rho}. \tag{8.41}$$

Either (8.40) or (8.41) provides the MHD generalization of Euler's equation. It is clear from (8.41) that the magnetic field introduces a pressure $B^2/2\mu_0$. The other magnetic term $(\mathbf{B}.\nabla)\mathbf{B}/\mu_0$ would be zero in a region of straight magnetic field lines. It arises only when magnetic field lines are bent and is of the nature of a tension force along magnetic field lines. We remind that reader of the discussion about the Maxwell stress tensor in Sect. 4.3 following (4.33). We can reach the same conclusions which we reached there.

8.5 Alfvén's Theorem of Flux Freezing

Let us now consider the physical significance of the induction Eq. (8.38), which is the central equation of MHD. Suppose the magnetic field inside the plasma has the typical value B, and the velocity field has the typical value v, whereas L is the typical length scale over which the magnetic or velocity fields vary significantly. Then the term $\nabla \times (\mathbf{v} \times \mathbf{B})$ in the induction Eq. (8.38) should be of order vB/L, while the other term $\eta\nabla^2\mathbf{B}$ in (8.38) should be of order $\eta B/L^2$. The ratio of these two terms is a dimensionless number known as the *magnetic Reynolds number* and is given by

$$\mathcal{R}_M \approx \frac{vB/L}{\eta B/L^2} \approx \frac{vL}{\eta}. \tag{8.42}$$

The important point to note here is that $\mathcal{R}_M$ goes as L, which is much larger for an astrophysical system than what it is for a laboratory plasma. In fact, it turns out that $\mathcal{R}_M$ is usually much smaller than 1 for laboratory plasmas and much larger than 1 for astrophysical systems. This means that $\eta\nabla^2\mathbf{B}$ is the dominant term on the RHS of (8.38) when we are dealing with laboratory plasmas, and $\nabla \times (\mathbf{v} \times \mathbf{B})$ is the dominant term when we are dealing with astrophysical plasmas. For laboratory plasmas, we can often write

$$\text{Laboratory:} \quad \frac{\partial \mathbf{B}}{\partial t} \approx \eta\nabla^2\mathbf{B}. \tag{8.43}$$

This is the vectorial version of the well-known *diffusion equation*, one of the standard equations of mathematical physics. This equation is not difficult to interpret. We see from (8.39) that η is essentially the inverse of conductivity σ, which means that η goes as the resistivity of the plasma. We know that the resistivity of a system makes currents in the system decay, and thereby magnetic fields produced by those currents also decay. The significance of (8.43) is that the magnetic field in the plasma diffuses away with time due to the resistivity, with the resistivity parameter η appearing as the diffusion coefficient. On the other hand, magnetic fields in astrophysical plasmas often evolve primarily due to the other term in (8.38), i.e. we can write

$$\text{Astrophysics:} \quad \frac{\partial \mathbf{B}}{\partial t} \approx \nabla \times (\mathbf{v} \times \mathbf{B}).$$

We now discuss the significance of this equation.

If the magnetic Reynolds number $\mathcal{R}_M$ of an astrophysical system is extremely large, then it is often justified to replace the approximation sign in the last equation by an equality sign, i.e.

$$\frac{\partial \mathbf{B}}{\partial t} = \nabla \times (\mathbf{v} \times \mathbf{B}). \tag{8.44}$$

When the magnetic field in the plasma evolves according to this equation, we can prove a very remarkable theorem called *Alfvén's theorem of flux freezing*. We first state this theorem before proving it.

Consider a surface S_1 inside a plasma at time t_1. The flux of magnetic field linked with this surface is $\int_{S_1} \mathbf{B} \cdot d\mathbf{S}$. At some future time t_2, the parcels of plasma which made up the surface S_1 at time t_1 will move away and will make up a different surface S_2. The magnetic flux linked with this surface S_2 at time t_2 will be $\int_{S_2} \mathbf{B} \cdot d\mathbf{S}$. The theorem of flux freezing states that

$$\int_{S_1} \mathbf{B} \cdot d\mathbf{S} = \int_{S_2} \mathbf{B} \cdot d\mathbf{S}$$

if $\mathbf{B}$ evolves according to (8.44). We write this more compactly in the form

$$\frac{d}{dt} \int_S \mathbf{B} \cdot d\mathbf{S} = 0, \tag{8.45}$$

where the Lagrangian derivative d/dt implies that we are considering the variation of the magnetic flux $\int_S \mathbf{B} \cdot d\mathbf{S}$ linked with the surface S as we follow the surface S with the motion of the plasma parcels constituting it.

Several steps necessary for proving this theorem were already worked out in Sect. 4.1.2, where we considered the rate of change of the magnetic flux $\int \mathbf{B} \cdot d\mathbf{S}$ linked with a moving circuit. Now we are considering a moving plasma surface. A little reflection should convince you the rate of change of magnetic flux can be treated in exactly the same manner in the present situation also, leading to (4.12).

Fig. 8.1 Illustration of flux
freezing. **a** A straight column
of magnetic field. **b**
Magnetic configuration after
bending the column. **c**
Magnetic configuration after
twisting the column

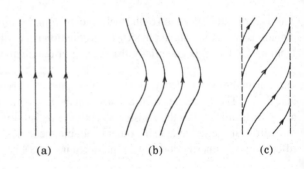

(a)   (b)   (c)

The last term in (4.12) was a line integral along the circuit, which will now be a line
integral along the boundary of the plasma surface. On converting this line integral
into a surface integral by Stokes's theorem (B.15), from (4.12) we get

$$\frac{d}{dt} \int_S \mathbf{B} \cdot d\mathbf{S} = \int \frac{\partial \mathbf{B}}{\partial t} \cdot d\mathbf{S} - \int [\nabla \times (\mathbf{v} \times \mathbf{B})] \cdot d\mathbf{S}. \tag{8.46}$$

If (8.44) holds, it is easy to see that the RHS of (8.46) is zero, which would imply
(8.45). This completes our proof that (8.44) leads to (8.45).

In astrophysical systems with high $\mathcal{R}_M$ within which (8.45) holds, we can imagine
the magnetic flux to be frozen in the plasma and to move with the plasma flows.
Suppose we have straight magnetic field lines going through a plasma column as
shown in Fig. 8.1a. If the plasma column is bent, then in the high $\mathcal{R}_M$ limit, the
magnetic field lines are also bent with it as shown in Fig. 8.1b. Only if the magnetic
field lines are bent with the plasma column, could the same magnetic flux continue
to pass through a cross section of the plasma column even after bending. On the
other hand, if one end of the plasma column is twisted, then the magnetic field lines
are also expected to be twisted as in Fig. 8.1c. As a result of the theorem of flux
freezing, the magnetic field in an astrophysical system can almost be regarded as a
plastic material which can be bent, twisted or distorted by making the plasma move
appropriately. This view of a magnetic field is radically different from that which
we normally encounter in laboratory situations, where the magnetic field appears as
something rather passive which we can switch on or off by sending a current through
a coil. In the astrophysical setting, the magnetic field appears to acquire a life of its
own.

We thus see that magnetic fields behave very differently in laboratory and astrophysical settings, due to the fact that the magnetic field evolves respectively according
to the two different equations (8.43) and (8.44) in these two situations. Alfvén [5]
coined the name *cosmical electrodynamics* to distinguish electrodynamics at cosmical scales from ordinary laboratory electrodynamics, although we start from the same
Maxwell equations and Ohm's law in both the cases. In the astrophysical setting, if
we know the initial configuration of the magnetic field and the nature of plasma flows,
we can almost guess on the basis of the flux-freezing theorem what the subsequent
magnetic field configuration is going to be (as we saw in Fig. 8.1). The human mind

is more attuned to thinking geometrically rather than thinking analytically. We may be able to solve an equation describing a process, but only when we are able to make a mental picture of how the process proceeds, we feel that we have understood the process. The beauty of cosmical electrodynamics is that the flux-freezing theorem allows us to make a mental picture of how the magnetic field evolves in an astrophysical plasma. An application of MHD to an astrophysical situation, where flux freezing is important, will be illustrated in Sect. 8.8.

8.6 Confining Plasmas with Magnetic Fields

We now consider an application of plasma physics which has the potential to change the course of human civilization. The huge energy needs of present-day civilization are mostly met by fossil fuels like coal and oil. Since the store of fossil fuels is limited, and they have undesirable effects such as the production of greenhouse gases, it is imperative to tap other sources of energy. One attractive possibility is the thermonuclear process, which generates energy in the interiors of stars. Deuterium happens to be the atom which seems most promising for this purpose. Since some molecules in seawater have deuterium atoms in place of hydrogen atoms, we have a huge source of deuterium atoms on the surface of the Earth. When two deuterium nuclei are brought together, they may fuse to produce a tritium or a helium nucleus, and a small fraction of mass is converted into energy during this process. Since deuterium nuclei are positively charged, two such nuclei would normally not come close to each other due to the Coulomb repulsion. Only when two such nuclei approach each other with sufficiently high relative velocity would it be possible to overcome the Coulomb barrier so that the thermonuclear reaction takes place. The easiest way of achieving this is to produce a sufficiently hot deuterium plasma. If the random velocities of the deuterium nuclei are sufficiently high due to the high temperature, then occasionally two nuclei may approach each other with high enough relative velocity to overcome the Coulomb barrier, leading to fusion.

Such a high-temperature deuterium plasma cannot be kept in ordinary material containers. The best option is to confine the hot plasma with the help of magnetic fields. Let us now use the MHD equations to determine whether a hot plasma can be kept in static equilibrium in which the various forces are in balance. For a static situation without any flow, we have $\mathbf{v} = 0$ so that it follows from (8.40) that

$$\rho \mathbf{F} - \nabla p + \frac{1}{\mu_0} (\nabla \times \mathbf{B}) \times \mathbf{B} = 0. \tag{8.47}$$

In a terrestrial laboratory, the body force $\mathbf{F}$ is nothing but the acceleration due to gravity $\mathbf{g}$. In the Earth's atmosphere, the pressure falls over heights of order kilometres, and the pressure gradient arising out of this fall balances $\mathbf{g}$. In a plasma experiment, however, we want to confine the plasma within a region having size of the order of metres. The pressure gradient in this situation has to be much larger than $\mathbf{g}$ and can only be balanced by the Lorentz force. We can neglect **g** in (8.47) for such a situation so that

$$\nabla p = \frac{1}{\mu_0} (\nabla \times \mathbf{B}) \times \mathbf{B}. \tag{8.48}$$

We now consider the confinement of a cylindrical column of plasma. We introduce cylindrical coordinates (r, θ, z) and assume cylindrical symmetry, i.e. assume that nothing varies in θ or z directions. Let us explore whether it is possible to confine this plasma column with the help of a magnetic field produced by sending a current $\mathbf{j} = j(r)\, \mathbf{e}_z$ along the axis of the column. It is easy to check that such a current will produce a magnetic field

$$\mathbf{B} = B_\theta(r)\, \mathbf{e}_\theta \tag{8.49}$$

in the θ direction, related to the current density by (8.34). In cylindrical coordinates with cylindrical symmetry, this equation becomes

$$\frac{1}{r} \frac{d}{dr} (r B_\theta) = \mu_0 j \tag{8.50}$$

on using the expression of the curl in cylindrical coordinates given by (C.3) with **B** as in (8.49). The other relevant equation for static equilibrium is obtained by substituting (8.49) in (8.48), which in a few easy steps left to the reader leads to

$$\frac{d}{dr} \left(p + \frac{B_\theta^2}{2\mu_0} \right) + \frac{B_\theta^2}{\mu_0 r} = 0 \tag{8.51}$$

on assuming the pressure $p(r)$ to be a function of r alone. We need to solve (8.50) and (8.51) to determine whether the plasma column can be confined with such a magnetic field.

We consider the simplest case of a constant current density through the plasma column. If j in (8.50) is constant, then we get

$$B_\theta = \frac{\mu_0}{2} jr$$

on using the boundary condition that B_θ has to be zero at $r = 0$ on the central axis of the plasma column. This can also be easily obtained from Ampere's law (3.15). Substituting this in (8.51) gives

$$\frac{dp}{dr} = -\frac{\mu_0}{2} j^2 r,$$

of which the solution is

$$p = p_0 - \frac{\mu_0 j^2 r^2}{4}, \tag{8.52}$$

Fig. 8.2 The profile of the azimuthal magnetic field B_θ and the pressure p inside a plasma column with a uniform current density j inside

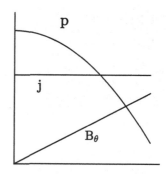

where p_0 is the gas pressure at the centre of the column. Figure 8.2 shows the current density, the magnetic field component B_θ and the pressure as a function of radius. It is to be noted that the pressure falls with radius, which we would expect on the basis of the result we derived towards the end of Sect. 4.3 that electric or magnetic field lines have tension along them. Due to this tension, any circular magnetic field line tries to shrink in size and squeezes the gas in its interior—an effect which is often referred to as the *pinching effect* of the magnetic field. We see from (8.52) that the pressure would become zero at a certain radius beyond which (8.52) ceases to hold, since a negative pressure is not physically admissible. If $r = a$ is the radius of the plasma column where the gas pressure falls to zero, we readily find from (8.52) that

$$a^2 = \frac{4p_0}{\mu_0 j^2}. \tag{8.53}$$

This raises the possibility that the pinching effect of the magnetic field may be able to confine a plasma column by making the pressure go to zero at a finite radius.

Is it possible that in the laboratory we confine a hot plasma column with the help of a magnetic field produced by sending a current along the axis of the column such that fusion reactions take place inside? An equilibrium configuration can be realized in a laboratory experiment only if it is stable, i.e. if any perturbations which may arise in the system are suppressed and do not grow. A formal stability analysis of this problem is complicated, and we shall not get into that mathematical theory here. On physical considerations, however, one can argue that the configuration we are discussing is actually *unstable*. Let us consider the two kinds of perturbations shown in Fig. 8.3a, b. Since magnetic field lines are crowded at the point P in Fig. 8.3a, it is obvious that the magnetic pressure is enhanced at that point, and this additional pressure will push the plasma column in such a way as to enhance the kink perturbation. In other words, the type of kink perturbation shown in Fig. 8.3a, once initiated, starts growing so that the system is unstable. This is called the *kink instability*. It is also easy to see that B_θ at point Q in Fig. 8.3b turns out to be larger than what B_θ just outside the column would have been if the column were not perturbed. The enhanced magnetic stress at Q will make the column still narrower there, triggering an instability. This is known as the *sausage instability*.

Fig. 8.3 Unstable
perturbations of a plasma
column having magnetic
fields produced by currents
flowing in the direction of
the axis. **a** Kink instability. **b**
Sausage instability

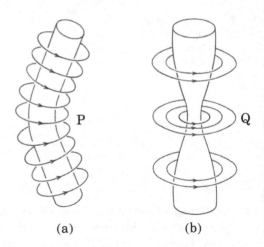

(a)   (b)

If there is a magnetic field along the axis of the plasma column (i.e. in the z direction), it tends to suppress both these types of instabilities. A kink perturbation bends the axial field lines, and the tension of the axial field clearly opposes the growth of the kink. In the case of a sausage perturbation, the axial field is compressed in the places where the column becomes narrower, and the enhanced pressure of the axial field opposes further narrowing of the column at that point. A rigorous analysis shows that $|B_z|$ has to be of the order of $|B_\theta|$ to suppress the instabilities.

While an axial magnetic field would suppress kink and sausage instabilities, does it make the plasma configuration stable? We have pointed out that MHD provides a macroscopic description of the plasma which is valid when certain criteria are satisfied. Even if a plasma configuration is stable according to MHD, other instabilities may arise from microscopic considerations. In fact, it has proved notoriously difficult to confine plasmas with magnetic fields due to various instabilities. It is necessary to spend some energy to set up a hot plasma configuration and to send currents through it to produce the necessary magnetic fields. The commercial production of energy is feasible only if the hot plasma remains confined for sufficient time for nuclear fusion reactions to generate more energy than what was fed into the system to set it up. When fusion research began in several countries soon after World War II, it was hoped that the commercial generation of energy would be possible in a few years. It was not realized initially that various instabilities would make it so difficult to confine the plasma, as a result of which the commercial generation of energy in this way has not been achieved yet. However, plasma physicists have learned over the years about many instabilities and how to suppress them. It is believed that commercial generation of energy will be possible with a sufficiently large plasma device. Such a large plasma device named International Thermonuclear Experimental Reactor (ITER) is now being constructed in southern France under an international collaboration. We hope that commercial generation of energy by thermonuclear fusion will be achieved when the ITER becomes operational in a few years.

8.7 MHD Waves in Magnetized Plasmas

When we discussed plasma oscillations and propagation of electromagnetic waves through plasmas in Sect. 8.3, we allowed ions and electrons to move differently (in fact, we considered ions to be at rest in our simplified treatment). It is possible to have other kinds of low-frequency waves (frequency $\ll \omega_p$) in a plasma in which electrons and ions remain together so that the plasma can be treated with the equations of MHD. We now present the theory of a very simple kind of wave of this nature which can be handled with MHD equations.

When a stretched string is disturbed, it can give rise to transverse waves propagating along it—driven by the tension in the string which provides the restoring force. We have pointed out that the magnetic force appearing in (8.41) has a magnetic tension part within it. We may expect the magnetic tension to drive a transverse wave along the magnetic field, just as the tension along a stretched string drives a wave along the string. To show that this indeed happens, we consider perturbing a uniform plasma containing a constant magnetic field $\mathbf{B}_0$. To simplify the treatment, we neglect the dissipative effect due to electrical resistivity given by the last term in (8.38). Let us write the perturbed magnetic field as $\mathbf{B} = \mathbf{B}_0 + \mathbf{B}_1(\mathbf{x}, t)$. The velocity field should be written as $\mathbf{v} = \mathbf{v}_1(\mathbf{x}, t)$, to make it clear that the velocity arises from the perturbation and does not have an unperturbed part. We substitute these in (8.38) and (8.40), keeping only the linear terms in the small perturbed quantities $\mathbf{v}_1$, $\mathbf{B}_1$ and throwing away the nonlinear terms. This gives

$$\frac{\partial \mathbf{v}_1}{\partial t} = \frac{1}{\mu_0 \rho}(\nabla \times \mathbf{B}_1) \times \mathbf{B}_0, \tag{8.54}$$

$$\frac{\partial \mathbf{B}_1}{\partial t} = \nabla \times (\mathbf{v}_1 \times \mathbf{B}_0). \tag{8.55}$$

Note that we have neglected the pressure gradient term by assuming the plasma continues to have uniform pressure even when it is perturbed. We shall later comment on this assumption.

By analogy with waves along stretched strings, we anticipate waves propagating along magnetic field lines. Let us take the perturbations $\mathbf{v}_1$, $\mathbf{B}_1$ to have solutions of the form $\exp[i(\mathbf{k}.\mathbf{x} - \omega t)]$ with $\mathbf{k}$ in the same direction as $\mathbf{B}_0$. Then we have to replace $\partial/\partial t$ by $-i\omega$ and ∇ by $i\mathbf{k}$ in (8.54) and (8.55). This gives

$$-\omega \mathbf{v}_1 = \frac{1}{\mu_0 \rho}(\mathbf{k} \times \mathbf{B}_1) \times \mathbf{B}_0 = \frac{1}{\mu_0 \rho}[(\mathbf{k}.\mathbf{B}_0)\mathbf{B}_1 - (\mathbf{B}_0.\mathbf{B}_1)\mathbf{k}], \tag{8.56}$$

$$-\omega \mathbf{B}_1 = \mathbf{k} \times (\mathbf{v}_1 \times \mathbf{B}_0) = (\mathbf{k}.\mathbf{B}_0)\mathbf{v}_1 - (\mathbf{k}.\mathbf{v}_1)\mathbf{B}_0. \tag{8.57}$$

Since we are expecting a transverse wave, let us look for a solution such that $\mathbf{v}_1$ and $\mathbf{B}_1$ are both perpendicular to $\mathbf{B}_0$ and $\mathbf{k}$ (which are in the same direction). For such a case, it is easy to see that the last terms in (8.56) and (8.57) will be zero. Then we are led to

$$-\omega \mathbf{v}_1 = \frac{1}{\mu_0 \rho} k B_0 \mathbf{B}_1,$$

$$-\omega \mathbf{B}_1 = k B_0 \mathbf{v}_1.$$

From these equations, we finally obtain

$$\omega^2 = k^2 \frac{B_0^2}{\mu_0 \rho}. \tag{8.58}$$

A little reflection will convince you that this corresponds to a wave propagating along the unperturbed magnetic field $\mathbf{B}_0$ with velocity given by

$$\mathbf{v}_A = \frac{\mathbf{B}_0}{\sqrt{\mu_0 \rho}}. \tag{8.59}$$

This type of wave is known as the *Alfvén wave*, and the velocity $\mathbf{v}_A$ is called *Alfvén velocity*. The existence of Alfvén waves has been demonstrated experimentally.

It may be noted that the analysis we have just presented was not of a very general nature. Anticipating a transverse wave propagating along the unperturbed magnetic field, we had taken $\mathbf{k}$ in the direction of $\mathbf{B}_0$, with $\mathbf{v}_1$ and $\mathbf{B}_1$ in the perpendicular direction. We also did not include pressure perturbations. In a more general situation, if we allow $\mathbf{v}_1$ to have a component in the direction of $\mathbf{k}$ (i.e. if the wave is partially longitudinal), then we have variations of density and consequently of pressure. To handle this situation, we have to include (8.22) and (8.29) in our analysis, with $\mathbf{k}$ inclined at any arbitrary direction with respect to $\mathbf{B}_0$. This leads to a much more complicated calculation, which we do not discuss here. It is well known that pressure perturbations in a gas can give rise to acoustic waves. The more complete analysis shows existence of waves which are mixtures of Alfvén waves and acoustic waves— known as *magnetoacoustic waves*.

8.8 Sunspots and Magnetic Buoyancy

Let us end our discussion with an illustration of how MHD is applied to astrophysics. As pointed out in Sect. 8.5, the magnetic Reynolds number $\mathcal{R}_M$ usually turns out to be much larger than 1 for astrophysical systems, and the magnetic field is frozen in the plasma. In fact, some of the most interesting theoretical predictions from MHD are for cases with high $\mathcal{R}_M$, which cannot be tested properly within laboratories where we usually have $\mathcal{R}_M \ll 1$. The Sun is our nearest large plasma body in which we can try to look for effects which are expected from MHD equations in the high-$\mathcal{R}_M$ situations. We often say that the Sun is our best laboratory for high-$\mathcal{R}_M$ MHD. A

Fig. 8.4 A magnetogram map of the Sun. The white and black colours indicate positive and negative magnetic polarities, respectively, whereas regions with weak magnetic fields are indicated in grey

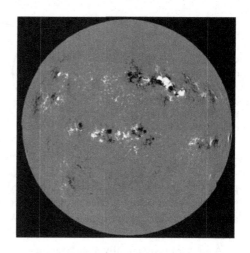

symbiotic relation exists between the study of the Sun and MHD. On the one hand, we observe phenomena in the Sun which require MHD for their explanation. On the other hand, the Sun is a convenient nearby plasma body in which we can look for effects predicted by MHD in the high-$\mathcal{R}_M$ limit. The explanation of solar phenomena by the application of MHD constitutes what is called *solar MHD*.

In 1908, Hale made the momentous discovery of the Zeeman splitting of spectral lines in the spectra of large sunspots, indicating that sunspots are regions of strong magnetic field of about 0.3 T ($\approx$ 5000 times the magnetic field around geomagnetic poles). This is the first time that astronomers found conclusive evidence for the existence of magnetic fields outside the Earth's environment. We now know that magnetic fields are ubiquitous in the astrophysical universe: many astrophysical bodies are found to have magnetic fields. The branch of MHD which deals with the question of how magnetic fields arise in astrophysical bodies is called *dynamo theory*. This theory is beyond the scope of this book. Readers desirous of reading a non-technical account of dynamo theory may turn to Choudhuri [4]. In the present account, we shall assume the existence of magnetic fields in the Sun and point out how MHD can be used for explaining various properties of sunspots.

We know that isolated magnetic monopoles cannot exist. Magnetic fields of both polarities have to be present in sunspots. Sunspots often occur in pairs, both of them appearing at approximately the same solar latitude. Two members of a sunspot pair are usually found to have opposite magnetic polarities. Figure 8.4 shows a magnetogram map of the Sun. A magnetogram measures the magnetic field at all points of the solar disk. Regions of positive and negative magnetic fields are respectively indicated by white and black colours, whereas regions without appreciable magnetic fields are shown in grey. A pair of sunspots at the same latitude with opposite polarities appear as a patch of white and a patch of black side by side in the magnetogram map. We see several such sunspot pairs in Fig. 8.4. In the northern hemisphere, the sunspot with positive polarity within the pair appears to right and the sunspot with negative

Fig. 8.5 The production of a strong toroidal magnetic field underneath the solar surface. **a** An initial field line of a dipolar magnetic field. **b** A sketch of the field line after it has been stretched by the faster rotation near the equatorial region

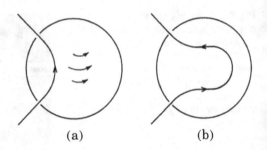

(a)   (b)

polarity to the left. This is seen to be reversed in the southern hemisphere. It may be pointed out that sunspots have cycles. The polarity patterns in both the hemispheres are reversed after approximately 11 years. We shall not get into a discussion of this cycle in this book.

A sunspot is essentially a dark region of concentrated magnetic field on the solar surface. The first question we address is this: why does the magnetic field get bundled up in a limited region which appears darker compared to the surroundings? Astrophysicists routinely construct theoretical models of the structures of stars. Models of the Sun suggest that the heat generated in the central region of the Sun is transported outward by convection in the outer layers of the Sun. A sunspot is, therefore, a bundle of magnetic flux sitting in a region where convection is taking place. To understand the formation of sunspots, we need to know how convection is affected by the presence of magnetic field. This subject is known as *magnetoconvection*. We have pointed out in Sects. 4.3 and 8.4.2 that a magnetic field has a tension force associated with it, which is expected to oppose fluid motions connected with convection. Detailed calculations of magnetoconvection suggest that if magnetic fields are present in a region of convection, they tend to be swept into confined regions within which convection is inhibited by magnetic tension, whereas the remaining regions become free from magnetic fields where convection can take place freely. Sunspots are merely regions within which magnetic fields are kept bundled up by convection. Since magnetic tension inhibits convection within a sunspot, heat transport is less efficient within a sunspot compared to the surroundings, leading to a cooler surface temperature of the sunspot. That is why a sunspot appears darker than the surroundings. A concentration of magnetic flux in a limited region with very little magnetic field in the surrounding region is referred to as a *magnetic flux tube*. A sunspot can be regarded as an example of this.

The magnetic field of the Earth is roughly like that of a magnetic dipole. Figure 8.5a shows a typical field line of a magnetic field with dipolar structure. The Earth rotates like a solid body. In contrast, it is found that the angular velocity of the Sun near the equatorial region is higher than that in the polar regions. Since we are dealing with a system having $\mathcal{R}_M \gg 1$ and the magnetic field would be frozen in the solar plasma according to Alfvén's theorem of flux freezing discussed in Sect. 8.5, we expect any dipolar magnetic field line of the Sun to be stretched out in the forward direction around the equatorial region, as shown in Fig. 8.5b. In other words, if we

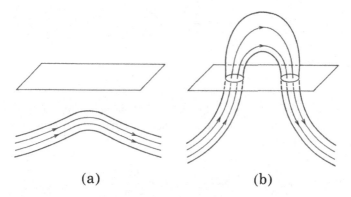

(a)   (b)

Fig. 8.6 Magnetic buoyancy of a flux tube. **a** A nearly horizontal flux tube under the solar surface.
b The flux tube after its upper part has risen through the solar surface

begin with a field line in the Sun as shown in Fig. 8.5a, after some time, it will look
like the field line as shown in Fig. 8.5b. Due to this effect, we expect the magnetic
field underneath the solar surface to have a strong component in the ϕ direction (if
we introduce spherical coordinates in the usual manner), which is usually called
the toroidal component. We note in Fig. 8.5b that this toroidal magnetic field has
opposite directions in the two hemispheres. Due to interactions with convection, this
toroidal magnetic field is expected to exist in the form of magnetic flux tubes.

We now come to the question of how a pair of sunspots with opposite polarity
lying at the same solar latitude arises. The most obvious explanation for this is
that there must be a magnetic flux tube underneath the solar surface aligned in the
toroidal direction, of which a part has come out through the solar surface as shown
in Fig. 8.6b. If the two sunspots in a pair are merely the two locations where this
strand of magnetic field intersects the solar surface, then we readily see that one
sunspot must have magnetic field lines coming out and that the other must have
field lines going in, implying that the two sunspots have opposite polarity. We have
already argued that the toroidal magnetic flux tubes will have oppositely directed
magnetic fields in the two hemispheres. This suggests that the sunspot pairs in the
two hemispheres should have opposite polarity patterns, as seen in the magnetogram
map shown in Fig. 8.4.

We finally come to the question how the magnetic configuration shown in Fig. 8.6b
arises. As we already discussed, magnetic fields exist in the form of toroidal magnetic
flux tubes, i.e. flux tubes in the horizontal direction with respect to the Sun. Neglecting
curvature, let us consider a nearly horizontal magnetic flux tube, as sketched in
Fig. 8.6a. There has to be a pressure balance between the inside and outside of the
magnetic flux tube across its surface. Let p_i be the gas pressure inside the magnetic
flux tube and p_e be the external pressure. We have seen in (8.41) that a magnetic
field gives rise to a pressure $B^2/2\mu_0$ wherever it exists. In order to have a pressure
balance across the bounding surface of the flux tube, we must have

$$p_e = p_i + \frac{B^2}{2\mu_0}. \tag{8.60}$$

It readily follows that

$$p_i < p_e. \tag{8.61}$$

This usually, though not always, implies that the internal density ρ_i is also less than the external density ρ_e. In the particular case when the temperatures inside and outside are both T, (8.60) leads to

$$R\rho_e T = R\rho_i T + \frac{B^2}{2\mu_0}, \tag{8.62}$$

where R is the gas constant appearing in the ideal gas equation of state $p = R\rho T$. It easily follows from (8.62) that

$$\frac{\rho_e - \rho_i}{\rho_e} = \frac{B^2}{2\mu_0 p_e}. \tag{8.63}$$

We thus see that the fluid in the interior of the flux tube is lighter and must be buoyant. In the limit of high $\mathcal{R}_M$, the magnetic field of the flux tube is frozen in the lighter fluid inside it. As a result, the flux tube with its internal magnetic field becomes buoyant as an entity and rises against gravity carrying the magnetic field. This very important effect known as *magnetic buoyancy* was discovered by E.N. Parker in 1955. Since (8.60) does not *always* imply that the interior of a flux tube is lighter, it is possible that one part of a flux tube becomes buoyant and not the other parts. Here, we shall not get into a discussion as to how this may come about. Suppose only the middle part of the flux tube shown in Fig. 8.6a has become buoyant. Then this middle part is expected to rise, eventually piercing through the surface and creating the configuration of Fig. 8.6b.

We thus find that we can combine the important MHD results of flux freezing, magnetic tension and magnetic pressure to explain many aspects of sunspots. It is due to flux freezing that the magnetic field inside the Sun would be stretched in the toroidal direction. The magnetic tension of the toroidal magnetic field in the environment of magnetoconvection suggests that the toroidal field is likely to exist in the form of magnetic flux tubes. Finally, the magnetic pressure inside these flux tubes would give rise to magnetic buoyancy, causing pairs of sunspots with opposite polarity to emerge on the solar surface.

Exercises

8.1 Show that the electromagnetic oscillations in a two-fluid plasma are damped if you include a collision term in the equation of motion of the electrons (8.9) so that it gets modified to

$$m_e \frac{\partial \mathbf{v}_e}{\partial t} = -e\mathbf{E} - m_e \frac{\mathbf{v}_e}{\tau}.$$

8.2 In the treatment of plasma oscillations given in Sect. 8.3, the motions of ions were neglected. Suppose ions with charge Ze and mass m_i also move in response to the electric field. Assuming the plasma to be a mixture of an electron fluid with velocity $\mathbf{v}_e$ and an ion fluid with velocity $\mathbf{v}_i$, show that the frequency of longitudinal plasma oscillations will be

$$\omega = \omega_p \sqrt{1 + \frac{Z m_e}{m_i}},$$

where ω_p is given by (8.14).

8.3 From the dispersion relation (8.18) of electromagnetic waves propagating through a plasma, show that the group velocity is

$$v_{gr} = c \sqrt{1 - \frac{\omega_p^2}{\omega^2}}.$$

A radio signal starting from a pulsar (a kind of star which emits radio pulses) and passing through the interstellar medium will reach an observer at a distance L in time

$$T_a = \int_0^L \frac{ds}{v_{gr}}.$$

Substitute the expression of v_{gr} in this and make a binomial expansion by assuming $\omega_p \ll \omega$. If signals with different frequencies started at the same time, show that they will be dispersed on reaching the observer, with the dispersion given by

$$\frac{dT_a}{d\omega} = -\frac{e^2}{\epsilon_0 m_e c \, \omega^3} \int_0^L N_e \, ds.$$

8.4 If there were free magnetic monopoles having magnetic charge g in our galaxy, they would have been accelerated to speed c by the galactic magnetic field. As a result of the galactic magnetic field B doing work on these monopoles, the energy of the magnetic field will decrease. Show that the decay time would be about

$$\tau \approx \frac{B}{2\mu_0 N g c},$$

where N is the number density of monopoles. Since our galaxy continues to have a magnetic field, you may assume the decay time to be larger than the lifetime of the galaxy, putting a limit on the quantity Ng (known as *Parker's limit*).

8.5 Consider a constant initial magnetic field $\mathbf{B} = B_0\,\mathbf{e}_y$ in a plasma of zero resistivity. Suppose a velocity field

$$\mathbf{v} = v_0\,e^{-y^2}\,\mathbf{e}_x$$

is switched on at time $t = 0$. Determine how the magnetic field evolves in time. Make a sketch of the magnetic field lines at some time after switching on the velocity field.

8.6 Suppose a plasma with a magnetic field is in equilibrium so that (8.48) is satisfied. If the gas pressure is negligible compared to the magnetic pressure, argue that the magnetic field has to obey the equation

$$\nabla \times \mathbf{B} = \alpha \mathbf{B},$$

where α is a scalar quantity which has to be constant along a magnetic field line, i.e. $\mathbf{B}.\nabla\alpha = 0$, but can vary from one field line to another. This type of magnetic field is known as a *force-free magnetic field*.

8.7 In an Alfvén wave, show that the average kinetic energy $\rho v_1^2/2$ is equal to the average magnetic energy density $B_1^2/2\mu_0$ associated with the perturbed magnetic field $\mathbf{B}_1$. What is the physical interpretation of this result?

8.8 Consider a horizontal magnetic flux tube with magnetic field B and radius of cross section a embedded in an isothermal atmosphere of perfect gas with constant gravity g. It is easy to check that $\Lambda = p/\rho g$ is a constant throughout an isothermal atmosphere. The flux tube rising due to magnetic buoyancy at speed v experiences a drag force per unit length given by

$$\frac{1}{2}C_D\rho v^2 a,$$

where C_D is a constant. Show that the flux tube eventually rises with an asymptotic speed

$$v_{\mathrm{A}}\left(\frac{\pi a}{C_D\Lambda}\right)^{1/2},$$

where $v_{\mathrm{A}} = B/\sqrt{\mu_0\rho}$.

References

1. Fermi, E.: Thermodynamics. Prentice-Hall (reprinted by Dover) (1937)
2. Reif, F.: Fundamentals of Statistical and Thermal Physics. McGraw Hill (1965)
3. Landau, L.D., Lifshitz, E.M.: The Classical Theory of Fields (Course of Theoretical Physics), vol.2, 4th edn. (1980)
4. Choudhuri, A. R.: Nature's Third Cycle: A Story of Sunspots. Oxford University Press (2015)
5. Alfvén, H.: Cosmical Electrodynamics. Oxford University Press (1950)

Appendix A
A Short Note on Gaussian Units

One inconvenient aspect of electromagnetic theory is that two systems of units—the Gaussian cgs units and the SI units—have been widely used, and the basic equations look different in these two systems. It is not essential that the basic equations should look different in two systems of units. For example, this is not the case with classical mechanics. Even though important quantities like the gravitational constant G have different numerical values in the cgs and mks units, the forms of the basic equations are the same in the two systems. However, the equations of electromagnetic theory have historically been written differently in the two systems.

Since the SI system of units is being used more widely nowadays, we have used this system throughout this book. However, as a working physicist is likely to encounter the Gaussian cgs system often while consulting older literature, it is still essential for a serious student of physics to form some notion about this system. We give a short introduction to the Gaussian cgs system here. It may be noted that the mechanical quantities (such as length, mass and force) in the Gaussian cgs system are expressed in cgs units (centimetre, gram, dyne), whereas they are expressed in mks units in the SI system (metre, kilogram, newton). We now explain how the electromagnetic quantities are expressed in the two different systems.

A.1 Electrostatic Units

Since the electrostatic unit (or esu) of charge is introduced through Coulomb's law, let us begin with a discussion of that. This law in SI units is given by (2.7). This equation suggests that the amplitude of the force between two charges q_1 and q_2 separated by a distance r is

$$F = \frac{1}{4\pi\epsilon_0} \frac{q_1 q_2}{r^2}. \tag{A.1}$$

© Springer Nature Singapore Pte Ltd. 2022
A. R. Choudhuri, *Advanced Electromagnetic Theory*, Lecture Notes in Physics 1009,
https://doi.org/10.1007/978-981-19-5944-8

We use the convention that positive F, which will result if the charges are of the same sign, corresponds to repulsion. In contrast, Coulomb's law in Gaussian units is written as

$$F_{cgs} = \frac{q_{1,es}\, q_{2,es}}{r_{cgs}^2}.$$ (A.2)

Here, the subscripts indicate that the charges have to be in esu, whereas the mechanical quantities like length and force are in cgs units.

We can take (A.2) as introducing the electrostatic unit of charge, which is defined in the following manner. An electrostatic unit of charge is such that two particles with unit charges kept at a distance of 1 cm repel each other with a force of 1 dyne. This may seem like a good definition of the unit of charge. The only problem is that this unit of charge turns out to be inconveniently small compared to the charges we encounter in everyday life. The SI coulomb unit of charge was introduced precisely in a way that it is convenient for our everyday use. The typical current that flows through a household wire is 5 A, which means that 5 C of charge flows through the wire every second. We have the following relation between coulomb and the esu of charge:

$$1\ C = 3 \times 10^9\ esu.$$ (A.3)

It may be noted that the numerical conversion factor given here is only approximate. We shall comment on its exact value in Sect. A.4. If a charge has the numerical value q C in SI and q_{es} in esu, then we have

$$q_{es} = 3.0 \times 10^9 q.$$

Since r and F in (A.1) have to be in metres and newtons, we also have

$$r_{cgs} = 10^2 r, \quad F_{cgs} = 10^5 F.$$

Substituting these conversion factors in (A.2), we get

$$F = 9 \times 10^9 \frac{q_1 q_2}{r^2}.$$ (A.4)

On comparing with (A.1), we easily conclude that $1/4\pi\epsilon_0$ has to have the numerical value 9×10^9 in SI units. The value of ϵ_0 given in (2.8) follows from such considerations.

By comparing (A.1) and (A.2), it is easy to conclude that the vectorial form of Coulomb's law in Gaussian cgs units is obtained by replacing $1/4\pi\epsilon_0$ by 1 in (2.7). Introducing the electric field in Gaussian units by writing the force on a charge as $\mathbf{F}_{cgs} = q_{es}\mathbf{E}_{cgs}$, we can easily see that the electric field due to a charge is given by an expression like (2.6) with 1 in place of $1/4\pi\epsilon_0$. This would imply the equations

$$\nabla_{cgs} \cdot \mathbf{E}_{cgs} = 4\pi\rho_{es},$$ (A.5)

$$\nabla_{cgs} \times E_{cgs} = 0. \tag{A.6}$$

These are the basic equations of electrostatics in Gaussian units and have to be compared with (1.8) and (1.9) in SI units.

We end our discussion of electrostatics by making a few comments about a dielectric medium. By repeating the arguments given in Sect. 2.14, we can introduce the polarization density P_{cgs} and show that the polarization charge density is given by $-\nabla_{cgs}.P_{cgs}$ in Gaussian units also. When we are dealing with a dielectric medium, we have to include this in (A.5), which would give

$$\nabla_{cgs} . E_{cgs} = 4\pi(\rho_{es} - \nabla_{cgs} . P_{cgs}).$$

This can be written as

$$\nabla_{cgs} . (E_{cgs} + 4\pi P_{cgs}) = 4\pi\rho_{es}. \tag{A.7}$$

This prompts us to introduce the electric displacement vector D_{cgs} in the following manner:

$$D_{cgs} = E_{cgs} + 4\pi P_{cgs}. \tag{A.8}$$

Using this, we can put (A.7) in the form

$$\nabla_{cgs} . D_{cgs} = 4\pi\rho_{es}. \tag{A.9}$$

Since the polarization density P_{cgs} is expected to be proportional to the electric field, (A.8) suggests that we can write

$$D_{cgs} = \kappa E_{cgs}. \tag{A.10}$$

Since it is clear from (A.8) that D_{cgs} and E_{cgs} must have the same unit, κ is obviously a dimensionless quantity. A little reflection should convince the reader that this is nothing different from the dielectric constant κ introduced in (2.103). Both being dimensionless quantities, their numerical values should be the same, irrespective of whether we use the Gaussian or the SI units.

A.2 Electromagnetic Units

We now introduce the electromagnetic unit of current which is defined through the Biot–Savart law. This law in SI units is given by (3.10). It implies that the amplitude of the force between two *parallel* current elements dl_1 and dl_2 carrying currents I_1 and I_2, respectively, and separated by a distance r is

$$dF = \frac{\mu_0}{4\pi} \frac{I_1 I_2 dl_1 dl_2}{r^2}. \tag{A.11}$$

This same law of interaction between current elements is written in Gaussian units as

$$dF_{cgs} = \frac{I_{1,em}\, I_{2,em}\, dl_{1,cgs}\, dl_{2,cgs}}{r_{cgs}^2}. \tag{A.12}$$

The subscripts of the currents indicate that they are in electromagnetic units. We can use (A.12) to define the electromagnetic unit of current in exactly the same way in which (A.2) was used for defining the esu. An electromagnetic unit of current is such that two current elements of length 1 cm carrying unit currents kept at a distance of 1 cm attract each other with a force of 1 dyne. We can then define the electromagnetic unit of charge as the amount of charge that flows in one second through a wire carrying a current of 1 electromagnetic unit. If the electromagnetic unit of charge is denoted as emu, then the electromagnetic unit of current has to be denoted as emu s^{-1}. We pointed out in Sect. 5.10.1 that the magnetic field is something like a relativistic effect. On this ground, we may expect that currents have to be sufficiently strong to produce appreciable forces. The electromagnetic unit of current, in fact, is related to ampere, the SI unit of current, in the following way:

$$1\,A = \frac{1}{10}\ \text{emu s}^{-1}. \tag{A.13}$$

If a current has the numerical value I A in SI and I_{em} emu s^{-1} in Gaussian units, then we have

$$I_{em} = I/10.$$

Using this conversion factor as well as the other conversion factors

$$r_{cgs} = 10^2 r, \quad dl_{cgs} = 10^2 dl, \quad dF_{cgs} = 10^5 dF,$$

we get from (A.12) that

$$dF = 10^{-7}\,\frac{I_1 I_2 dl_1 dl_2}{r^2}. \tag{A.14}$$

On comparing with (A.11), we can conclude that $\mu_0/4\pi$ has to have the numerical value 10^{-7} in SI units. The value of μ_0 given in (3.11) arises from these considerations.

As in the case of electrostatics, it is easy to conclude that the vectorial form of the Biot–Savart law in Gaussian units is obtained by replacing $\mu_0/4\pi$ by 1 in (3.10). The expression (3.9) for the force on a current due to a magnetic field is also used for introducing the magnetic field in Gaussian units when the current is in electromagnetic units and the mechanical quantities in cgs units. It is easy to check that the magnetic field due to a current in Gaussian units is given by an expression like (3.8) with 1 in place of $\mu_0/4\pi$. A little work then shows that the basic equations of magnetostatics should be

$$\nabla_{cgs} \cdot \mathbf{B}_{cgs} = 0, \tag{A.15}$$

$$\nabla_{cgs} \times \mathbf{B}_{cgs} = 4\pi \mathbf{j}_{em}. \tag{A.16}$$

These equations have to be compared with (1.10) and (1.11), which are the basic equations of magnetostatics in SI units.

Just as we made a few comments about a dielectric medium in our discussion of electrostatic units, we now make a few comments about a (para)magnetic medium. By introducing the magnetic polarization density $\mathbf{M}_{cgs}$, we can repeat the arguments of Sect. 3.7 to show that the polarization current density is given by $\nabla_{cgs} \times \mathbf{M}_{cgs}$. When considering a magnetic medium, we have to include this in (A.16) so that

$$\nabla_{cgs} \times \mathbf{B}_{cgs} = 4\pi (\mathbf{j}_{em} + \nabla_{cgs} \times \mathbf{M}_{cgs}).$$

This can be written as

$$\nabla_{cgs} \times (\mathbf{B}_{cgs} - 4\pi \mathbf{M}_{cgs}) = 4\pi \mathbf{j}_{em}. \tag{A.17}$$

We can introduce the other magnetic field vector $\mathbf{H}_{cgs}$ through the following relation:

$$\mathbf{B}_{cgs} = \mathbf{H}_{cgs} + 4\pi \mathbf{M}_{cgs}. \tag{A.18}$$

Using this, we can put (A.17) in the form

$$\nabla_{cgs} \times \mathbf{H}_{cgs} = 4\pi \mathbf{j}_{em}. \tag{A.19}$$

In a paramagnetic substance, the polarization density $\mathbf{M}_{cgs}$ is proportional to the magnetic field, and we can write

$$\mathbf{B}_{cgs} = \kappa_m \mathbf{H}_{cgs}. \tag{A.20}$$

It is easy to see from (A.18) that κ_m is a dimensionless quantity independent of the system of units and is the same as the permeability κ_m defined in (3.51).

A.3 Lorentz Force Equation

When working with Gaussian units, we can use electrostatic units when dealing with electrostatics and electromagnetic units when dealing with magnetostatics. In other words, it is not necessary to mix the two kinds of units when we consider static situations. However, it becomes essential to mix them in dynamical situations. Let the ratio of the emu of charge to the esu of charge be denoted by c. If q_{es} and q_{em} are the values of a charge in esu and emu, then we have

$$q_{em} = \frac{q_{es}}{c}. \tag{A.21}$$

It is obvious that the values of current in the two units should also satisfy the same relation

$$\mathbf{j}_{em} = \frac{\mathbf{j}_{es}}{c}. \tag{A.22}$$

From (A.2) and (A.12), it is easy to show that the charge in esu and emu has the dimensions $[M^{1/2}L^{3/2}T^{-1}]$ and $[M^{1/2}L^{1/2}]$ respectively. It then follows from (A.21) that c should have the dimension $[LT^{-1}]$. We conclude from (A.3) and (A.13) that $c = 3 \times 10^{10}$ cm s^{-1}. We shall show in Sect. A.4 that c turns out to be the speed of light, but we do not assume this a priori.

We have already pointed out that the force exerted by a magnetic field on a current is given by an expression of the form (3.9), when all the quantities are in cgs or electromagnetic units. This suggests that the force on a moving charge should be

$$q_{em}\mathbf{v}_{cgs} \times \mathbf{B}_{cgs} = \frac{q_{es}}{c}\mathbf{v}_{cgs} \times \mathbf{B}_{cgs}$$

making use of (A.21). Adding the electric force to this, the full Lorentz force equation in Gaussian units for a charge moving in an electromagnetic field is

$$\mathbf{F}_{cgs} = q_{es}\left(\mathbf{E}_{cgs} + \frac{\mathbf{v}_{cgs}}{c} \times \mathbf{B}_{cgs}\right). \tag{A.23}$$

Since we know how to relate velocity, force and charge in SI and Gaussian units, it is easy to use (A.23) to determine out how the electric and magnetic fields are related in the two unit systems. The SI unit of electric field (volts per metre) is related to its Gaussian unit (dynes per esu) as follows:

$$1 \text{ V m}^{-1} = \frac{1}{3} \times 10^{-4} \text{ dyne esu}^{-1}. \tag{A.24}$$

The relation between the SI unit of the magnetic field (tesla) and its Gaussian unit (gauss) is

$$1 \text{ T} = 10^4 \text{ gauss}. \tag{A.25}$$

We now comment on one fact which is obvious from (A.21) and (A.22): the electromagnetic units of charge and current are much larger than the electrostatic units, if c is the speed of light. This is not surprising if we keep in mind that the magnetic field is a relativistic effect, as stressed in Sect. 5.10.1. Hence, a current that produces an appreciable magnetic field has to be much larger than a typical current following from electrostatic considerations. We pointed out towards the end of Sect. 5.10.1 that, even when introducing SI units, we face the problem that the currents which we encounter in everyday life would correspond to very large charges if the screening by opposite charges did not take place.

A.4   Maxwell's Equations

Let us now determine how the equations of electrostatics and magnetostatics in Gaussian units have to be extended for situations involving time variations. We can proceed as in Sect. 4.1 to determine the time derivative terms.

We can write (A.19) as

$$\nabla_{cgs} \times \mathbf{H}_{cgs} = \frac{4\pi}{c} \mathbf{j}_{es}, \tag{A.26}$$

making use of (A.22). If the charge conservation equation

$$\frac{\partial \rho_{es}}{\partial t} + \nabla_{cgs} \cdot \mathbf{j}_{es} = 0 \tag{A.27}$$

has to be satisfied, then (A.26) must be extended to have the form

$$\nabla_{cgs} \times \mathbf{H}_{cgs} = \frac{4\pi}{c} \mathbf{j}_{es} + \frac{1}{c} \frac{\partial \mathbf{D}_{cgs}}{\partial t}. \tag{A.28}$$

When we take the divergence of this equation, it is easy to check that (A.27) with (A.9) would ensure that the RHS is zero, which is required for consistency.

To determine the other time derivative term, we turn to the discussion of Sect. 4.1.2. When we write (4.9) in Gaussian units, we have to divide the $\mathbf{v} \times \mathbf{B}$ term by c because of (A.23). It easily follows from (4.13) that this will make k appearing in the discussion of Sect. 4.1.2 equal to $1/c$, which would lead to the equation

$$\nabla_{cgs} \times \mathbf{E}_{cgs} = -\frac{1}{c} \frac{\partial \mathbf{B}_{cgs}}{\partial t}. \tag{A.29}$$

It should be clear that (A.9) and (A.15), along with (A.28) and (A.29), constitute Maxwell's equations in Gaussian units.

We now want to argue that c appearing in (A.28) and (A.29) is nothing but the speed of light. To show this, we apply Maxwell's equations to a uniform medium within which the charge and current densities are zero. Making use of (A.10), (A.20) and keeping in mind that κ and κ_m are constant for a uniform medium, we get

$$\nabla_{cgs} \cdot \mathbf{E}_{cgs} = 0,$$

$$\nabla_{cgs} \cdot \mathbf{B}_{cgs} = 0,$$

$$\nabla_{cgs} \times \mathbf{E}_{cgs} = -\frac{1}{c} \frac{\partial \mathbf{B}_{cgs}}{\partial t},$$

$$\frac{1}{\kappa_m} \nabla_{cgs} \times \mathbf{B}_{cgs} = \frac{\kappa}{c} \frac{\partial \mathbf{E}_{cgs}}{\partial t}.$$

It is easy to combine these equations to arrive at

$$\frac{\partial^2 \mathbf{E}_{cgs}}{\partial t^2} = \frac{\kappa \kappa_m}{c^2} \nabla^2_{cgs} \mathbf{E}_{cgs},$$ (A.30)

with an exactly similar equation for $\mathbf{B}_{cgs}$. It is obvious that the speed of electromagnetic waves in Gaussian units is given by same expression (4.41) which gives the speed in SI units (keep in mind that the numerical value of c is different in the two systems of units). In a vacuum, the speed of light or any other electromagnetic wave comes out to be c. Note that when we write down Maxwell's equations in Gaussian cgs units, c has to be the speed of light *in cgs units*.

Maxwell's original derivation of the electromagnetic wave in 1865 was carried out using Gaussian units. What he showed was that the speed of light was equal to the ratio of the emu of charge to the esu of charge—as indicated by c in (A.21). Since the esu and the emu of charge can be established from electrical and magnetic measurements on the basis of (A.2) and (A.12), Maxwell's famous result meant that the speed of light could be determined from purely electrical and magnetic measurements.

In view of the fact that the ratio of the emu to esu is c, let us revisit the conversion factors in (A.3) and (A.13). The current convention is to take (A.13) to be an exact relation, which implies that the value of $\mu_0/4\pi$ appearing in the formulae used in SI units is exactly equal to 10^{-7}, as indicated in (3.11). From (A.13), we get

$$1 \text{ C} = \frac{1}{10} \text{ emu} = \frac{c}{10} \text{ esu}$$ (A.31)

on using the relation between emu and esu. If we use the best measured value of the speed of light c currently available and keep five significant digits of that value in cgs units, then it follows from (A.31) that

$$1 \text{ C} = 2.9979 \times 10^9 \text{ esu.}$$ (A.32)

Obviously, (A.3) is a very good approximation. If the esu of charge is defined through (A.2), then a measured value of c fixes the coulomb, the SI unit of charge, in accordance with (A.31). When giving the value of ϵ_0 in (2.8), we noted that this value is also fixed by the speed of light c. These various considerations are certainly interrelated.

A.5   A Few Important Results

Since Maxwell's equations in Gaussian units look somewhat different from the way they look in SI units, we expect that various results derived from Maxwell's equations would also look different in Gaussian units. From the knowledge of Maxwell's equations in Gaussian units, readers should be able to determine how different results of electromagnetic theory derived in this book would look in Gaussian units. We present a discussion of only a few very important selected results.

Let us consider the discussion of the energy of the electromagnetic field presented in Sect. 4.2. The rate of work done by the electromagnetic field per unit volume is $\mathbf{j}_{es}.\mathbf{E}_{cgs}$ in Gaussian units. Keeping this in mind, if we repeat the derivation in Sect. 4.2 using Maxwell's equations written in Gaussian units, then we end up with the equation

$$\frac{\partial \mathcal{E}_{mech,cgs}}{\partial t} = - \oint \frac{c}{4\pi} (\mathbf{E}_{cgs} \times \mathbf{H}_{cgs}).\, d\mathbf{S} - \int \frac{\partial}{\partial t} \left(\frac{\mathbf{H}_{cgs} \cdot \mathbf{B}_{cgs}}{8\pi} + \frac{\mathbf{E}_{cgs} \cdot \mathbf{D}_{cgs}}{8\pi} \right) dV$$

$$(A.33)$$

in place of (4.17). This suggests that the Poynting vector representing the energy flux is given by

$$\mathbf{N}_{cgs} = \frac{c}{4\pi} (\mathbf{E}_{cgs} \times \mathbf{H}_{cgs}), \tag{A.34}$$

whereas the energy density of the electromagnetic field is given by

$$u_{cgs} = \frac{\mathbf{E}_{cgs} \cdot \mathbf{D}_{cgs}}{8\pi} + \frac{\mathbf{H}_{cgs} \cdot \mathbf{B}_{cgs}}{8\pi}. \tag{A.35}$$

These can be compared with (4.19) and (4.20) giving the expressions of the Poynting vector and the electromagnetic energy density in SI units.

We have discussed in Sect. 4.10 how Maxwell's equations in free space lead to two inhomogeneous wave equations. We now point out how this discussion may be modified in Gaussian units. Since κ and κ_m are 1 for free space, it follows from (A.10) and (A.20) that we can replace $\mathbf{D}_{cgs}$ and $\mathbf{H}_{cgs}$ by $\mathbf{E}_{cgs}$ and $\mathbf{B}_{cgs}$ when we use Gaussian units in free space. To obtain the inhomogeneous wave equations, we first have to introduce the scalar and vector potentials. It follows from (A.15) that we can introduce the vector potential

$$\mathbf{B}_{cgs} = \nabla_{cgs} \times \mathbf{A}_{cgs} \tag{A.36}$$

in the same way as in (4.128). Substituting in (A.29), we get

$$\nabla_{cgs} \times \left(\mathbf{E}_{cgs} + \frac{1}{c} \frac{\partial \mathbf{A}_{cgs}}{\partial t} \right) = 0.$$

This suggests that we can introduce the scalar potential as follows:

$$\mathbf{E}_{cgs} = -\nabla_{cgs} \Phi_{cgs} - \frac{1}{c} \frac{\partial \mathbf{A}_{cgs}}{\partial t}, \tag{A.37}$$

which can be compared with (4.130). We now have to substitute (A.36) and (A.37) into the Eqs. (A.9) and (A.28) with source terms and use the Lorentz gauge condition

$$\nabla_{cgs} \cdot \mathbf{A}_{cgs} + \frac{1}{c} \frac{\partial \Phi_{cgs}}{\partial t} = 0, \tag{A.38}$$

which is somewhat different from (4.135). This leads to the inhomogeneous wave equations

$$\left(\nabla_{cgs}^2 - \frac{1}{c^2} \frac{\partial^2}{\partial t^2} \right) \Phi_{cgs} = -4\pi \rho_{es}, \tag{A.39}$$

$$\left(\nabla_{cgs}^2 - \frac{1}{c^2} \frac{\partial^2}{\partial t^2} \right) \mathbf{A}_{cgs} = -\frac{4\pi}{c} \mathbf{j}_{es}. \tag{A.40}$$

We have discussed solutions of inhomogeneous wave equations in detail in Chaps. 6 and 7. Comparing (A.39) and (A.40) with (4.138) and (4.139), it is easy to determine how the solutions obtained in SI units have to be modified to give solutions in Gaussian units. For example, a comparison of (A.39) and (4.138) suggests that we merely have to replace $1/\epsilon_0$ in the solution of (4.138) in SI units by 4π to obtain the solution of (A.39) in Gaussian units. This can be done with (7.2) to give the expression of the radiation electric field due to an accelerated charge in Gaussian units. Once we have the radiation electric field, it is straightforward to apply the expression (A.34) of the Poynting vector to calculate the energy flux in Gaussian units. By integrating over different directions, we can get the total energy of radiation emitted per unit time by the charged particle, following the procedure of Sect. 7.2. The final result is the Gaussian version of Larmor's formula, which is

$$P_{cgs} = \frac{2 q_{es}^2 \dot{v}_{cgs}^2}{3 c^3}. \tag{A.41}$$

This has to be compared with (7.22). This example illustrates how we have to modify electromagnetic results given in SI units to arrive at the corresponding results in Gaussian cgs units.

Following our discussion, readers should be able to arrive at the Gaussian cgs version of any result derived in this book. In fact, it will be an instructive exercise for the readers to determine how the cyclotron frequency given by (5.124), the Debye length given by (8.5) and the plasma frequency given by (8.14) are modified in the Gaussian cgs units.

Appendix B
Useful Vector Relations

We assume that the readers of this book are familiar with the dot and cross products of vectors and the gradient, divergence and curl operators. Here, we collect some important relations involving the products and the operators, which are used throughout the book.

B.1 General Identities

For products of three vectors **A**, **B** and **C**, we have the following identities:

$$\mathbf{A}.(\mathbf{B} \times \mathbf{C}) = \mathbf{B}.(\mathbf{C} \times \mathbf{A}) = \mathbf{C}.(\mathbf{A} \times \mathbf{B}), \tag{B.1}$$

$$\mathbf{A} \times (\mathbf{B} \times \mathbf{C}) = (\mathbf{A}.\mathbf{C})\mathbf{B} - (\mathbf{A}.\mathbf{B})\mathbf{C}. \tag{B.2}$$

Let ϕ, ψ be two scalar fields and **A**, **B** two vector fields. The different operators acting on products satisfy the following relations:

$$\nabla(\phi\psi) = \phi\nabla\psi + \psi\nabla\phi, \tag{B.3}$$

$$\nabla.(\psi\mathbf{A}) = \mathbf{A}.\nabla\psi + \psi\nabla.\mathbf{A}. \tag{B.4}$$

$$\nabla \times (\psi\mathbf{A}) = \psi\nabla \times \mathbf{A} - \mathbf{A} \times \nabla\psi, \tag{B.5}$$

$$\nabla(\mathbf{A}.\mathbf{B}) = \mathbf{A} \times (\nabla \times \mathbf{B}) + \mathbf{B} \times (\nabla \times \mathbf{A}) + (\mathbf{A}.\nabla)\mathbf{B} + (\mathbf{B}.\nabla)\mathbf{A}, \tag{B.6}$$

$$\nabla.(\mathbf{A} \times \mathbf{B}) = \mathbf{B}.(\nabla \times \mathbf{A}) - \mathbf{A}.(\nabla \times \mathbf{B}), \tag{B.7}$$

$$\nabla \times (\mathbf{A} \times \mathbf{B}) = \mathbf{A}(\nabla.\mathbf{B}) - \mathbf{B}(\nabla.\mathbf{A}) + (\mathbf{B}.\nabla)\mathbf{A} - (\mathbf{A}.\nabla)\mathbf{B}. \tag{B.8}$$

© Springer Nature Singapore Pte Ltd. 2022
A. R. Choudhuri, *Advanced Electromagnetic Theory*, Lecture Notes in Physics 1009,
https://doi.org/10.1007/978-981-19-5944-8

Two operators operating in succession satisfy the relations:

$$\nabla \times \nabla \psi = 0, \tag{B.9}$$

$$\nabla.(\nabla \times \mathbf{A}) = 0, \tag{B.10}$$

$$\nabla.\nabla \psi = \nabla^2 \psi, \tag{B.11}$$

$$\nabla \times (\nabla \times \mathbf{A}) = \nabla(\nabla.\mathbf{A}) - \nabla^2 \mathbf{A}. \tag{B.12}$$

B.2 Integral Relations

If V is a volume bounded by a closed surface S, with $d\mathbf{S}$ taken positive in the outward direction,

$$\oint_S \mathbf{A}.\,d\mathbf{S} = \int_V (\nabla.\mathbf{A})\,dV, \tag{B.13}$$

$$\oint_S \psi\,d\mathbf{S} = \int_V (\nabla\psi)\,dV, \tag{B.14}$$

where (B.13) is Gauss's theorem.

If an open surface S is bounded by a contour C, of which $d\mathbf{l}$ is the line element, then Stokes's theorem states that

$$\oint_C \mathbf{A}.\,d\mathbf{l} = \int_S (\nabla \times \mathbf{A}).\,d\mathbf{S}. \tag{B.15}$$

Appendix C
Formulae and Equations in Cylindrical and Spherical Coordinates

C.1 Vector Formulae in Cylindrical Coordinates

If ψ is a scalar field and $\mathbf{A} = A_r \mathbf{e}_r + A_\theta \mathbf{e}_\theta + A_z \mathbf{e}_z$ is a vector field, then

$$\nabla \psi = \frac{\partial \psi}{\partial r} \mathbf{e}_r + \frac{1}{r} \frac{\partial \psi}{\partial \theta} \mathbf{e}_\theta + \frac{\partial \psi}{\partial z} \mathbf{e}_z, \tag{C.1}$$

$$\nabla . \mathbf{A} = \frac{1}{r} \frac{\partial}{\partial r}(r A_r) + \frac{1}{r} \frac{\partial A_\theta}{\partial \theta} + \frac{\partial A_z}{\partial z}, \tag{C.2}$$

$$\nabla \times \mathbf{A} = \left(\frac{1}{r} \frac{\partial A_z}{\partial \theta} - \frac{\partial A_\theta}{\partial z} \right) \mathbf{e}_r + \left(\frac{\partial A_r}{\partial z} - \frac{\partial A_z}{\partial r} \right) \mathbf{e}_\theta + \frac{1}{r} \left[\frac{\partial}{\partial r}(r A_\theta) - \frac{\partial A_r}{\partial \theta} \right] \mathbf{e}_z, \tag{C.3}$$

$$\nabla^2 \psi = \frac{1}{r} \frac{\partial}{\partial r} \left(r \frac{\partial \psi}{\partial r} \right) + \frac{1}{r^2} \frac{\partial^2 \psi}{\partial \theta^2} + \frac{\partial^2 \psi}{\partial z^2}. \tag{C.4}$$

C.2 Vector Formulae in Spherical Coordinates

If ψ is a scalar field and $\mathbf{A} = A_r \mathbf{e}_r + A_\theta \mathbf{e}_\theta + A_\phi \mathbf{e}_\phi$ is a vector field, then

$$\nabla \psi = \frac{\partial \psi}{\partial r} \mathbf{e}_r + \frac{1}{r} \frac{\partial \psi}{\partial \theta} \mathbf{e}_\theta + \frac{1}{r \sin \theta} \frac{\partial \psi}{\partial \phi} \mathbf{e}_\phi, \tag{C.5}$$

$$\nabla . \mathbf{A} = \frac{1}{r^2} \frac{\partial}{\partial r}(r^2 A_r) + \frac{1}{r \sin \theta} \frac{\partial}{\partial \theta}(\sin \theta A_\theta) + \frac{1}{r \sin \theta} \frac{\partial A_\phi}{\partial \phi} \tag{C.6}$$

A. R. Choudhuri, *Advanced Electromagnetic Theory*, Lecture Notes in Physics 1009,
https://doi.org/10.1007/978-981-19-5944-8

$$\nabla \times \mathbf{A} = \frac{1}{r \sin \theta} \left[\frac{\partial}{\partial \theta} (\sin \theta A_\phi) - \frac{\partial A_\theta}{\partial \phi} \right] \mathbf{e}_r + \left[\frac{1}{r \sin \theta} \frac{\partial A_r}{\partial \phi} \right.$$

$$\left. - \frac{1}{r} \frac{\partial}{\partial r} (r A_\phi) \right] \mathbf{e}_\theta + \frac{1}{r} \left[\frac{\partial}{\partial r} (r A_\theta) - \frac{\partial A_r}{\partial \theta} \right] \mathbf{e}_\phi \tag{C.7}$$

$$\nabla^2 \psi = \frac{1}{r^2} \frac{\partial}{\partial r} \left(r^2 \frac{\partial \psi}{\partial r} \right) + \frac{1}{r^2 \sin \theta} \frac{\partial}{\partial \theta} \left(\sin \theta \frac{\partial \psi}{\partial \theta} \right) + \frac{1}{r^2 \sin^2 \theta} \frac{\partial^2 \psi}{\partial \phi^2}. \tag{C.8}$$

Suggestions for Further Reading

This is not meant to be an exhaustive bibliography of all books on electromagnetic theory. I am mainly listing those books which I personally have found useful over the years.

Early Classics

Apart from laying the foundation of modern electromagnetic theory, Maxwell wrote a classical exposition of the subject in two volumes (Maxwell [1]) advocating the field approach over the action-at-a-distance approach, which was the standard approach followed in electromagnetic theory textbooks at that time. The vector notation had not yet been invented in Maxwell's time. All the vectorial equations, including the expressions of gradient, divergence and curl, were written in long scalar notation at that time. Maxwell's book, as well as the other early textbooks to follow the Maxwellian approach such as the book by Jeans [2], followed the long scalar notation.

The book which introduced Maxwellian electrodynamics in Germany was by Föppl [3], which later became Abraham and Föppl [4], to be translated into English after another transformation to Abraham and Becker [5]. This was one of the first electromagnetic theory textbooks to develop the subject in vectorial notation and became the standard textbook of the subject from which at least two generations of physicists learned the subject.

Two great physicists of that era who wrote well-known series of basic physics textbooks—Max Planck and Arnold Sommerfeld—devoted a volume to electromagnetic theory (Planck [6]; Sommerfeld [7]). As electromagnetic theory started occupying a place of pride in the graduate physics curricula of American universities (along with quantum mechanics), several voluminous textbooks were written by professors of American universities. The most well known among these early books were Harnwell [8], Stratton [9] and Smythe [10].

After the pattern of teaching electromagnetism in physics departments somewhat stabilized after the Second World War, textbooks at two distinct levels began to appear. The intermediate-level textbooks were designed for teaching electromagnetic theory in the first or second year of college, whereas the more advanced textbooks

© Springer Nature Singapore Pte Ltd. 2022
A. R. Choudhuri, *Advanced Electromagnetic Theory*, Lecture Notes in Physics 1009,
https://doi.org/10.1007/978-981-19-5944-8

had advanced undergraduate or graduate students in mind. We now list some of the most widely used textbooks at both these levels.

Intermediate-Level Textbooks

Perhaps the most outstanding textbook at this level for many years has been the book by the Nobel-winning physicist Purcell [11], which was Volume 2 of the celebrated *Berkeley Physics Course*. It is now updated to [12]. A special mention should be made of a book of deep insight by another great physicist: Volume 2 of the *Feynman Lectures* (Feynman, Leighton and Sands [13]). The other well-known textbooks at this level are Reitz, Milford and Christy [14] and Griffiths [15].

Graduate-Level Textbooks

One of the earliest textbooks which is still useful is Panofsky and Phillips [16]. It is a compact book and is a favourite of this author. The present book is perhaps more strongly influenced by this book than any other book. The textbook which dominated the year-long teaching of electromagnetic theory in American universities for several decades is the voluminous book by Jackson [17, 1st edn. 1962]. The more recent voluminous textbooks are Garg [18] and Zangwill [19]. One compact book suitable for a one-semester course is by Melia [20]. Two volumes in the famous Landau–Lifshitz *Course of Theoretical Physics* are devoted to electromagnetism (Landau and Lifshitz [21]; Landau, Lifshitz and Pitaevsky [22]).

Plasma Physics and MHD

Since some of the standard textbooks on electromagnetic theory do not cover plasma physics and MHD, we give some references on these subjects. A well-known textbook on plasma physics is Chen [23]. The pioneering textbook on MHD by Cowling [24] can still be read profitably for its clarity. Although written primarily for astrophysics students, Choudhuri [25] can be consulted for an introduction to plasma physics and MHD.

References

1. Maxwell, J.C.: A Treatise on Electricity and Magnetism, vols. I & II. Clarendon Press (reprinted by Dover), Oxford (1873)
2. Jeans, J.H.: Mathematical Theory of Electricity and Magnetism. Cambridge University Press (1925)
3. Föppl, A.: Einführung in die Maxwellsche Theorie der Elektrizität. Teubner, Leipzig (1894)
4. Abraham, M., Föppl, A.: Theorie der Elektrizität: Einführung in die Maxwellsche Theorie der Elektrizität. Teubner, Leipzig (1904)
5. Abraham, M., Becker, R.: The Classical Theory of Electricity and Magnetism. Blackie and Son (1932)
6. Planck, M.: Theory of Electricity and Magnetism. Macmillan (1932)
7. Sommerfeld, A.: Electrodynamics: Lectures on Theoretical Physics, vol. III. Academic Press (1952)
8. Harnwell, G.P.: Principles of Electricity and Electromagnetism. McGraw-Hill (1938)

9. Stratton, J.A.: Electromagnetic Theory. McGraw Hill (1941)
10. Smythe, W.R.: Static and Dynamic Electricity. McGraw Hill (1950)
11. Purcell, E.M.: Electricity and Magnetism. McGraw Hill (1965)
12. Purcell, E.M., Morin, D.J.: Electricity and Magnetism, 3rd edn. Cambridge University Press (2013)
13. Feynman, R.P., Leighton, R.B., Sands, M.: The Feynman Lectures on Physics, vol. II. Mainly Electromagnetism and Matter. Addison-Wesley (1964)
14. Reitz, J.R., Milford, F.J., Christy, R.W.: Foundations of Electromagnetic Theory, 4th edn. Addison-Wesley (2008)
15. Griffiths, D.J.: Introduction to Electrodynamics, 4th edn. Cambridge University Press (2017)
16. Panofsky, W.H.F., Phillips, M.: Classical Electricity and Magnetism, 2nd edn. Addison-Wesley (1962)
17. Jackson, J.D.: Classical Electrodynamics, 3rd edn. Wiley (1999)
18. Garg, A.: Classical Electromagnetism in a Nutshell. Princeton University Press (2012)
19. Zangwill, A.: Modern Electrodynamics. Cambridge University Press (2013)
20. Melia, F.: Electrodynamics. University of Chicago Press (2001)
21. Landau, L.D., Lifshitz, E.M.: The Classical Theory of Fields: Volume 2 (Course of Theoretical Physics), 4th edn. Butterworth-Heinemann (1980)
22. Landau, L.D., Lifshitz, E.M., Pitaevsky, L.P.: Electrodynamics of Continuous Media: Volume 8 (Course of Theoretical Physics), 2nd edn. Butterworth-Heinemann (1984)
23. Chen, F.F.: Introduction to Plasma Physics and Controlled Fusion. Volume 1: Plasma Physics, 2nd edn. Springer (2006)
24. Cowling, T.G.: Magnetohydrodynamics, 2nd edn. Adam Hilger (1976)
25. Choudhuri, A.R.: The Physics of Fluids and Plasmas: An Introduction for Astrophysicists. Cambridge University Press (1998)

Index

© Springer Nature Singapore Pte Ltd. 2022
A. R. Choudhuri, *Advanced Electromagnetic Theory*, Lecture Notes in Physics 1009,
https://doi.org/10.1007/978-981-19-5944-8

Printed in the United States
by Baker & Taylor Publisher Services